2014
RECENT ADVANCES IN FINANCIAL ENGINEERING

Proceedings of the
TMU Finance Workshop 2014

2014
RECENT ADVANCES IN FINANCIAL ENGINEERING

Proceedings of the
TMU Finance Workshop 2014

Tokyo Metropolitan University (TMU)
Akihabara Satellite Campus
6 – 7 November 2014

editors

Masaaki Kijima
Tokyo Metropolitan University, Japan

Yukio Muromachi
Tokyo Metropolitan University, Japan

Takashi Shibata
Tokyo Metropolitan University, Japan

NEW JERSEY • LONDON • SINGAPORE • BEIJING • SHANGHAI • HONG KONG • TAIPEI • CHENNAI • TOKYO

Published by

World Scientific Publishing Co. Pte. Ltd.
5 Toh Tuck Link, Singapore 596224
USA office: 27 Warren Street, Suite 401-402, Hackensack, NJ 07601
UK office: 57 Shelton Street, Covent Garden, London WC2H 9HE

British Library Cataloguing-in-Publication Data
A catalogue record for this book is available from the British Library.

RECENT ADVANCES IN FINANCIAL ENGINEERING 2014
Proceedings of the TMU Finance Workshop 2014

ISBN 978-981-4730-76-1

In-house Editor: Alisha Nguyen

Typeset by Stallion Press
Email: enquiries@stallionpress.com

Printed in Singapore

PREFACE

Tokyo Metropolitan University (TMU) organized the *TMU Finance Workshop 2014* at the Akihabara Satellite Campus of TMU on November 6 and 7, 2014. This annual workshop is a successor of the Daiwa International Workshop (2004 to 2008), the KIER-TMU International Workshop (2009 to 2010), and the International Workshop on Finance (2011 to 2012). The TMU Finance Workshop 2014 was thus the 10th anniversary workshop.

The workshop has been designed as the forum to exchange new ideas in finance (including mathematical finance and financial engineering) and served as a bridge between academic researchers and practitioners. To these ends, the speakers have shared various interesting ideas, information on new methods, and their up-to-date research results. In the 2014 workshop, we invited 13 leading scholars, including one keynote speaker, from various countries, for the two-day workshop resulting in many fruitful discussions.

This book contains ten papers presented at the workshop. The editors made the final selection based on recommendations by referees. In these papers, the latest concepts, methods, and techniques related to current topics in finance are proposed and reviewed. We hope that the reader of this book will obtain a great deal about cutting-edge research in finance.

We would like to express our utmost gratitude to those who contributed their papers. Many thanks to all the anonymous reviewers who evaluated the papers submitted. Financial support from the JSPS KAKENHI (Grant Numbers 26242028 and 26285071), the TMU Program for Enhancing the Quality of University Education, and the Credit Pricing Corporation is greatly appreciated.

We owe much gratitude to Satoshi Kanai for helping us in preparing the LaTeX documents. We are grateful to Alisha Nguyen, Agnes Ng, and the editorial committee members of World Scientific Publishing Co. for their kind assistance and support in publishing this book.

Tokyo, July 2015

Masaaki Kijima, Tokyo Metropolitan University
Yukio Muromachi, Tokyo Metropolitan University
Takashi Shibata, Tokyo Metropolitan University

TMU Finance Workshop 2014

Date
November 6 and 7, 2014

Place
Akihabara Satellite Campus, Tokyo Metropolitan University,
Tokyo, Japan

Organizer
Graduate School of Social Sciences, Tokyo Metropolitan University

Supported by
JSPS KAKENHI A (Grant Number 26242028)
JSPS KAKENHI B (Grant Number 26285071)
TMU Program for Enhancing the Quality of University Education
Credit Pricing Corporation

Program Committee
Masaaki Kijima, Chair, Tokyo Metropolitan University
Yukio Muromachi, Tokyo Metropolitan University
Takashi Shibata, Tokyo Metropolitan University

Program

November 6 (Thursday), 2014

Opening Address

Yukio Muromachi (Tokyo Metropolitan University)

(Chair: Hoi Ying Wong)

13:10–14:10 Jordan Stoyanov (Newcastle University)
Plenary talk: "Probability distributions in stochastic financial models"

14:10–14:50 Mikhail Zhitlukhin (Steklov Mathematical Institute)
"General reward-to-variability performance measures"

(Chair: Sebastian Jaimungal)

15:20–16:00 Hoi Ying Wong (Chinese University of Hong Kong)
"Ambiguous correlation in asset pricing"

16:00–16:40 Michael Tehranchi (University of Cambridge)
"Asymptotics of implied volatility far from maturity"

16:40–17:20 Kentaro Kikuchi (Shiga University)
"Quadratic Gaussian joint pricing model for stocks and bonds: Theory and application to Japanese data"

November 7 (Friday), 2014

(Chair: Chi Chung Siu)

10:00–10:40 Sebastian Jaimungal (University of Toronto)
"A mean field game approach to optimal execution"

10:40–11:20 Ryan Donnelly (École Polytechnique Fédéral de Lausanne)
"Volume imbalance and algorithmic trading"

(Chair: Mikhail Zhitlukhin)

13:00–13:40 Kazutoshi Yamazaki (Kansai University)
"Cash management and control band policies for spectrally one-sided Lévy processes"

13:40–14:20 Tadao Oryu (Kyoto University)
"An excursion-theoretic approach to regulator's bank reorganization problem"

14:40–15:00 Tsz King Chung (Tokyo Metropolitan University)
"Optimal timing for short covering of an illiquid security"

(Chair: Michael Tehranchi)

15:30–16:10 Chi Chung Siu (University of Technology, Sydney)
"A class of nonzero-sum stochastic differential investment and reinsurance games"

16:10–16:50 Daisuke Yoshikawa (Hokkai-Gakuen University)
"An equilibrium approach to indifference pricing with model risk"

16:50–17:30 Xue Cui (Tokyo Metropolitan University)
"The effect of reversibility on investment timing and quantity with random acquisition"

Closing Address

Takashi Shibata (Tokyo Metropolitan University)

CONTENTS

Moment Properties of Probability Distributions Used in Stochastic Financial Models

Jordan Stoyanov*

School of Mathematics & Statistics, Newcastle University, U.K.
stoyanovj@gmail.com

Our goal in this paper is to look at stochastic financial models and especially on those properties of the involved distributions which are expressed in terms of the moments. The questions discussed and the results presented reveal the role which the moments play in the analysis of distributions. Interesting conclusions can be derived in both cases when available are finitely many moments and when we know all moments.

Among the results included in the paper are sharp lower and/or upper bounds for option prices in terms of a finite number of moments. However the main attention is paid to the determinacy of distributions by their moments. While any light tailed distribution is uniquely determined by its moments, the uniqueness may fail for heavy tailed distributions. And, here is the point. Heavy tailed distributions are essentially involved in stock market modelling, and most of them are non-unique in terms of the moments. This is why the phenomenon *non-uniqueness of distributions* deserves a special attention.

We treat distributions on the positive half-line used to describe, e.g., stock prices and option prices, and distributions on the whole real line describing log-returns. For reader's convenience we give a brief and unified general picture of existing results about uniqueness and non-uniqueness of distribution in terms of their moments. Some of the results are classical and well-known. In several cases we provide here new arguments along with presenting new recent results on the moment determinacy of random variables and their non-linear transformations and also of stochastic processes which are solutions to SDEs.

Key words: probability distributions, moments, uniqueness, non-uniqueness, Cramér's condition, Carleman's condition, Krein's condition, Hardy's condition, Lin's condition, Stieltjes class, rate of growth of moments, option price bounds, Black-Scholes model, stochastic differential equations, stationarity

*The final version of this paper was completed after I moved recently to Ljubljana, Slovenia. I would like to thank the Mathematics Department, University of Ljubljana for the hospitality and the provided status of Guest-Professor.

1. Introduction

The evolution of stock prices, derivative prices, log-returns, etc., is diverse, dynamic and random. This is why for modelling we use a variety of stochastic processes with discrete or continuous time, with trajectories which are continuous or have jumps, and with specific distributional properties. The distributions are essential for describing financial instruments. In many cases the questions of interest and hence the answers are given in terms of the moments.

Notice, the moments are regarded as, and they indeed are, the simplest characteristics of a distribution. This is why we pay much attention to those properties of a distribution which are expressed in terms of the moments.

It is known and easy to imagine that, e.g., we cannot identify a distribution by knowing only a finite number of its moments. Hence the question is: Can we determine a distribution by knowing all moments? For many distributions the answer is "yes" and they are called *M-determinate*. However for quite popular distributions used in stochastic financial models the answer is "no". These are *M-indeterminate* distributions. The basics are summarized in Section 2.

It is well-known that the log-normal distribution is non-unique in terms of the moments. Why is this so, how to show the non-uniqueness and how this property affects conclusions involving the log-normal distribution? For reader's convenience we provide in Section 3 the necessary details.

In Section 4 we present a brief, systematic and unified picture of the most useful and workable conditions in the area of moment determinacy of probability distributions. Some of the results are classical and well-known, others are quite recent. We apply these tools in Section 5 to some distributions frequently used in stochastic modelling. We answer completely the question about the moment determinacy of all power transformations (usually called Box-Cox transformations) of the exponential and the normal distributions. New arguments are given to prove previously known statements.

The goal of Section 6 is to present interesting and non-trivial results which are based on knowing only a finite number of moments. The upper and/or lower bounds for option prices are important not only in the context of stochastic finance but in general. These are universal sharp bounds for probability distributions.

Moment determinacy of the solutions of stochastic differential equations are discussed in Sections 7 and 8. We describe examples of SDEs showing different kind of moment determinacy of their solutions. We also present a few new results providing explicit constructions of stochastic processes with prescribed moments.

The list of References, papers and books, is extensive. Most items are published over the last years, some are new and available only online. Useful comments are given at the end of the paper.

Readers who wish to become better familiar with the topic 'determinacy of probability distributions' may use the present paper as a gentle introduction. We have included the most important ideas, old and very new results, techniques and illustrations. Useful comments are provided in many places in the text. Outlined are challenging conjectures and open questions.

2. Basic Notions and Notations

We assume that all random variables and stochastic processes considered in this paper are defined on an underlying probability space $(\Omega, \mathcal{F}, \mathbf{P})$. For some purposes we may need this to be a 'usual filtered probability space'.

We use $\mathbf{E}[X]$ to denote the expectation of the random variable X with respect to the probability measure $\mathbf{P}$. Also we use $\tilde{\mathbf{E}}_t$ for the time-t expectation with respect to a martingale probability measure.

We write $X \sim F$, if the distribution function of X is F. We use f for the density of X, and of F, $f = F'$, if it exists. If X is such that $\mathbf{E}[|X|^k] < \infty$ for all $k = 1, 2, \ldots$, then $m_k = \mathbf{E}[X^k]$ is the *moment of order* k and $\{m_k, k = 1, 2, \ldots\}$ the *moment sequence* of X, and of F.

Clearly, the moments are simple characteristics of a distribution and it is remarkable that important properties of F and X can be derived or expressed only in terms of the moments. Notice, the moment sequence $\{m_k, k = 1, 2, \ldots\}$ is a discrete and countable object. This is just a sequence of numbers. If X is a positive random variable, i.e. its values are in $\mathbb{R}_+ = [0, \infty)$, the moments of X increase and tend to infinity. (Exceptions are random variables with values in $[0, 1]$, which is an easy case.) If X takes values in $\mathbb{R} = (-\infty, \infty)$, then the even order moments increase and tend to infinity.

It is non-trivial that based on such a discrete information, the moment sequence $\{m_k,\ k = 1, 2, \ldots\}$, we can derive essential properties or even exactly identify the unknown distribution $F(x),\ x \in \mathbb{R}$. Notice, F is obviously an infinite and uncountable object. (In Functional Analysis, this is equivalent to the non-trivial problem of finding exactly an unknown function f (density or arbitrary function) by knowing the values at zero of all derivatives of $\hat{f}$, the Fourier transform of f.)

If the distribution function F is the only one with the moment sequence $\{m_k\}$, we say that F is ***M-determinate***, or that F is uniquely determined by its moments. Otherwise, F is ***M-indeterminate***, or non-unique in terms of the moments. In the latter case there must be at least one distribution function, say G, such that G has the same moments as F, however $G \neq F$. There is a deep analytical and non-trivial result which can be extracted from works by C. Berg and collaborators; for references, see, e.g., [89].

Fundamental Theorem. *Suppose that F is M-indeterminate. Then there are infinitely many absolutely continuous distributions and infinitely many purely discrete distributions all having the same moments as F.*

Examples 3.2 and 7.3 are a partial but good illustration of this result.

It is well-known that there are conditions which are necessary and sufficient for uniqueness of a distribution by its moments. These are important theoretical results. However, from practical point of view they are uncheckable and hence not so useful. Thus a great attention was paid to find conditions which are only sufficient, or only necessary, but which are easy to check. In Section 3 we give a compact summary of what is available nowadays.

The moment problem has a specific name depending on the support of the distribution, supp(F), or the range of values of the random variable:

supp(F): $[0, 1]$ (*Hausdorff moment problem*);
$\mathbb{R}_+ = [0, \infty)$ (*Stieltjes moment problem*);
$\mathbb{R} = (-\infty, \infty)$ (*Hamburger moment problem*).

Most important for the M-(in)determinacy is the asymptotic behavior of its tail(s). We will be also looking at another property of a distribution, the infinite divisibility. Most of the distributions used in stochastic financial models are heavy tailed and infinitely divisible. The interplay between these two properties is quite delicate. Some illustrations are given in the next sections.

What to say about distributions used in financial models? It looks we have a lot of 'freedom'. And indeed a variety of distributions are involved. Since stock prices and derivatives prices are positive, we use for them random variables or stochastic processes with values in $\mathbb{R}_+$. Respectively, the log-returns are random variables with values in $\mathbb{R}$. The specific choice of a model depends on characteristics observed for the financial phenomenon we want to study. The models themselves are always interesting because their study brings to us new and challenging theoretical questions. However important is that some models, as intended, are 'good' for describing adequately aspects of the reality in the financial world.

3. First Illustrations

We start with the normal distribution $\mathcal{N}$ and the log-normal distribution $Log\,\mathcal{N}$ both widely used in stochastic financial modelling. Consider two random variables,

$$Z \sim \mathcal{N} \text{ and } S \sim Log\,\mathcal{N}.$$

We are interested in the M-determinacy of Z, S and their non-linear transformations. Look first at something classic.

Example 3.1. There are two coefficients, β_1 and β_2, *skewness* and *kurtosis*, introduced for any random variable $X \sim F$ with finite the first four moments and defined as follows:

$$\beta_1 = \frac{\mu_3}{\sigma^3}, \quad \beta_2 = \frac{\mu_4}{\sigma^4}.$$

Here $\sigma^2 = \mathbf{E}[(\xi - m_1)^2]$ and $\mu_j = \mathbf{E}[(\xi - m_1)^j]$ for $j = 3, 4;\ m_1 = \mathbf{E}[\xi]$. The role of these two coefficients when 'describing' the shape of a distribution is well-known in statistical theory and practice.

For a random variable $Z \sim \mathcal{N}(0, 1)$, the first four moments are $0, 1, 0, 3$, hence Z has skewness $\beta_1 = 0$ and kurtosis $\beta_2 = 3$.

Suppose that from a large data set you have calculated the empirical moments of order up to four and then found that $\hat{\beta}_1 \approx 0$ and $\hat{\beta}_2 \approx 3$. Are you willing to accept that your data come from a normal distribution?

It is known that the answer is 'no'. It is useful and instructive, however, to have a couple of examples. Indeed, consider a purely discrete random variable Y:

$$\mathbf{P}[Y = \pm\sqrt{3}] = \frac{1}{6}, \ \mathbf{P}[X = 0] = \frac{2}{3}.$$

It is easy to see that the first four moments of Y are $0, 1, 0, 3$, the same as the first four moments of Z. (Even more, Z and Y have coinciding all odd order moments, they are just all equal to zero.) Hence Y has skewness 0 and kurtosis 3.

There are purely discrete random variables with infinitely many values in $\mathbb{R}$, not just three as above, such that the skewness is 0 and the kurtosis is 3.

The next illustration involves absolutely continuous distributions. Consider the family $DG(a)$ of double-gamma distributions with parameter a, $a > 0$. A random variable $Y \sim DG(a)$ if its density is

$$g(x) = \frac{1}{2\Gamma(a)}|x|^{a-1}\mathrm{e}^{-|x|}, \ x \in \mathbb{R}.$$

For a particular value of a, $a = \frac{1}{2}(\sqrt{13}+1)$, we find that $\beta_1 = 0$ and $\beta_2 = 3$.

Hence there are infinitely many distributions, continuous and discrete, all different from the normal, and all having the same skewness 0 and kurtosis 3.

After the above we may think that if knowing all moments we are in a better position allowing us to identify uniquely a distribution. In general, this is not the case as shown in the next example involving the log-normal distribution.

Example 3.2. The density f of the random variable $X \sim Log\,\mathcal{N}(0,1)$ is

$$\text{(1)} \qquad f(x) = \frac{1}{\sqrt{2\pi}}\frac{1}{x}\exp\left[-\frac{1}{2}(\ln x)^2\right], \ x > 0; \ \ f(x) = 0, \ x \leq 0.$$

It can be shown that all positive integer order moments of X are finite and

$$m_k = \mathbf{E}[X^k] = \mathrm{e}^{k^2/2}, \ k = 1, 2, \ldots.$$

Interestingly, the order k can also be any negative integer, and we have the relation $\mathbf{E}[X^k] = \mathbf{E}[X^{-k}]$. This follows from the fact that X^{-1} is a well-defined random variable with the same $Log\,\mathcal{N}$ distribution as X.

Let us define two infinite sets of random variables, call them *Stieltjes classes*:

$$\mathbf{S}_c = \{X_\varepsilon, \ \varepsilon \in [-1,1]\}, \ X_\varepsilon \sim f_\varepsilon, \ f_\varepsilon(x) = f(x)\,[1 + \varepsilon\sin(2\pi\ln x)], \ x \in \mathbb{R};$$

$$\mathbf{S}_d = \{Y_a, \ a > 0\}, \ \mathbf{P}[Y_a = a\mathrm{e}^n] = a^{-n}\mathrm{e}^{-n^2/2}/A, \ n = 0, \pm 1, \pm 2, \ldots.$$

Here A is the normalizing constant. In the notations $\mathbf{S}_c$ and $\mathbf{S}_d$, $\mathbf{S}$ stands for Stieltjes, c for continuous and d for discrete.

Amazing property: With $X_0 = X \sim Log\,\mathcal{N}(0,1)$, for any $\varepsilon \in [-1,1]$ and any $a > 0$, the following relations hold:

$$\mathbf{E}[X_\varepsilon^k] = \mathbf{E}[Y_a^k] = \mathbf{E}[X^k] = \mathrm{e}^{k^2/2}, \quad \text{for } k = 1, 2, \ldots .$$

Conclusion: The log-normal distribution is M-indeterminate! We see explicitly that there are so 'many' other distributions all with the same moments. Notice, the family $\mathbf{S}_c$ consists of *absolutely continuous* variables, while $\mathbf{S}_d$ consists of *purely discrete* variables. Then, obviously, we can take mixtures.

The above explicit constructions are two different ways to show that $Log\,\mathcal{N}$ is M-indeterminate. There are at least four other methods.

Another immediate conclusion is that in the Black-Scholes model, the stock price S_t being log-normally distributed, is M-indeterminate for any $t \in (0, T]$.

Two other important properties are that $Log\,\mathcal{N}$ is unimodal and infinitely divisible. It is easy to see that in the Stieltjes class $\mathbf{S}_c$ above, for any $\varepsilon \neq 0$, f_ε has infinitely many modes.

Conjecture 3.1. It is known that for $\varepsilon = 1$, X_1 is not infinitely divisible. We conjecture that for any $\varepsilon \in (-1, 1)$ and $\varepsilon \neq 0$, f_ε is not infinitely divisible.

Open Question 3.1. Suppose that F is an absolutely continuous distribution function on $\mathbb{R}_+$ with all moments finite. Let the following two 'properties' of F hold: (i) $m_k(F) = \int_0^\infty x^k\,\mathrm{d}F(x) = \mathrm{e}^{k^2/2}$, $k = 1, 2, \ldots$; (ii) F is unimodal. Prove that $F = Log\,\mathcal{N}$. Otherwise give a counterexample.

It is worth mentioning that of interest is also a class, denoted by $LSN(\lambda)$, and called *logarithmic skew-normal distributions*. Here λ can be any real number. We say that $X \sim LSN(\lambda)$ if its density is

$$f_\lambda(x) = \varphi(x)\,\Phi(\lambda x), \; x \in \mathbb{R},$$

where φ and Φ are the standard normal density and distribution function. If $\lambda = 0$, we get the usual $Log\,\mathcal{N}$ distribution. The class $LSN(\lambda)$ is used in stochastic modelling, it is wider than $Log\,\mathcal{N}$. However, any random variable $X_\lambda \sim LSN(\lambda)$ is M-indeterminate. Available are also explicit Stieltjes classes; see [66].

4. Classical and Recent Conditions

For a reader's convenience we give below in a compact form essential and workable conditions for either uniqueness or non-uniqueness of a distribution in terms of the moments. Historically these conditions appeared at a different time and in a different form. Sometime they were 'ready' to be used, sometime needed to be extracted from the original purely analytical forms and reformulated and given in the language of modern probability and statistics.

4.1 Cramér's condition

We start with a classical result which appeared in the first decades of 20th century and described by H. Cramér in the middle of 30-ties.

Let X be a random variable with values in $\mathbb{R}$ such that its moment generating function exists, i.e., for some $t_0 > 0$,

$$M(t) = \mathbf{E}[e^{tX}] < \infty \quad \text{for all} \quad t \in (-t_0, t_0) \quad (\textbf{\textit{Cramér's condition}}).$$

In such a case we say that the distribution of X has *light tails*, or that X has an exponential moment.

Statement 4.1. *Let Cramér's condition hold for the random variable X. Then all moments of positive integer order of X are finite and X is M-determinate.*

An easy consequence is that any random variable with values in a bounded interval is M-determinate. This is the Hausdorff case.

If there is no moment generating function, $M(t) = \mathbf{E}[e^{tX}] = \infty$ for all $t \neq 0$, we say that the distribution of X has *heavy tails.* Distributions such as Pareto and Student, frequently used in financial modelling, not only do not have moment generating functions, but they are 'very heavy', only a finite number of their moments are finite. If, however, a heavy tailed distribution has all moments finite, it is then reasonable to ask about its moment determinacy. There are two possibilities, either X is M-determinate, or it is M-indeterminate.

Note that Cramér's condition is the strongest sufficient condition for uniqueness. There are heavy tailed distributions, i.e., no Cramér's condition, which are M-determinate. Later we illustrate this.

4.2 Hardy's condition

The next condition, applicable to positive random variables, is based on two old papers by G. H. Hardy published in 1917/1918. In its precise modern form, as given below, it appeared explicitly in the literature only recently; see [94].

Statement 4.2. *Consider a random variable $X > 0$, $X \sim F$, where F is an arbitrary distribution function. Assume that $\sqrt{X}$ has a moment generating function (equivalently, Cramér's condition holds for $\sqrt{X}$), namely: for some $t_0 > 0$,*

$$\mathbf{E}[e^{t\sqrt{X}}] < \infty \quad \textit{for all} \quad t \in [0, t_0) \qquad (\textbf{\textit{Hardy's condition}}).$$

Then X has all moments finite, i.e. $m_k = \mathbf{E}[X^k] < \infty$, $k = 1, 2, \ldots$, and moreover, F is the only distribution function with the moment sequence $\{m_k, k = 1, 2, \ldots\}$.

This condition is sufficient for uniqueness. It is weaker than Cramér's condition. In fact, we have the following general statement: If X is a random variable with values in $\mathbb{R}_+$ of $\mathbb{R}$, and Cramér's condition holds for X, then X^2 is

M-determinate. Notice, the condition is on X, and the conclusion is for its square X^2.

4.3 Carleman's condition

Consider a random variable X with distribution function F and denote its moment sequence by $\{m_k,\ k \in \mathbb{N}\}$. If the range of values of X, i.e. supp(F), is bounded, X is M-determinate by Cramér's condition. Thus, assume that supp(F) is unbounded. In the Hamburger case supp(F) $= \mathbb{R}$ or is its subset, while in the Stieltjes case supp(F) $= \mathbb{R}_+$ or is its subset. We use the moments $\{m_k\}$ to define an infinite series, called usually *Carleman's series*:

$$\mathrm{C} = \begin{cases} \sum_{k=1}^{\infty} \frac{1}{(m_{2k})^{1/2k}} & \text{(Hamburger case)}, \\ \sum_{k=1}^{\infty} \frac{1}{(m_k)^{1/2k}} & \text{(Stieltjes case)}. \end{cases}$$

Statement 4.3a. *The condition* $\mathrm{C} = \infty$, *called* ***Carleman's condition***, *implies that X, and also F, is M-determinate.*

This result, established by T. J. Carleman in 1926, became a very useful tool, 'easy' to apply and as soon as $\mathrm{C} = \infty$, to conclude that a distribution is M-determinate. Carleman's condition uses explicitly all moments. Since important is the divergence of the Carleman's series, this can be derived by knowing only the asymptotic behavior of the moments m_k for large k. Sometime this small 'trick' helps to establish the determinacy property.

Note, however, that Carleman's condition is only sufficient for the uniqueness. If a distribution is M-indeterminate, then necessarily $\mathrm{C} < \infty$. There are M-determinate distributions with Carleman's series $\mathrm{C} < \infty$; see [93].

We need also a result which is converse to Statement 4.3. For details see [81].

Statement 4.3b. *Suppose that Carleman's condition is not satisfied:* $C < \infty$. *Assume that F is absolutely continuous and there is* $x_0 > 0$ *such that its density* $f = F'$ *is positive for* $x > x_0$ *and the function* $-\ln f(e^x)$ *is convex for* $y > y_0$, *where* $y_0 = \ln x_0$. *This condition together with* $C < \infty$ *imply that X is M-indeterminate.*

4.4 Krein's condition

Suppose now that X is an absolutely continuous random variable with distribution function G, its density is $g = G'$, and that the moments of all orders are finite. Another assumption is that X takes values in an unbounded domain, and that in this domain $g(x) > 0$. In the Hamburger case the domain is $\mathbb{R}$. In the Stieltjes case the domain is $\mathbb{R}_+$ if g is bounded on $\mathbb{R}_+$; however, if g is unbounded near zero, we take (x_0, ∞), for some $x_0 > 0$, to avoid the 'singularity' of g. We define the following normalized logarithmic integral (called also *Krein's integral*):

$$\mathrm{K}[g] = \begin{cases} \int_{-\infty}^{\infty} \frac{-\ln g(y)}{1+y^2}\mathrm{d}y & \text{(Hamburger case)}, \\ \int_{x_0}^{\infty} \frac{-\ln g(y^2)}{1+y^2}\mathrm{d}y & \text{(Stieltjes case)}. \end{cases}$$

Statement 4.4a. *The condition* $\mathrm{K}[g] < \infty$, *called* ***Krein's condition***, *implies that* X *is M-indeterminate.*

Krein's condition is only sufficient for non-uniqueness of a distribution. If a distribution is M-determinate, and its density is positive, then necessarily the Krein's integral is divergent. Statement 4.4a in the Hamburger case is one of the famous results obtained by M. G. Krein in 1944. Both cases, Hamburger and Stieltjes, were further studied and extended in works by N. I. Akhiezer, H. Pedersen, E. Slud, G. D. Lin and A. G. Pakes; details can be found in [65, 81, 89, 93].

We also need a result which is converse to Statement 4.4a.; see [65].

Statement 4.4b. *Suppose that the density* g *of the random variable* X *is as above and that Krein's integral is divergent:* $\mathrm{K}[g] = \infty$. *In the Hamburger case we require* g *to be symmetric. Let in both Hamburger and Stieltjes cases the following* ***Lin's condition*** *be satisfied: for some* $x_0 > 0$, g *is differentiable for* $x > x_0$ *and the ratio* $\mathrm{L}(x) := -x\,g'(x)/g(x)$ *is ultimately monotone increasing and tending to infinity as* $x \to \infty$. *The two conditions,* $\mathrm{K}[g] = \infty$ *and Lin's condition, imply that* X *is M-determinate.*

4.5 Explicit Stieltjes classes

The way we illustrated in Section 3 the non-uniqueness of the log-normal distribution gives a clear idea of the term *Stieltjes class*. Look at the sets $\mathbf{S}_c$ and $\mathbf{S}_d$. Each one is a parametrized infinite family of different random variables, hence different distributions, all having the same moments as one initial distribution. The idea of constructing such a family goes back to T. J. Stieltjes (1894), who was the first to describe explicit distributions supported by the positive half-line which are M-indeterminate.

We deal, as in [91], with an absolutely continuous random variable X with distribution function G and density g assuming that all moments are finite.

Suppose that we have found a function $h = (h(x),\ x \in \mathbb{R})$, a sign function, such that $|h(x)| \leq 1$ for all x and the product $g(x)h(x),\ x \in \mathbb{R}$, has 'vanishing moments' in the sense that $\int x^k g(x)h(x)\,\mathrm{d}x = 0,\ k = 0, 1, 2, \ldots$. The function h is called a *perturbation* and the ***Stieltjes class*** based on the density g and the perturbation h is defined as follows:

$$\mathbf{S} = \mathbf{S}(g, h) = \{g_\varepsilon(x) : \ g_\varepsilon(x) = g(x)[1 + \varepsilon\, h(x)];\ x \in \mathbb{R},\ \varepsilon \in [-1, 1]\}.$$

Statement 4.5. *For any* $\varepsilon \in [-1, 1]$, *the function* g_ε *is a density, hence there is a random variable, say* X_ε *such that* $X_\varepsilon \sim G_\varepsilon$, *where* $G'_\varepsilon = g_\varepsilon$. *While the distributions in* $\mathbf{S}$ *are obviously different, all moments of* X_ε *are the same as those of* $X = X_0$:

$$\mathbf{E}[X_\varepsilon^k] = \mathbf{E}[X^k] \ \textit{for all } k = 1, 2, \ldots \textit{ and any } \varepsilon \in [-1, 1].$$

To have written an explicit Stieltjes class is equivalent to the statement that all members in this class are M-indeterminate. However, to construct a Stieltjes class is a delicate analytical problem. If we are successful, this would confirm the claim that we deal with distributions which are M-indeterminate.

The distribution function G, also its density g we started with, play the role of a center in the class **S**. Important is that all these distributions are moment equivalent. Clearly, we can use different perturbations h and the same density g. The most important is the 'vanishing moments' property for the product hg. This property is equivalent to the fact that given g, we are looking for a function h which in the Hilbert space $L_2[g]$ (g is the 'weight function') is orthogonal to the set of all polynomials. If there is no such h, the Stieltjes class is trivial, it consists of only one element, G itself, then G is M-determinate. Further details about Stieltjes classes can be found in [91, 82, 80, 79].

4.6 Rate of growth of the moments

Intuitively, the tails of a distribution are related to the growth of its moments. If the tails are heavier, the moments grow 'faster'. We will see that there is a critical boundary of growing of the moments which separates the distributions into two groups, M-determinate and M-indeterminate. We treat separately positive random variables and variables with values in $\mathbb{R}$. We provide a little more details since the results are recent and available only online; see [67, 95, 68].

Stieltjes case. For a random variable X with values in $\mathbb{R}_+$ and moments $m_k = \mathbf{E}[X^k]$, $k = 1, 2, \ldots$, we consider the ratio of two consecutive moments:

$$\Delta_k = \frac{m_{k+1}}{m_k}.$$

We can easily check that Δ_k increases in k, $k \geq k_0$ for some fixed integer $k_0 \geq 1$. Suppose that there is a number $\gamma > 0$ such that

$$\Delta_k = O((k+1)^\gamma) \text{ as } k \to \infty.$$

The number γ is called a ***rate of growth of the moments*** of X.

Statement 4.6a. *If the rate $\gamma \leq 2$, then X is M-determinate.*

The natural question to ask is about the role, or sharpness of the value $\gamma = 2$. The answer is given by the next result.

Statement 4.6b. *The value $\gamma = 2$ is the best possible constant for which X is M-determinate. There is a positive random variable, Y, with finite moments such that $\Delta_k = O((k+1)^{2+\delta})$ for some $\delta > 0$ and Y is M-indeterminate.*

The question now is about the moment determinacy of X if the rate is $\gamma > 2$. We may suggest that such an X is M-indeterminate. We need however an additional condition and here is one possible answer.

Statement 4.6c. *Suppose Y is a positive random variable with finite moments and its rate of growth of the moments is $\gamma > 2$. Suppose in addition that Y is absolutely continuous with smooth density g such that Lin's condition is satisfied: for some $y_0 > 0$, the ratio $-yg'(y)/g(y)$, $y > y_0$ is monotone increasing and tending to infinity as $y \to \infty$. Then Y is M-indeterminate.*

Hamburger case. Similarly to the above, we can study random variables with values in $\mathbb{R}$ and finite moments. We work now with the even order moments. First we define the rate of growth γ as follows:

$$\Delta_{2(k+1)} = \frac{m_{2(k+1)}}{m_{2k}} = O((k+1)^{\gamma}) \text{ as } k \to \infty.$$

We can formulate three statements which are analogous to those in the Stieltjes case. We do not give details here and leave this task to the reader with a reference to another recent paper; see [95].

Remark 4.1. The conditions and the results described above provide us with diverse tools allowing, in principle, to analyze the moment determinacy of any distribution. However there are cases when to find the answer is not so straightforward. It is always useful to work with specific distributions frequently used in statistical theory and practice.

Remark 4.2. Relevant questions and results can be found in many papers, see, e.g., [39, 40, 41, 52, 72, 77, 84, 92, 96].

5. Details about the Exponential and the Normal Distributions

Let us use the knowledge from Section 4 and give all details about the moment determinacy of two random variables, $X \sim Exp(1)$ and $Z \sim \mathcal{N}(0, 1)$, and also of their power transformations.

5.1 Exponential distribution

The random variable $X \sim Exp(1)$ has density e^{-x}, $x > 0$ and moments $m_k(X) = k!$, $k = 1, 2, \ldots$. Since X satisfies Cramér's condition, X is M-determinate. Another way is to check that Carleman's condition holds, hence again X is M-determinate. The third proof uses the Krein-Lin result, Statement 4.4b. The forth proof is to apply Statement 4.6a and see that X has a rate of growth of the moments $\gamma = 1$.

What about X^2, X^3, or any power, X^r, for real r? In each of these cases we can write explicitly the density and use it in the analysis. The answer is well-known: X^2 is heavy tailed, but is M-determinate. This is true because, e.g., its moments $m_k(X^2) = (2k)!$, $k = 1, 2, \ldots$ satisfy Carleman's condition. We can also refer to Hardy's condition: X satisfies Cramér's condition, hence X^2 is M-determinate. At least two more ways can be suggested to prove the claim.

In general, X^r is M-determinate for any real $r \in [0, 2]$ and M-indeterminate for all $r > 2$. In particular, X^3 is M-indeterminate, i.e. $r = 3$ is the smallest integer power such that X^r obeys this property. The moments are $m_k(X^3) = (3k)!$, $k = 1, 2, \ldots$, they grow 'very fast'. Since the density f of X^3 is

$$f(x) = \frac{1}{3}\,x^{-2/3}\,\mathrm{e}^{-x^{1/3}},\ x > 0,$$

Krein's condition (in the Stieltjes case) is satisfied, so indeed X^3 is M-indeterminate.

We can make one step more. Namely, to write an explicit *Stieltjes class* for the cube $Y = X^3$ of $X \sim Exp(1)$:

$$\mathbf{S}(f,h) = \{f_\varepsilon = f[1 + \varepsilon h], \varepsilon \in [-1,1]\}.$$

Here f is the density of X^3 and the perturbation is $h(x) = \sin\left(\frac{\pi}{6} - \sqrt{3}x^{1/3}\right)$, $x > 0$.

Notice, in $\mathbf{S}(f,h)$, the function f_ε, for each $\varepsilon \in [-1,1]$, is a density of a random variable, say Y_ε, and we have $\mathbf{E}[Y_\varepsilon^k] = m_k = (3k)!$, $k = 1, 2, \ldots$. These are the same as the moments of X^3.

Open Question 5.1. Find a discrete random variable with values in $\mathbb{R}_+$ such that its moments are $\{(3k)!,\ k = 1, 2, \ldots\}$.

The Fundamental Theorem, see Section 2, states only the existence of such distributions. We do not know any, yet.

Conjecture 5.1. While it is known that $Y = X^3$ for $X \sim Exp$ is infinitely divisible, we conjecture here that for any $\varepsilon \neq 0$ in the above Stieltjes class, the random variable Y_ε is not infinitely divisible.

Note finally that the density of X^3 is unimodal, while the density f_ε of Y_ε has infinitely many modes for any $\varepsilon \neq 0$.

6. Normal Distribution

What can we say about the M-determinacy of a random variable $Z \sim \mathcal{N}(0,1)$ and then of its powers Z^2, Z^3, Z^4 and $|Z|^r$ for arbitrary real r?

The results given in Section 4 allow to find the answer. Let us just list what is true: Z and Z^2 are M-determinate (by Cramér's condition, and/or Hardy's condition, or to use the existence of the moment generating function of $Z^2 = \chi_1^2$, or the rate of growth of the moments).

More interesting is the next observation. Since Z^2 has light tail, this means that Z^4 satisfies Hardy's condition, so Z^4 is M-determinate. Notice, to use twice Cramér's condition and twice Hardy's condition is the shortest and most beautiful way to show that Z^4, where $Z \sim \mathcal{N}$, is M-determinate.

However, Z^3 turns to be M-indeterminate. Once again: the M-indeterminate Z^3 is 'between' the M-determinate Z^2 and Z^4. Even more surprising is that the absolute value $|Z^3|$ is M-determinate. Each of these can be proved in several ways. We see the 'small' difference: the cube Z^3 is a Hamburger case, there are two heavy tails, while $|Z^3|$ is a Stieltjes case with only one heavy tail.

The next fact to mention here is that for any real $r > 4$, $|Z|^r$ is M-indeterminate. Just write the density of $|Z|^r$, this is a Stieltjes case, and apply Krein's condition. Moreover, available are explicit Stieltjes classes for Z^3 and $|Z|^r$, $r > 4$.

Remark 5.2. We can write explicit Stieltjes classes for Z^3 and state an open question and a conjecture as done for the exponential distribution, see the previous sub-section. Similarly for $|Z|^r$, $r > 4$.

7. Relations between Stock Prices and Option Prices

The M-indeterminacy of the log-normal distribution is a fact implying that we have to be careful when studying models of Black-Scholes type and using the moments. Besides the geometric Brownian motion, several other heavy tailed distributions, including stable distributions, are essentially used to model different market ingredients. And, not surprisingly, most of them are M-indeterminate.

Any financial model is based on objects with specific distributions. Some of the questions we ask are in terms of the moments. A standard assumption is that there is no arbitrage opportunity in the market. So all expectations of random quantities will be taken with respect to a martingale probability measure.

Let us write a short list of possible and reasonable questions arising when studying any stochastic financial model. Typically, there is one or more underlying assets in such model which is described by a stochastic process S_t, $t \in [0, T]$.

• Direct question: In the pricing of an European contingent claim, there is a payoff f that is payable at the claim's maturity T. In this context, what are the best possible bounds for the price of the claim based on knowing the first n moments of the stock price at maturity?

• Converse question: We do not know the stock price and available to us are observable option prices. Can we derive conclusions about the stock price dynamics?

• Given observable option prices, can we find bounds for the prices of other derivatives on the same stock?

• How to answer the previous questions if there are transaction costs?

• Can we price options by knowing all moments of the stock price at maturity?

The specific results given below are variations of results due to several authors. The results are spread in the literature; see, e.g., [5, 6, 15, 21, 22, 44, 46, 50, 61, 62, 63, 71] and more references therein.

7.1 Upper bound for call option price based on two moments

Suppose that the stock price process S_t, $0 \le t \le T$ is quite 'general' and such that we do not known its distributions including the distribution at maturity T. We deal with an European call option with strike K. We assume that in the market the bank interest is $r = 0$ and there is no arbitrage, so if Q is a martingale probability measure, we write just $\mathbf{E}$ instead of $\mathbf{E}_Q$. The only information available to the traders is that at maturity T the random variable S_T has finite second moment and explicitly known are the first two moments

$$m_1 = \mathbf{E}[S_T] \text{ and } m_2 = \mathbf{E}[S_T^2].$$

Since the fair price of the European call is $C(K) := \mathbf{E}[(S_T - K)^+]$, the question of interest is: *What is the maximal value of* $C(K)$*?*

Equivalently, what is the maximum value of the expectation $\mathbf{E}[(S_T - K)^+]$ taken over all random variables S_T with fixed the first two moments m_1 and m_2. Clearly, such an upper bound will depend on K, m_1, m_2.

Statement 6.1. *Under the above conditions we have the following:*

$$\mathbf{E}[(S_T - K)^+] \le \begin{cases} \frac{1}{2}\left(m_1 - K + \sqrt{m_2 - 2m_1K + K^2}\right), & \text{if } K \ge m_2/(2\,m_1), \\ m_1 - Km_1^2/m_2, & \text{otherwise.} \end{cases}$$

The remarkable feature is that this bound is valid for all square integrable stock price processes as soon as fixed are the first two moments of S_T.

7.2 Other sharp bounds based on higher order moments

Assume now that the interest rate is $r > 0$ and let the time-t expected payoff of the European call be denoted by $\tilde{\mathbf{E}}_t[(S_T - K)^+]$; the time-$t$ neutral option price will be equal to $\mathrm{e}^{-r(T-t)}\mathbf{E}[(S_T - K)^+]$.

Notice first that we can write an upper bound for $\tilde{\mathbf{E}}_t[(S_T - K)^+]$, similar to the above one in Statement 6.1. We only need instead of m_1 and m_2 to use

$$m_1(t) = \tilde{\mathbf{E}}_t[S_T] \text{ and } m_2(t) = \tilde{\mathbf{E}}_t[S_T^2].$$

Statement 6.2a. *Let for some* $p \ge 1$ *the pth order moment of* S_T *be finite. Then*

$$\tilde{\mathbf{E}}_t[(S_T - K)^+] \le \begin{cases} (\mathbf{E}[S_T^p])^{1/p} - K, & \text{if } K \le \frac{p-1}{p}(\mathbf{E}[S_T^p])^{1/p}, \\ (\mathbf{E}[S_T^p])^{1/p}\,\frac{1}{p}\,(\frac{p-1}{pK})^{p-1}, & \text{otherwise.} \end{cases}$$

In the next result we provide a two-sided bound for the time-t option price. We use the quantity $\sigma(t)$, where $\sigma^2(t) = m_2(t) - m_1^2(t)$, the time-$t$ variance of S_T.

Statement 6.2b. *Suppose we know the time-t tail probability* $c_t := \mathbf{P}_t[S_T > K]$. *Then the following two-sided relation holds:*

$$(m_1(t) - K)\,c_t \le \tilde{\mathbf{E}}_t[(S_T - K)^+] \le (m_1(t) - K)\,c_t + \sigma_t\,\sqrt{c_t - c_t^2}.$$

Remark 6.2. It is important to mention that the bounds in Statements 6.2a and 6.2b are sharp. There are specific distributions for which equality is attained in each case. Hence these statements and other once available in the literature provide *universal bounds* for the expectation of truncated random variables. See the references mentioned before.

Clearly, instead of upper bounds we can look from the very beginning for the lower bound $\min \mathbf{E}[(S_T - K)^+]$ for the European call option. The minimum is taken over all random variables S_T with fixed the first two moments (or, which is the same, with fixed the first moment and the variance). Similarly, we can find upper bound or lower bound of European put option, written on the same stock and having the same strike and maturity. Available are bounds for European type of options at maturity T or at a current time $t < T$.

In most cases, the payoff of options of European type is a *piece-wise linear function*. This is the property allowing to use, e.g., linear programming techniques.

Available in the literature are diverse bounds when known about S_T are, e.g., the first three moments, m_1, m_2, m_3, or the first n moments, $m_1, \ldots, m_n$. More delicate are the non-linear problems arising in portfolio selection. The goal is for a non-linear function $u(\cdot)$ to optimize $\mathbf{E}_Q[u(X)]$ over all random variables with fixed two, three or larger number of moments. Useful references are [6, 33, 61, 62, 63].

7.3 Bounds for call and put option prices

Suppose that the market is as before: the distribution of the stock price S_T at maturity T is unknown, traded in the market are an European call option and European put option with the same strike K and maturity T. Either the interest rate is $r = 0$ or $r > 0$. Both cases are of interest and they are treated in the literature. The first and main step is to analyze models with interest rate $r = 0$. A standard assumption is that there is no arbitrage in the market, so all calculations are performed with respect to a martingale probability measure, the expectations are denoted by $\tilde{\mathbf{E}}$. The call and put option prices are

$$C(K) = \tilde{\mathbf{E}}[(S_T - K)^+] \quad \text{and} \quad P(K) = \tilde{\mathbf{E}}[(K - S_T)^+].$$

Statement 6.3. *Suppose that all positive and all negative order moments of the stock price S_T at maturity T are finite, i.e., $\mathbf{E}[S_T^p] < \infty$ and $\mathbf{E}[S_T^{-q}] < \infty$ for any real $p > 0$ and $q > 0$.*

(a) For any fixed p and all strikes $K > 0$, we have the call price upper bound:

$$C(K) \le \frac{B_0\, \mathbf{E}[S_T^{p+1}]}{p+1} \left(\frac{p}{p+1}\right)^p \frac{1}{K^p}.$$

(b) For any fixed q and all strikes K, the put price upper bound is

$$P(K) \le \frac{B_0\, \mathbf{E}[S_T^{-q}]}{q+1} \left(\frac{q}{q+1}\right)^q K^{q+1}.$$

Remark 6.3. We can use the put-call parity relation and in case (a) easily derive lower bound for $P(K)$ in terms of positive $(p+1)$th order moment of S_T, while in case (b) write a lower bound for $C(K)$ via qth order negative moment of S_T.

In most of the models considered in the literature, but not always, the stock price process at any time has all positive and all negative order moments finite.

This is true for any S_T satisfying Cramér's condition, i.e., for light tailed distributions. However this is true also for several heavy tailed distributions. Such an example is the geometric Brownian motion, in which case we have explicit expressions for the moments, positive and negative. This allows to find two-sided bounds for the call option price and for the put option price. We leave this to the reader.

The last comment here is about the role of the strike K in the above bounds. It turns out some bounds 'work' well for small K, others are good for large K.

The upper bound for $C(K)$, see (a), is 'useful' for large K. If $\mathbf{E}[S_T^{p+1}] < \infty$ for some $p > 0$, then $C(K) = O(K^{-p})$ as $K \to \infty$.

The upper bound for $P(K)$, see (b), is 'useful' for small K. If $\mathbf{E}[S_T^{-q}] < \infty$ for some $q > 0$, then $P(K) = O(K^{q+1})$ as $K \to 0$.

Take $p \downarrow 0$ in (a) to arrive at the familiar bound $C(K) \leq B_0\mathbf{E}[S_T]$.

Take $q \downarrow 0$ in (b), find another familiar bound $P(K) \leq B_0 K$.

8. Moment Determinacy of the Solutions of SDEs

In this section we discuss the moment determinacy property of a stochastic process $X = (X_t,\ t \in [0, T])$ which is the solution of the following Itô's type SDE:

$$\mathrm{d}X_t = a(t, X_t)\mathrm{d}t + \sigma(t, X_t)\mathrm{d}W_t,\ X_0,\ t \in [0, T], \ \text{ or } \ t \geq 0.$$

As usual $W = (W_t, t \geq 0)$ is a standard Brownian motion, the initial condition X_0 is either a constant or a random variable independent of W. Under some 'general conditions' on the drift $a(\cdot)$, the diffusion $\sigma^2(\cdot)$ and the initial position X_0, this SDE has a unique solution such that X_t, at any time t, has all moments finite. Since we are basically interested in the moments, it is enough for us to consider uniqueness in a weak sense of the solutions of SDEs.

The initial value X_0 does play a role for the finiteness of the moments of X_t for $t > 0$. It is easy to take $X_0 = \mathit{const}$. If, however, X_0 is a random variable, we assume that all its moments are finite and that X_0 is M-determinate. If X_0 is M-indeterminate, this property will be inherited by the solution X_t, $t > 0$, which case does not seem interesting. Thus in what follows we work with the drift and the diffusion coefficients and if necessary, we specify X_0.

Question 7.1. When are the one-dimensional and the finite-dimensional distributions of the process X uniquely determined by their marginal or multi-indexed moments?

It may happen that a SDE has a unique weak solution which, however, is non-unique in terms of the moments. This should not be surprising since these are quite different properties.

According to a recent result in [56], if all marginal one-dimensional distributions of the process X are M-determinate, then any finite-dimensional distribution of X is M-determinate. Hence enough is to give details just for the random variable X_t, where $t > 0$ is fixed. We assume generally that X_0, $a(\cdot)$, and $\sigma^2(\cdot)$ are 'nice' in the sense that there is a unique weak solution of the corresponding SDE.

Let us exhibit four examples covering a range of SDEs with different moment determinacy and different integrability property.

Example 7.1. If $a(\cdot)$ and $\sigma^2(\cdot)$ are 'nice' and $\sigma^2(\cdot)$ is uniformly bounded, then X_t is M-determinate, X_t^2 is M-determinate, however X_t^3 is M-indeterminate. We note first that the moment properties of the solution X_t in this case are the 'same' as those of the standard Brownian motion W_t. Then we easily derive the above determinacy from the properties of the powers of a normally distributed random variable, see Section 5. The same moment determinacy properties for X_t, X_t^2 and X_t^3 are valid for any Gaussian process. In particular, such statements hold for Ornstein-Uhlenbeck type processes (particular linear SDEs)

$$\mathrm{d}X_t = (\alpha_0 + \alpha_1 X_t)\mathrm{d}t + \sigma \,\mathrm{d}W_t,$$

where $\alpha_0, \alpha_1, \sigma$ are constants.

Another case, which is more general, is instead to require $\sigma^2(\cdot)$ to be uniformly bounded, to assume that $a(\cdot)$ and $\sigma^2(\cdot)$ are such that X_t satisfies Cramér's condition, i.e., X_t to have an exponential moment. Then X_t is M-determinate. Moreover, if $a(\cdot)$ and $\sigma^2(\cdot)$ are such that the random variable $X_T^* = \sup_{0\le t\le T} X_t$ is exponentially integrable, then X_t is M-determinate for any $t \in [0, T]$.

The exponential integrability was intensively studied for several classes of SDEs, martingales, Lévy processes; see, e.g., [26, 43, 51, 69]. One of the consequences in any such a case is that the distributions involved are M-determinate. Conclusions can also be derived for the absolute value $|X_t|$ and its powers $|X_t|^r$. We have to see which of the tools in Section 4 is best to use.

Example 7.2. Consider the familiar linear SDE (*Black-Scholes Model, GBM*):

$$\mathrm{d}S_t = \mu S_t \,\mathrm{d}t + \sigma S_t \,\mathrm{d}W_t, \; S_0 = s_0 = const > 0, \; t \in [0, T], \;\text{ or }\; t \ge 0.$$

There is a unique weak solution S_t, $t \in [0, T]$, where

$$S_t \sim Log\,\mathcal{N}(a_t, b_t^2), \; a_t = (\mu - \tfrac{1}{2}\sigma^2)t, \; b_t^2 = \tfrac{1}{2}\sigma^2 t.$$

Hence all moments of S_t, $t > 0$, are finite, they can be written explicitly and the conclusion is that S_t for $t > 0$ is M-indeterminate as we know from Section 3.

Here is another interesting fact. While the moment non-uniqueness implies that S_t is not exponentially integrable, we can claim that even a 'smaller' random variable, $\sqrt{S}_t$, is also not exponentially integrable. This follows by Hardy's condition, discussed in Section 4. Alternative proofs can also be given.

Example 7.3. Let us present a simple example of a *non-linear* SDE with unique non-trivial weak solution such that X_t is M-indeterminate for each $t > 0$.

Take this one:

$$dX_t = 3\,X_t^{1/3}\,dt + 3\,X_t^{2/3}\,dW_t,\ X_0 = 0,\ t \geq 0.$$

(We ignore the trivial solution $X_t = 0$ for all $t \geq 0$.) How to show the moment non-uniqueness of X_t? First, by using Itô's formula we find that the unique weak solution is $X_t = W_t^3$. Second, since $W_t \sim \mathcal{N}$, we refer to Section 5 and the fact that the cube of a normally distributed random variable is M-indeterminate.

Example 7.4. Suppose that X_0, $a(\cdot)$ and $\sigma^2(\cdot)$ are such that X_t does not satisfy Cramér's condition, i.e., X_t is heavy tailed (no exponential moment). We still assume that all moments of X_t are finite. There are two possibilities. The first one, X_t to be M-indeterminate, was illustrated in Examples 7.2 and 7.3. The second possibility is to choose the drift and the diffusion such that the unique (non-trivial) weak solution to the corresponding SDE to be M-determinate, even having heavy tail(s). To see that this is possible, consider the following SDE:

$$dX_t = dt + 2\,X_t^{1/2}\,dW_t,\ X_0 = 0,\ t \geq 0.$$

This SDE has a unique non-trivial weak solution with the property that for any $t > 0$, X_t has all moments finite, and it is M-determinate. The reader can easily see that this follows from the fact that $X_t = W_t^2$ and referring to Section 5.

Remark 7.1. Different ideas can be exploited to characterize the moment determinacy of the solutions of general SDEs. For example, an answer can be found if we know properties of the density $p(s, x; t, y)$, the solution of the forward Kolmogorov PDE. Clearly, the drift and the diffusion are essentially involved. We fix the first three arguments s, x, t and analyze $p(\cdot)$ as a function of y, and especially the asymptotic behavior of $p(\cdot)$ as $y \to \infty$. Then we can use Krein's condition or the converse to Krein's condition together with Lin's condition and make a conclusion about the moment determinacy of X_t. It is interesting and relevant to study also the moment determinacy of X_t as $t \to \infty$.

In the literature available are upper bounds for the moments $\mathbf{E}[X_t^{2k}]$, $\mathbf{E}[X_t^{2k+1}]$ or $\mathbf{E}[|X_t|^k]$. However they are not so useful since they lead to converging Carleman's series. Hence, no conclusion can be drawn about the moment determinacy.

If we look again at the examples above we see that the diffusion coefficient $\sigma^2(\cdot)$ is most important for the moment determinacy of the solution of a SDE. If we write formally $|\sigma(x)| \approx c|x|^\gamma$, we see that for some γ the solution X_t is M-determinate, for others it is M-indeterminate.

It is desirable to have answers to questions as above for stochastic process which are solutions to SDEs of the following type:

$dX_t = (\alpha_0 + \alpha_1\,X_t)\,dt + (\beta_0 + \beta_1\,X_t)\,dW_t,\ X_0;$

$dX_t = (\alpha_0 + \alpha_1\,X_t)\,dt + \sigma\,X_t^\gamma\,dW_t,\ X_0;$

$dr_t = (\alpha_0 + \alpha_1\,r_t + \alpha_2\,r_t^2 + \alpha_3/r_t)\,dt + \sqrt{\beta_0 + \beta_1\,r_t + \beta_2\,r_t^\gamma}\,dW_t,\ r_0.$

Such results would be of independent interest, however they can be involved when using SDEs to model stock price dynamics, interest rates, etc.

These problems and related topics are currently under study (a joint project with M. Zhitlukhin).

9. More Challenging Questions and Some New Results

Let us ask a few questions in a very general form.

Question 8.1. Do you believe that there is only one SDE such that at any time $t \geq 0$ its solution $X_t, t \geq 0$ has the moments $\mathbf{E}[X_t^k] = k!,\ k = 1, 2, \ldots$?

Question 8.2. Do you believe that there is only one SDE such that at any time $t \geq 0$ its solution has the moments $\mathbf{E}[X_t^k] = (2k)!,\ k = 1, 2, \ldots$?

Question 8.3. Do you believe that there are infinitely many SDEs such that all solutions at any time $t \geq 0$ have the same moments $\{(3k)!,\ k = 1, 2, \ldots\}$?

After seeing the numbers $k!$, $(2k)!$, $(3k)!$, the reader may identify them as the moments of the random variables ξ, ξ^2, ξ^3, where $\xi \sim Exp(1)$. Since we want these to be the moments at any time $t \geq 0$, this suggests that the processes must be stationary. However, the stationarity of a process requires the marginal distributions to be time-shift invariant and the process to have a specified correlation structure. Hence, in each of the above three cases we have to provide the complete description of the process we want to construct. Below c is a constant, $c > 0$.

Proposition 8.1. *There is only one SDE with explicit drift and diffusion coefficients such that it has a unique weak solution $X = (X_t, t \geq 0)$ which is a stationary diffusion Markov process with correlation function e^{-ct} and at any time $t \geq 0$ the moments of X_t are $\mathbf{E}[X_t^k] = k!,\ k = 1, 2, \ldots$. Moreover, X_t is exponentially integrable.*

Proposition 8.2. *There is only one SDE with explicit coefficients such that it has a unique weak solution $X = (X_t, t \geq 0)$ which is a stationary diffusion Markov process with correlation function e^{-ct} and at any time $t \geq 0$ the moments of X_t are $\mathbf{E}[X_t^k] = (2k)!,\ k = 1, 2, \ldots$. Moreover, in this case X_t is not exponentially integrable.*

Proposition 8.3. *There is a parametrized infinite family of SDEs, $\{SDE^{(\varepsilon)}, \varepsilon \in [-1, 1]\}$, with explicit coefficients such that their solutions are stationary diffusion Markov processes $\{X^{(\varepsilon)},\ \varepsilon \in [-1, 1]\}$ each with correlation function e^{-ct} and at any time $t \geq 0$ all $X_t^{(\varepsilon)}$ have the same moments $\mathbf{E}[(X_t^{(\varepsilon)})^k] = (3k)!,\ k = 1, 2, \ldots$. None is exponentially integrability.*

Sketch of Proofs. As mentioned above, we will use our knowledge about the moment determinacy of $\xi \sim Exp(1)$ and its powers ξ^2 and ξ^3.

However, we need to know how to construct a stochastic process with a specified distributional and correlation structure. We provide first the necessary details.

Let F be an absolutely continuous distribution function with density $f = F'$ and X a random variable, $X \sim F$. Denote by (a, b) the range of values of X, i.e., the support of F: $\mathrm{supp}(F) = (a, b)$. Assume that f is continuous, bounded and strictly positive in the domain (a, b) which can be a finite or infinite interval. One requirement more is that F has a finite second moment, hence finite variance. We use the usual notation m_1 for the first moment of F.

Thus, given F and f and taking any number $c > 0$, we define a 'new' function, $v = (v(x),\ x \in (a, b))$, as follows:

$$v(x) = \frac{2\,c}{f(x)} \int_a^x (m_1 - u) f(u)\,\mathrm{d}u = \frac{2\,c}{f(x)} \left(m_1\, F(x) - \int_a^x u\, f(u)\,\mathrm{d}u \right).$$

It can be checked that the function v is strictly positive in (a, b), hence its square-root, $\sqrt{v(x)}$, is a well-defined function.

Theorem. (See [7]) *Suppose that F, f and v are as above, W is a standard Brownian motion which is independent of the random variable X_0 and $c > 0$ is an arbitrary number. Then the following SDE*

$$dX_t = -c(X_t - m_1)\,dt + \sqrt{v(X_t)}\,dW_t,\ X_t|_{t=0} = X_0,\ t \geq 0$$

has a unique weak solution $X = (X_t, t \geq 0)$ which is a diffusion Markov process. Moreover, X is ergodic with ergodic density f. If the density of X_0 is f, the process X is stationary with correlation function e^{-ct}, $t \geq 0$.

This theorem gives a clear idea of what we need to do now. We take $F = Exp(1)$, its support is $(a, b) = (0, \infty) = \mathbb{R}_+$. Hence we are ready to give the main details in each of the above three cases, Propositions 8.1, 8.2 and 8.3.

Details for Case 1. We work with ξ, $m_k(\xi) = k!$, $f(x) = \mathrm{e}^{-x}$ and assume that $X_0 \sim Exp(1)$. We have all ingredients to calculate $v(x)$. Hence we can write explicitly just one SDE and conclude that its unique weak solution $X = (X_t, t \geq 0)$ is exactly the desired stationary and diffusion Markov process X. The correlation function is e^{-ct}, $t \geq 0$; X_t has kth order moment equal to $k!$, and X_t is M-determinate. The exponential integrability of X_t follows from the Cramér's condition for $\xi \sim Exp(1)$.

Details for Case 2. Now we need ξ^2 for which $m_k(\xi^2) = (2k)!$. Take $X_0 \sim f$, where $f(x) = \frac{1}{2}\, x^{-1/2}\mathrm{e}^{-x^{1/2}}$, $x > 0$ is the density of ξ^2. Use this f to calculate v, write explicitly the SDE and produce the unique weak solution with the desired properties. One difference from Case 1 is that here X_t is not exponentially integrable. This is because ξ^2 is heavy tailed, no Cramér's condition.

Details for Case 3. Here we work with the cube ξ^3. The moment of order k is $m_k(\xi^3) = (3k)!$ and the density to be used next is $f(x) = \frac{1}{3}\, x^{-2/3}\mathrm{e}^{-x^{1/3}}$, $x > 0$. Take this f as a density of X_0, calculate v, write the SDE and get a stationary diffusion Markov process X with all required properties.

However, as we know, ξ^3 is M-indeterminate. Hence we can construct a Stieltjes class $\mathbf{S}(f,h) = \{f_\varepsilon = f[1+\varepsilon h],\ \varepsilon \in [-1,1]\}$ based on the density f of ξ^3 and the perturbation $h(x) = \sin\left(\frac{\pi}{6} - \sqrt{3}x^{1/3}\right),\ x > 0$; see also Section 5. For any fixed $\varepsilon \in [-1,1]$ we use the density f_ε and calculate v_ε. Thus we obtain a SDE$^{(\varepsilon)}$ whose unique weak solution $X^{(\varepsilon)}$ is a stationary diffusion Markov process. Take now $\varepsilon \in [-1,1]\}$ and get the desired infinite family of processes. The drift and the diffusion coefficients are different for different ε. However, at any time $t \geq 0$, the solution of any of these SDEs has the moments $\{(3k)!\}$. The non-exponential integrability is obvious because ξ^3 is 'too heavy tailed'.

Remark 8.1. We have chosen the moments $k!$, $(2k)!$, $(3k)!$, $k = 1, 2, \ldots$ not just for curiosity. On one side it was convenient to use our findings in the previous sections. On the other side, more important is that these numbers show exactly where are the boundaries between the following three groups of SDEs:

SDEs with M-determinacy and exponential integrability;

SDEs with M-determinacy but no exponential integrability;

SDEs with M-indeterminacy and hence no exponential integrability.

We have a general result about a SDE which involves the moment sequence $\{m_k\}$ of an arbitrary absolutely continuous distribution F. We do not give details.

10. Final Comments

10.1 Briefly on relevant topics not discussed here

The role of the underlying stochastic process is essential for any valuation of financial instruments. However, sometime we do not know the stock price process and instead we have in our disposal a 'huge' data set of observable option prices. Based on these we want to 'find', 'identify' the underlying process. This is called an *Inverse Problem*. These are difficult theoretical problems and they are computationally heavy; see [11, 70, 85] and more references therein.

We have discussed in this paper only a few popular distributions each used in one or another way in stochastic financial models. Many other distributions are also exploited in the literature. To mention just a few: skew-normal and skew-Student distributions, polynomial logistic, negative inverse Gaussian, generalized gamma, generalized hyperbolic; see, e.g., [3, 4, 8, 9, 10, 14, 18, 25, 27, 30, 31, 33, 34, 36, 57, 58, 64, 74, 75, 86].

Special attention deserve the *explosion phenomenon* of solutions of SDEs, see [2, 42, 43, 45], and also the *stochastic volatility processes*, see [21, 45, 59, 88].

There are several methods based on moments and used to analyse stochastic models. This includes variations of the classical Pearson's method of moments. It turns out that the moments of distributions can be involved in option pricing when following semi-definite optimization methods. Pricing of options based on moments is treated in [1, 24, 32, 38, 60].

An interesting class of problems is to construct a stochastic process with prescribed specific properties. For example, we may require the process to have a special martingale structure, or to follow given one-dimensional and/or multi-dimensional distributions, or to have a given sequence of moments. Sometime we want the process to be a solution to a SDE. This kind of questions and results for general Lévy processes can be found in [7, 12, 19, 28, 29, 35, 37, 47, 49, 53, 73]. The results in Section 8 fall into this category.

10.2 Comments on the References

Well-known are the classical works by N. I. Akhiezer, M. G. Krein, J. A. Shohat and J. D. Tamarkin, B. Simon and C. Berg. They are not included here.

The items included into the list below are chosen to reflect, in one or another way, the fact that the moments of distributions are involved and/or used essentially in the analysis of stochastic financial models. Some papers contain the term 'moments' in the title and/or deal explicitly with the moment determinacy of distributions. More than a half of the papers contain 'moments' in the abstract and essentially in the main text. And some papers deal with specific distributions without mentioning the term 'moments' at all. However, when reading such papers one may get the feeling that implicitly the moments are there. Or, as one may say: 'The rabbit is behind the bush!'.

Most important is that all papers and books in the References list are indeed relevant and essential in the area of stochastic financial modelling.

Acknowledgement

I am grateful to the Organizers of the 'Finance Workshop' (Tokyo Metropolitan University, 6–7 November 2014) for inviting me to attend the event and deliver a talk. I enjoyed a series of fruitful discussions with Professors Takashi Shibata, Yukio Muromachi and Masaaki Kijima and with many young and enthusiastic PhD students, post-docs and visitors at TMU. The present paper is an extended version of the material reported at the Workshop.

It was also a pleasure to deliver seminar and/or colloquium talks at Kyoto University (Professor Kouji Yano), Osaka University (Professor Masaaki Fukasawa) and Tokyo City University (Professor Shuya Kanagawa).

I would like to acknowledge with thanks the support from the above institutions and the support for my visit to Japan from Emeritus Fellowship provided by Leverhulme Trust (UK).

My thanks are addressed also to G. D. Lin, M. Zhitlukhin, M. Zervos, D. Hobson, the Referee and the Editors for their relevant suggestions and comments well used when preparing the final text of the paper for which I am solely responsible.

References

1. Albanese C., Osseiran A. (2007). Moment methods for exotic volatility derivatives. Available online: arXiv:0710.2991v1 [math.PR] 16 Oct 2007.
2. Andersen L. B. G., Piterbarg V. V. (2007). Moment explosions in stochastic volatility models. *Finance & Stochastics* **11**, 29–50.
3. Barndorff-Nielsen O. E. (1997). Normal IG distributions in stochastic volatility modelling. *Scandinavian Journal of Statistics* **24**, 1–13.
4. Basrak B., Davis R. A., Mikosch T. (2002). Regular variation of GARCH processes. *Stochastic Processes & Applications* **99**, 95–115.
5. Beiglböck M., Henri-Labordére P., Penkner F. (2013). Model independent bounds for option prices: a mass transport approach. *Finance & Stochastics* **17**, 477–501.
6. Bertsimas D., Popescu I. (2002) On the relation between option and stock prices: a convex optimization approach. *Operations Research* **50**, 358–374.
7. Bibby B. M., Skovgaard I. M., Sorensen, M. (2005). Diffusion-type models with given marginal distribution and auto-correlation function. *Bernoulli* **11**, 191–220.
8. Bingham N. H., Kiesel, R. (2001). Semi-parametric modelling in finance: theoretical foundations. *Quantitative Finance* **1**, 1–10.
9. Biró T. S., Rosenfeld, R. (2007). Microscopic origin of non-Gaussian distributions of financial returns. Available: arXiv:0705.4112v2 [q-fin.ST] 6 Jul 2007.
10. Blasi F., Scarlatti S. (2012). From normal vs skew-normal portfolios: FSD and SSD rules. *Journal of Mathematical Finance* **2**, 90–95.
11. Bouchouev I., Isakov V. (1997). The inverse problem of option pricing. *Inverse Problems* **13**, 11–17.
12. Campi L. (2004). Arbitrage and completeness in financial markets with given *N*-dimensional distributions. *Decisions & Economics Finance* **27**, 57–80.
13. Carr P., Madan D. (2009). Saddlepoint methods for option prices. *Journal of Computational Finance* **13**, 49–61.
14. Chen J. T., Gupta A. K., Troskie C. G. (2003). The distribution of stock returns when the market is up. *Communications in Statistics: Theory & Methods* **32** , 1541–1558.
15. Chen L., He S., Zhang S. (2011). Tight bounds for some risk measures, with applications to robust portfolio selection. *Operations Research* **59**, 847–865.
16. Chin S., Dufresne D. (2012). A general formula for option prices in a stochastic volatility model. *Applied Mathematical Finance* **19**, 313–340.
17. Concuera J. M., Nualart D., Schoutens W. (2005). Completion of a Lévy market by power-jump assets. *Finance & Stochastics* **9**, 109–127.
18. Cont R., Tankov P. (2004). *Financial Modelling with Jump Processes.* Chapman & Hall/CRC, London.
19. Cox A. M. G., Hobson D., Obloj J. (2011). Time homogeneous diffusions with a given marginals at a random time. *ESAIM Probability Statistics* **15**, 511–524.
20. Cuchiero C., Keller-Ressel M., Teichmann J. (2012). Polynomial processes and their applications to mathematical finance. *Finance & Stochastics* **16**, 711–740.
21. Davis R. A., Mikosch T. (2009). Probabilistic properties of stochastic volatility models. In: Andersen T. G., Davis R. A., Kreiss J.-P., Mikosch T. (eds.) *Handbook of Financial Time Series*, pp. 255–267. Springer, New York.
22. De La Peña V. H., Ibragimov R., Jordan S. (2004). Option bounds. *Journal of Applied Probability* **41A**, 145–156.

23. De Schepper, A., Heijnen, B. (2007). Distribution-free option pricing. *Insurance: Mathematics and Economics* **40**, 179–199.
24. Demos A. (2002). Moments and dynamic structure of a time-varying parameter stochastic volatility in mean model. *Econometrics Journal* **5**, 345–357.
25. Dragulescu A. A., Yakovenko V. M. (2002). Probability distribution of returns in the Heston model with stochastic volatility. *Quantitative Finance* **2**, 443–453.
26. Duffie D., Filipović D., Schachermeiyer W. (2003). Affine processes and applications in finance. *Annals of Applied Probability* **13**, 98–1053.
27. Eberlein, E., Keller, U. (1995). Hyperbolic distributions in finance. *Bernoulli* **1**, 281–299.
28. Ekström E., Hobson D. (2011). Recovering a time-homogeneous stock price process from perpetual option prices. *Annals of Applied Probability* **21**, 1102–1135.
29. Ekström E., Hobson D., Janson S., Tysk J. (2013). Can time-homogeneous diffusions produce any distribution? *Probability Theory and Related Fields* **155**, 493–520.
30. Eling M. (2014). Fitting asset returns to skewed distributions: Are the skew-normal and skew-Student good models? *Insurance: Mathematics and Economics* **59**, 45–56.
31. Embrechts P., Klüppelberg C., Mikosch T. (1997). *Modelling Extremal Events for Insurance and Finance.* Springer, Berlin.
32. Eriksson B., Pistorius M. (2011). Method of moments approach to pricing double barrier contracts in polynomial jump-diffusion model. *International Journal of Theoretical & Applied Finance* **14**, 1139–1158.
33. Fabozzi F. J., Rachev S. T., Menn C. (2005). *Fat-Tailed and Skewed Asset Return Distributions: Implications for Risk Management Portfolio Selections and Option Pricing.* Wiley, New York.
34. Fajardo J., Farias A. (2004). Generalized hyperbolic distributions and Brazilian data. *Brazilian Review of Econometrics* **24** 249–271.
35. Filipović D., Larsson M. (2014). Polynomial preserving diffusions and applications to finance. Available online: arXiv:1404.0989v3 [math.PR] 11 Aug 2014.
36. Fischer M. J. (2014). *Generalized Hyperbolic Secant Distributions: With Applications to Finance.* Springer, Berlin.
37. Föllmer H., Wu C.-T., Yor M. (2000). On weak Brownian motion of arbitrary order. *Annales Institut H. Poincaré: Probabilité et Statistique* **36**, 447–487.
38. Fusai G., Tagliani A. (2002). An accurate valuation of Asian options using moments. *International Journal of Theoretical & Applied Finance* **5**, 147–169.
39. Gallant A. R., Tauchen G. (1996). Which moments to match? *Econometric Theory* **12**, 657–681.
40. Gavriliadis, P. N. (2008). Moment information for probability distributions, without solving the moment problem. I. Where is the mode? *Communications in Statistics: Theory & Methods* **37**, 671–681.
41. Gavriliadis P. N., Athanassoulis G. A. (2009). Moment information for probability distributions, without solving the moment problem, II: Main-mass, tails and shape approximation. *Journal of Computational & Applied Mathematics* **229**, 7–15.
42. Gerhold S. (2011). Moment explosion in the LIBOR market model. *Statistics & Probability Letters* **81**, 560–562.
43. Glasserman P., Kim K.-K. (2010). Moment explosions and stationary distributions in affine diffusion models. *Mathematical Finance* **20**, 1–33.

44. Grundy B. D. (1991). Option prices and the underlying asset's return distribution. *Journal of Finance* **46**, 1045–1069.
45. Gulisashvili A. (2012). *Analytically Tractable Stock Price Models.* Springer, New York.
46. Gotoh J., Konno H. (2002). Bounding option prices by semidefinite programming: a cutting plane algorithm. *Management Science* **48**, 665–678.
47. Hamza K., Klebaner F. C. (2007). A family of non-Gaussian martingales with Gaussian marginals. *Journal of Applied Mathematics & Stochastic Analysis* **2007**, article ID 92723, 19 pp.
48. Hobson D. (1998). Robust hedging of the lookback option. *Finance & Stochastics* **2**, 329–347.
49. Hobson D. (2015). Mimicking martingales. Available: arXiv:1505.03709v1 [math.PR] 14 May 2015.
50. Jarrow R., Rudd A. (1982). Approximate option valuation for arbitrary stochastic processes. *Journal of Financial Economics* **10**, 347–369.
51. Jeablanc M., Yor M., Chesney M. (2009). *Mathematical Methods for Financial Markets.* Springer, London.
52. Kagan A., Nagaev S. (2001). How many moments can be estimated from a large sample? *Statistics & Probability Letters* **55**, 99–105.
53. Keller-Ressel M., Papapantoleon A. (2013). The affine LIBOR models. *Mathematical Finance* **23**, 627–658.
54. Kijima, M. (2012). *Stochastic Processes with Financial Applications.* 2nd ed. Chapman & Hall/CRC, London.
55. Kijima M., Muromachi Y. (2008). An extension of the Wang transform derived from Bühlmanns economic premium principle for insurance risk. *Insurance: Mathematics and Economics* **42**, 887-896.
56. Kleiber C., Stoyanov, J. (2013). Multivariate distributions and the moment problem. *Journal of Multivariate Analysis* **113**, 7–17.
57. Koutras V. M., Drakos K., Koutras M. V. (2014). A polynomial logistic distribution and its applications in finance. *Communications in Statistics: Theory & Methods* **43**, 2045–2065.
58. Küchler U., Neumann K. (1999). Stock returns and hyperbolic distributions. *Mathematical & Computer Modelling* **29**, 1–15.
59. Kulik R., Soulier Ph. (2013). Estimation of limiting conditional distributions for heavy tailed long memory stochastic volatility process. *Extremes* **16**, 203–239.
60. Lasserre J. B., Prieto-Rumeau T., Zervos M. (2006). Pricing a class of exotic options via moments and SDP relaxations. *Mathematical Finance* **16**, 469–494.
61. Lee R. W. (2004). The moment formula for implied volatility at extreme strikes. *Mathematical Finance* **14**, 469–480.
62. Lee R. W. (2004). Option pricing by transform methods: extensions, unification and error control. *Journal of Computational Finance* **7**, 51–86.
63. Levy H. (1985). Upper and lower bounds of put and call option value: stochastic dominance approach. *Journal of Finance* **40**, 1197–1217.
64. Li H., Melnikov A. (2014). Polynomial extensions of distributions and their applications in actuarial and financial modeling. *Insurance: Mathematics and Economics* **55**, 250–260.

65. Lin G. D. (1997). On the moment problem. *Statistics & Probability Letters* **35**, 85–90. Addendum: ibid **50**(2000), 205.
66. Lin G. D., Stoyanov J. (2009). The logarithmic skew-normal distributions are moment indeterminate. *Journal of Applied Probability* **46**, 909–916.
67. Lin G. D., Stoyanov J. (2014). Moment determinacy of powers and products of non-negative random variables. *Journal of Theoretical Probability*. In press. Available on-line: doi 10.1007/s10959-014-0546-z
68. Lin G. D., Stoyanov J. (2015). On the moment determinacy of products of non-identically random variables. *Probability and Mathematical Statistics*. In press. Available online: arXiv:1406.1654v2 [math.PR] 12 Nov 2014.
69. Lions P.-L., Musiela M. (2007). Correlations and bounds for stochastic volatility models. *Annales Institut H. Poincaré. Analysis* **24**, 1–16.
70. Lishang J., Youshan T. (2001). Identifying the volatility of underlying assets from option prices. *Inverse Problems* **17**, 137–155.
71. Lo A. W. (1987). Semi-parametric upper bounds for option prices and expected payoffs. *Journal of Financial Economics* **19**, 373–387.
72. Lykov K. V. (2012). New uniqueness conditions for the classical moment problem. *Mathematical Notes* **92**, 797–806.
73. Madan D., Yor M. (2002). Making Markov martingales meet marginals: with explicit constructions. *Bernoulli* **8**, 509–536.
74. Malevergne Y., Pisarenko V., Sornette D. (2005). Empirical distributions of stock returns: between the stretched exponential and the power law? *Quantitative Finance* **5**, 379–401.
75. Maslov S., Zhang Y.-C. (1999). Probability distribution of drawdowns in risky investments. *Physics A* **262**, 232–241.
76. Merton R C, Samuelson P A (1974). Fallacy of the log-normal approximation to optimal portfolio decision-making over many periods. *Journal of Financial Economics* **1**, 67–94.
77. Mnatsakanov R., Hakobyan A. (2009). Recovery of distributions via moments. *IMS Lecture Notes - Monograph Series* **57**, 252–265.
78. Ninomiya S., Victoir N. (2008). Weak approximation of SDEs and application to derivative pricing. *Applied Mathematical Finance* **15**, 107-121.
79. Ostrovska S. (2014). Constructing Stieltjes classes for M-indeterminate absolutely continuous probability distributions. *ALEA Latin American Journal of Probability & Mathematical Statistics* **11**, 253–258.
80. Ostrovska S., Stoyanov J. (2005). Stieltjes classes for M-indeterminate powers of inverse Gaussian distributions. *Statistics & Probability Letters* **71**, 165–171.
81. Pakes A. G. (2001). Remarks on converse Carleman and Krein criteria for the classical moment problem. *Journal of the Australian Mathematical Society* **71**, 81–104.
82. Pakes A. G. (2007). Structure of Stieltjes classes of moment-equivalent probability laws. *Journal of Mathematical Analysis & Applications* **326**, 1268–1290.
83. Rolski T., Schmidli H., Schmidt V., Teugels J. (1999). *Stochastic Processes for Insurance and Finance*. Wiley, New York.
84. Rasz S., Tari A., Telek M. (2006). A moments based distribution bounding method. *Mathematical & Computational Modelling* **43**, 1367–1382.

85. Rodrigo M. R., Mamon R. S. (2007). Recovery of time-dependent parameters of a Black-Scholes-type equation: an inverse Stieltjes moment approach. *Journal of Applied Mathematics* **2007**, Article ID 62098, 8 pp.
86. Satchell S., Knight J. (2000). *Return Distributions in Finance.* Elsevier, Amsterdam.
87. Shephard N., Andersen T. G. (2009). *Stochastic Volatility: Origins and Overview.* Springer, London.
88. Stein M. S., Stein C. S. (1991). Stock price distributions with stochastic volatility: an analytic approach. *Review of Financial Studies* **4**, 727–752.
89. Stoyanov J. (2000). Krein condition in probabilistic moment problems. *Bernoulli* **6**, 939–949.
90. Stoyanov J. (2002). Moment problems related to the solutions of SDEs. *Lecture Notes in Control and Information Sciences* (Springer) **280**, 459–469.
91. Stoyanov J. (2004). Stieltjes classes for moment indeterminate probability distributions. *Journal of Applied Probability* **41A**, 669–685.
92. Stoyanov J. (2012). Inference problems involving determinacy of distributions. *Communications in Statistics: Theory & Methods* **41**, 2864–2878.
93. Stoyanov J. (2013). *Counterexamples in Probability*, 3rd edn. Dover, New York.
94. Stoyanov J., Lin G. D. (2012). Hardy's condition in probabilistic moment problems. *Theory of Probability and Its Applications* **57**, 811–820. [SIAM edition **57**(2013), 699–708.]
95. Stoyanov J., Lin G. D., DasGupta A. (2014). Hamburger moment problem for powers and products of random variables. *Journal of Statistical Planning and Inference* **154**, 166–177.
96. Tagliani A., Milev M. (2013). Laplace transform and finite difference method for the Black-Scholes equation. *Applied Mathematics & Computations* **220**, 649–658.
97. Tehranchi M. R. (2013). On the uniqueness of martingales with certain prescribed marginals. *Journal of Applied Probability* **50**, 557–575.
98. Tian R. (2008). Moment Problems with Applications to Value-At-Risk and Portfolio Management. Georgia State University. Dissertations, Paper 21.

An Equilibrium Approach to Indifference Pricing with Model Uncertainty*

Mark H.A. Davis[1] and Daisuke Yoshikawa[2]

[1]Department of Mathematics, Imperial College London, London SW7 2AZ, United Kingdom. Email: mark.davis@imperial.ac.uk
[2]Department of Business Administration, Hokkai-Gakuen University, 4-1-40 Asahimachi, Toyohira Ward, Sapporo 062-8605, Japan. Email: yoshikawa@ba.hokkai-s-u.ac.jp

Utility indifference pricing is an effective method for investors to construct a strategy in an incomplete market. In fact, if an investor can trade a random endowment under the criteria shown by utility indifference pricing, they can devise financial contracts that are optimized according to their preferences. However, because it does not have the direct implication of equilibrium, the value of the random endowment given by indifference pricing is not necessarily the same as the market price. In this study, we attempt to derive the equilibrium of random endowment under the framework of indifference pricing. However, letting the utility function be of exponential type means that any trade involving random endowment will not appear in equilibrium. Thus, we show that non-zero trade in equilibrium appears by introducing uncertainty in a model, which is one of the sources of market incompleteness.

Key words: Utility indifference pricing, Equilibrium, Model uncertainty, Maxmin expected utility

1. Introduction

A classical problem in finance is that investors cannot necessarily hedge all risks. To fix this problem, the theory on incomplete market has been developed over 30 years. This work has enjoyed excellent results; mean-variance hedging and local-risk minimization are representative pricing standards in an incomplete

*Send all correspondence to Daisuke Yoshikawa, Department of Business Administration, Hokkai-Gakuen University, 4-1-40 Asahimachi, Toyohira Ward, Sapporo 062-8605, Japan. Email:yoshikawa@ba.hokkai-s-u.ac.jp.

market, of which Schweizer (1999) [28] gives us a good summary. They give us a criteria to minimize risk in an incomplete market. Various pricing measures have also been developed as methods to minimize risks. They include minimal martingale measure (Föllmer and Schweizer (1991) [8]), variance-optimal martingale, p-optimal martingale (Grandits (1999) [14]) and the minimal entropy martingale (Frittelli (2000) [10], Goll and Rüschendorf (2001) [12] and Monoyios (2007) [22]).

These methods can greatly contribute to investors who seek to minimize risk in an incomplete market, if the criteria related to each method has affinities with each investor's trading policy. This implies that although one method might be good for one investor, it might not be good for another. In this sense, each of these methods are not uniformly applicable for various types of investors because of the diversity of investors' preferences.

Utility indifference pricing provides a solution for this. In fact, if an investor can trade a random endowment at a better price than the one shown in indifference pricing, (s)he can optimize the risk in her(his) portfolio. Even if the price shown in indifference pricing might be different for each investor, the framework of indifference pricing itself is common for each investor. Therefore, this method is applicable for all types of investors with various preferences.

In addition to the wide applicability discussed above, there is economic justification in utility indifference pricing. The reason for this is that the utility indifference pricing is based on the principle that every investor maximizes their expected utility[1]. We can enjoy various benefits from utility indifference pricing, such as those mentioned by Becherer (2003) [1], Sircar and Zariphopoulou (2004) [30], Monoyios (2004) [21], Hugonnier and Kramkov (2004) [16], Hugonnier, et al. (2005) [17], Davis (2006) [5], Monoyios (2008) [23] and Biagini et al. (2011) [2].

The framework of indifference pricing requires a calculation of utility maximization in two cases: First is utility maximization where the investor holds random endowment. Second is utility maximization where the investor does not hold random endowment. If the expected utility in the former case is greater than that in the latter, it is rational for the investor to hold the random endowment, and vice versa. The utility indifference price is the price that equates the expected utility including random endowment with the expected utility that does not include random endowment.

Consider the situation wherein the investor wants to sell the random endowment. In this case, if the investor can sell the random endowment at a price higher than the utility indifference price, then it becomes beneficial for the investor. In this sense, the utility indifference price will be the minimum price of the random endowment for the investor as a seller.

---

[1] A rigorous formulation of utility maximization in an incomplete market is given by Kramkov and Schachermayer (1999) [20] and Schachermayer (2000) [27].

Conversely, when the investor wants to buy the random endowment, the utility indifference price will be the maximum price of the random endowment. The name utility 'indifference' price is derived from this context.

Although utility indifference pricing shows an optimal way to minimize the risk of each investor, it is not the market price. This is because the price shown by utility indifference pricing does not imply an equilibrium price in the market of the random endowment. In this study, we will show the equilibrium using utility indifference pricing. For this purpose, the procedure is divided into three parts: First, we show the equilibrium in the normal framework of utility indifference pricing. Second, we try to extend the framework by introducing uncertainty into the model. Third, a further extension is attempted by embedding a risk management policy in the framework, in addition to uncertainty.

More precisely, this paper is constructed as follows: Section 2 gives an overview of utility indifference pricing, where the utility function is given as an exponential type[2]. Section 3 describes the equilibrium of random endowment according to utility indifference pricing. In Section 4, we extend utility indifference pricing by introducing uncertainty and a risk management policy in the model. In Section 5, to show numerical examples, we apply the results described in Sections 3 and 4 to a more concrete model, which is related to the model shown by Davis (2006) [5][3].

2. Model and an Overview of Utility Indifference Pricing

We consider the mathematical framework in given probability space $(\Omega, \mathcal{F}, \mathbb{F}, P)$, where $\mathbb{F} := (\mathcal{F}_t)_{0\le t\le T}$, $\mathcal{F} := \mathcal{F}_T$ and $\mathcal{F}_0$ is trivial. Stochastic process $S \in \mathbb{R}^d$ is defined as semimartingale, and the expected value of S is given by the probability measure P. Consider the $\mathcal{F}_T$-measurable random variable (or 'random endowment') B which will generate some payoff at time T. The random endowment B is assumed bounded from below (Delbaen et al. (2002) [6]). We assume that for some fixed $\alpha, \epsilon \in (0, \infty)$,

$$\mathbb{E}[e^{(\alpha+\epsilon)B}] < \infty \text{ and } \mathbb{E}[e^{-\epsilon B}] < \infty.$$

This assumption guarantees that B is Lebesgue integrable for all martingale measures for which the relative entropy is able to be defined (Becherer (2003) [1]). Utility indifference price is defined as the value or price of the random endowment B which equates the maximized expected utility of the terminal wealth without the random endowment and the maximized expected utility of the terminal

---

[2]HARA utility function or CRRA type utility are also available to use and it is also one of the representative way; e.g. Zariphopoulou (2001) [31], Monoyios (2010) [24]

[3]Davis (2006) [5] uses the model in which there are two assets; the one is tradable asset and the other is untradeable one.

wealth including the random endowment. We define $X^{x,q} = x + pq + \int_0^T \theta_t^\top dS_t$, where S can be interpreted as a discount price process of tradable assets. Then, the terminal wealth with random endowment is $X^{x,q} - Bq$, where q is an amount of B, $x + pq$ is initial capital (x is money market account, p implies a price of the random endowment B), $\theta := \{\theta_t;\ t \in [0, T]\} \in \Theta$ is $\mathbb{R}^d$-valued admissible trading strategy[4], Θ is set of S-integrable and predictable processes. We also define a set $\mathcal{X}_0$ as follows,

$$\mathcal{X}_0 := \{X \geq 0 : X \in L^1(Q) \text{ for all } Q \in \mathcal{M} \text{ and } U(X) \in L^1(P)\},$$

where the utility function $U : \mathbb{R} \to \mathbb{R}$ as a C^1-strictly increasing, strictly concave function satisfying the following conditions:

$$\lim_{r\to\infty} U'(r) = 0, \qquad \lim_{r\to-\infty} U'(r) = \infty,$$

$\mathcal{M}$ is a set of Θ-martingale measures satisfying $H(Q|P) < \infty$, where $H(Q|P)$ is relative entropy of Q with respect to P, which is defined as follows,

$$H(Q|P) := \begin{cases} \int \frac{dQ}{dP} \ln \frac{dQ}{dP} dP & Q \ll P \\ +\infty & \text{otherwise.} \end{cases}$$

Note that relative entropy is always non negative (c.f. Theorem 1.4.1 of Ihara (1993) [18]). On this set $\mathcal{X}_0$, the maximized expected utility with the random endowment is described as follows,

$$\sup_{X^{x,q}-Bq\in\mathcal{X}_0} \mathbb{E}\left[U(X^{x,q} - Bq)\right], \tag{1}$$

By definition, positive q implies the investor with the above expected utility selling the random endowment B, and negative q implies buying the random endowment B.

The maximized expected utility without random endowment is described as follows,

$$\sup_{X^{x,0}\in\mathcal{X}_0} \mathbb{E}\left[U(X^{x,0})\right]. \tag{2}$$

If p is the utility indifference price, then it satisfies

$$\sup_{X^{x,0}\in\mathcal{X}_0} \mathbb{E}\left[U(X^{x,0})\right] = \sup_{X^{x,q}-Bq\in\mathcal{X}_0} \mathbb{E}\left[U(X^{x,q} - Bq)\right]. \tag{3}$$

When $q > 0$, we call price p the utility indifference sell price, because it implies that the investor with the utility $U(\cdot)$ sells the random endowment at the price p.

[4]More precisely, $\theta \in L(S)$ and for some constant $c \in \mathbb{R}$, $\int_0^t \theta_s^\top dS_s \geq c,\ t \in [0, T]$.

This definition implies that the utility indifference sell price is the minimum price of random endowment B. The reason is that, if the market price of the random endowment B is less than utility indifference sell price, then the investor does not have the ground to sell the random endowment. It is natural to define the utility indifference buy price p which satisfies (3) for $q < 0$. Hereafter, we describe the utility indifference price as $p(B; q)$. To be clarify the sell side and buy side, $p(B; q)$ is described as $p^s(B; q)$ which is the utility indifference sell price and implies $q > 0$, and $p^b(B; q')$ which is the utility indifference buy price and implies $-q' = q < 0$: i.e.

$$p(B;q) = \begin{cases} p^s(B;q), & \text{for } q > 0 \\ p^b(B;q'), & \text{for } -q' = q < 0. \\ p^* & \text{for } q = 0. \end{cases}$$

Now, we specify the utility function as $U(x) = -e^{-\gamma x}$, where $\gamma \in \mathbb{R}_+$ is risk-aversion. Then, the dual function is given by $V(y) = \frac{y}{\gamma}\left(\ln\left(\frac{y}{\gamma}\right) - 1\right)$. Consider the problem (2) which is the left hand side of (3). Using the function $V(\cdot)$, (2) is rewritten as follows,

$$\text{(4)} \qquad \sup_{X^{x,0} \in \mathcal{X}_0} \mathbb{E}\left[U(X^{x,0})\right] = \inf_{\eta \in \mathbb{R}_+, Q \in \mathcal{M}} \left\{ \mathbb{E}\left[V\left(\eta \frac{dQ}{dP}\right)\right] + \eta x \right\}.$$

Hereafter, we assume that

$$\mathcal{M} \neq \emptyset.$$

The above equation (4) is calculated continuously,

$$\begin{aligned} &\inf_{\eta \in \mathbb{R}_+, Q \in \mathcal{M}} \left\{ \mathbb{E}\left[V\left(\eta \frac{dQ}{dP}\right)\right] + \eta x \right\} \\ &= \inf_{\eta \in \mathbb{R}_+, Q \in \mathcal{M}} \left\{ \frac{\eta}{\gamma} \ln\left(\frac{\eta}{\gamma}\right) \mathbb{E}\left[\frac{dQ}{dP}\right] + \frac{\eta}{\gamma} \mathbb{E}\left[\frac{dQ}{dP} \ln\left(\frac{dQ}{dP}\right)\right] - \frac{\eta}{\gamma} \mathbb{E}\left[\frac{dQ}{dP}\right] + \eta x \right\} \\ &= \inf_{\eta \in \mathbb{R}_+} \left\{ \frac{\eta}{\gamma} \ln\left(\frac{\eta}{\gamma}\right) - \frac{\eta}{\gamma} + \eta x + \frac{\eta}{\gamma} \inf_{Q \in \mathcal{M}} H[Q|P] \right\} \\ &= \inf_{\eta \in \mathbb{R}_+} \left\{ \frac{\eta}{\gamma} \ln\left(\frac{\eta}{\gamma}\right) - \frac{\eta}{\gamma} + \eta x + \frac{\eta}{\gamma} H[Q^0|P] \right\}. \end{aligned}$$

On line 3, we use $\eta \in \mathbb{R}_+$. We write the solution of $\inf_{Q \in \mathcal{M}} H[Q|P]$ as $Q^0 \in \mathcal{M}$ which we call minimal entropy martingale measure (MEMM). The solution of $\inf_{\eta \in \mathbb{R}} \left\{ \frac{\eta}{\gamma} \ln\left(\frac{\eta}{\gamma}\right) - \frac{\eta}{\gamma} + \eta x + \frac{\eta}{\gamma} H[Q^0|P] \right\}$ is given by $\eta^0 = \gamma e^{-\gamma\left(x + \frac{1}{\gamma} H[Q^0|P]\right)}$. Then, the maximized expected utility is described as follows,

$$\text{(5)} \qquad \sup_{X^{x,0} \in \mathcal{X}_0} \mathbb{E}\left[U(X^{x,0})\right] = -e^{-\gamma\left(x + \frac{1}{\gamma} H[Q^0|P]\right)}.$$

Likewise, we can solve the utility maximization problem (1) which is the right hand side of (3).

$$\sup_{X^{x,q}-Bq\in\mathcal{X}_0} \mathbb{E}\left[U(X^{x,q}-Bq)\right]$$

$$(6)\qquad = \inf_{\eta\in\mathbb{R}_+}\left\{\frac{\eta}{\gamma}\ln\left(\frac{\eta}{\gamma}\right)-\frac{\eta}{\gamma}+\eta(x+pq)+\frac{\eta}{\gamma}\inf_{Q\in\mathcal{M}}\left\{H[Q|P]-\gamma q\mathbb{E}^Q[B]\right\}\right\}.$$

Note that an existence of the solution of $\inf_{Q\in\mathcal{M}}\left\{H[Q|P]-\gamma q\mathbb{E}^Q[B]\right\}$ is shown by Proposition 2.2 of Becherer (2003) [1]. Then, the maximized expected utility is given such that,

$$(7)\qquad \sup_{X^{x,q}-qB\in\mathcal{X}_0}\mathbb{E}\left[U(X^{x,q}-qB)\right]=-e^{-\gamma\left(x+pq+\frac{1}{\gamma}\inf_{Q\in\mathcal{M}}\left\{H[Q|P]-\gamma q\mathbb{E}^Q[B]\right\}\right)}.$$

This result is consistent with the main result of Delbaen et al (2002) [6].

Remark 2.1. By definition of dual function, it holds for any $X^{x,q}$ that,

$$\mathbb{E}\left[U(X^{x,q}-Bq)\right]\le\inf_{\eta\in\mathbb{R}_+,Q\in\mathcal{M}}\left\{\mathbb{E}\left[V\left(\eta\frac{dQ}{dP}\right)+\eta\frac{dQ}{dP}\left(x+pq+\int_0^T\theta_t^\top dS_t-Bq\right)\right]\right\}.$$

The equality (6) holds for supremum of left hand side of the above equation (c.f. Davis (2006)[5]). Similarly, the equality (4) holds.

□

By definition of the utility indifference price and (5)(7), $p(B;q)$ satisfies,

$$-e^{-\gamma\left(x+\frac{1}{\gamma}H[Q^0|P]\right)}=-e^{-\gamma\left(x+p(B;q)q+\frac{1}{\gamma}\inf_{Q\in\mathcal{M}}\left\{H[Q|P]-\gamma q\mathbb{E}^Q[B]\right\}\right)}.$$

Therefore, utility indifference price is given by,

$$p(B;q)=\frac{1}{\gamma q}H[Q^0|P]-\frac{1}{q}\inf_{Q\in\mathcal{M}}\left\{\frac{1}{\gamma}H[Q|P]-q\mathbb{E}^Q[B]\right\}.$$

For $q>0$, utility indifference sell price is,

$$(8)\qquad p^s(B;q)=\sup_{Q\in\mathcal{M}}\left\{\mathbb{E}^Q[B]-\frac{1}{\gamma q}\left(H(Q|P)-H(Q^0|P)\right)\right\},$$

and utility indifference buy price is,

$$p^b(B;q)=\inf_{Q\in\mathcal{M}}\left\{\mathbb{E}^Q[B]+\frac{1}{\gamma q}\left(H(Q|P)-H(Q^0|P)\right)\right\}.$$

Before proceeding to the discussion of equilibrium, we consider upper and lower bound of utility indifference price. By the previous discussion, utility indifference sell and buy price is given as follows,

$$p^s(B;q) = \sup_{Q\in\mathcal{M}} \left\{ \mathbb{E}^Q[B] - \frac{1}{\gamma q}\left(H(Q|P) - H(Q^0|P)\right)\right\}$$

$$\leq \sup_{Q\in\mathcal{M}} \left\{ \mathbb{E}^Q[B] \right\}. \tag{9}$$

$$p^b(B;q) = \inf_{Q\in\mathcal{M}} \left\{ \mathbb{E}^Q[B] + \frac{1}{\gamma q}\left(H(Q|P) - H(Q^0|P)\right)\right\}$$

$$\geq \inf_{Q\in\mathcal{M}} \left\{ \mathbb{E}^Q[B] \right\}. \tag{10}$$

Although these bounds seem simple and straightforward, it might be too wide. It is available to make it narrower. We will show it in Corollary 3.1.

3. The Equilibrium in the Market

In this section, we consider the equilibrium in the market of random endowment B. First, we give a definition of the equilibrium (c.f. Mas-Collel *et al.* (1995) [25]).

Definition 3.1. Let an economy specify the investors' preference which is described by the utility function $U := \{U_i(\cdot); U_i(x) := -e^{-\gamma_i x},\ i = 1,\cdots,I+J\}$. An allocation $q^s := \{q^s_i, i = 1,\cdots,I\}$, $q^b := \{q^b_j, j = 1,\cdots,J\}$ and a price p of the random endowment B constitutes a price equilibrium if there is an assignment such that

1. **Offer price condition**: For any investor with utility function $\{U_i,\ i = 1,\cdots,I\}$, when the investor sells q^s_i-units of the random endowment, (p, q^s_i) is preferred to all other allocations $(p, (q^s_i)')$; that is, an expected utility corresponding to the allocation (p, q^s_i) is larger than another expected utility corresponding to the allocation $(p, (q^s_i)')$.

2. **Bid price condition**: For any investor with utility function $\{U_{I+j},\ j = 1,\cdots,J\}$, when the investor buys q^b_j-units of the random endowment, (p, q^b_j) is preferred to all other allocations $(p, (q^b_j)')$; that is, an expected utility corresponding to the allocation (p, q^b_j) is larger than another expected utility corresponding to the allocation $(p, (q^b_j)')$.

3. **Market cleared condition**: $\sum_{i=1}^{I} q^s_i = \sum_{j=1}^{J} q^b_j$.

Theorem 3.1. *If investors in the market of random endowment B act according to utility maximization, there is an equilibrium price p^*, such that*

$$p^* = \mathbb{E}^{Q^0}[B].$$

Before proving this Theorem, we show a few Lemmas. Here, we define a set of Radon-Nikodým derivatives of equivalent martingale measure for physical measure P;

$$\mathcal{N} := \left\{ \frac{dQ}{dP}(\omega) \middle| \, \omega \in \Omega, Q \in \mathcal{M} \right\}.$$

We also define a set of terminal wealths for all admissible strategies Θ,

$$\mathcal{G} := \left\{ G(\theta) \middle| \theta \in \Theta, G(\theta) = \int_0^T \theta_t^\top dS_t < \infty \right\}.$$

In the following Lemma, we discuss the form of MEMM Q^0 in our context. The uniqueness of MEMM is shown by Frittelli (2000)[10] and Schweizer (2010)[29].

Lemma 3.1. *For MEMM Q^0, Radon-Nikodým derivative $\frac{dQ^0}{dP}(\omega)$ is given as follows,*

$$\frac{dQ^0}{dP} = \frac{e^{-\gamma G(\theta^0)}}{\int e^{-\gamma G(\theta^0)} dP},$$

where γ is a risk-aversion and θ^0 is the optimizer of the problem (2).

Proof MEMM is given as the solution of the problem

$$\inf_{Q \in \mathcal{M}} \{H(Q|P)\}. \tag{11}$$

Since $Q \in \mathcal{M}$, for any admissible strategy $\theta \in \Theta$, it holds that

$$\int_\Omega G(\theta)(\omega) \tfrac{dQ}{dP}(\omega) dP(\omega) = 0, \text{ for all } \tfrac{dQ}{dP} \in \mathcal{N}.$$

It means that $\frac{dQ}{dP} \in \mathcal{N}$ has Θ-martingale property and this martingale property is the constraint of the optimization problem. Therefore, we can rewrite the optimization problem (11) as follows,

$$\inf_{\frac{dQ}{dP} \in \mathcal{N}, \lambda \in \Lambda} \left\{ \int_\Omega \frac{dQ}{dP}(\omega) \ln \frac{dQ}{dP}(\omega) dP(\omega) + \int_\Omega \langle \lambda, G(\Theta) \rangle (\omega) \frac{dQ}{dP}(\omega) dP(\omega) \right\}, \tag{12}$$

where $G(\Theta)$ is defined for all admissible strategies Θ, Λ is the set of which elements consist of real numbers and have the same dimension as $G(\Theta)$, and we used Frechét differentiability of the relative entropy functional $\int_\Omega \frac{dQ}{dP}(\omega) \ln \frac{dQ}{dP}(\omega) dP(\omega)$ and the continuous linear functional $\int_\Omega G(\theta)(\omega) \frac{dQ}{dP}(\omega) dP(\omega)$ in order to apply the Lagrange multiplier method (Favretti (2005) [7], Cochrane and Magcgregor

(1978) [4] and Botelho (2009) [3]). For some $\frac{dQ'}{dP} \in \mathcal{N}$, $\epsilon \in \mathbb{R}_+$ and any function $\eta(\omega)$ such that $\int_\Omega \eta(\omega)dP(\omega) = 0$, we set as follows,

$$\frac{dQ}{dP}(\omega) = \frac{dQ'}{dP}(\omega) + \epsilon\eta(\omega).$$

Let λ^* be the optimizer of the problem (12). By a variational method using $\frac{dQ}{dP}(\omega) = \frac{dQ'}{dP}(\omega) + \epsilon\eta(\omega)$, the first order condition of the optimization is deduced as follows,

$$\int_\Omega \eta(\omega)\left\{\ln\frac{dQ}{dP}(\omega) + 1 + \langle\lambda^*, G(\Theta)\rangle(\omega)\right\}dP(\omega) = 0.$$

Therefore, for any $\omega \in \Omega$, $\ln\frac{dQ}{dP}(\omega) + 1 + \langle\lambda^*, G(\Theta)\rangle(\omega) = \text{constant}$. Since it is required that $\int \frac{dQ}{dP}dP = 1$, $\frac{dQ}{dP}$ has the form,

$$\frac{dQ}{dP}(\omega) = \frac{e^{-\langle\lambda^*, G(\Theta)\rangle(\omega)}}{\int_\Omega e^{-\langle\lambda^*, G(\Theta)\rangle(\omega)}dP(\omega)}.$$

By the martingale property of $Q \in \mathcal{M}$, dQ/dP,

$$\int_\Omega G(\theta')(\omega)\frac{dQ}{dP}(\omega)dP(\omega) = \int_\Omega G(\theta')(\omega)\frac{e^{-\langle\lambda^*, G(\Theta)\rangle(\omega)}dP(\omega)}{\int_\Omega e^{-\langle\lambda^*, G(\Theta)\rangle(\omega)}dP(\omega)} = 0 \text{ for all } \theta' \in \Theta.$$

Since $\langle\lambda^*, G(\Theta)\rangle$ is given by the linear combination of λ^* and $G(\Theta)$, it holds $\langle\lambda^*, G(\Theta)\rangle \in \mathcal{G}$. That is, for some $\theta^* \in \Theta$, it holds that $\langle\lambda^*, G(\Theta)\rangle = G(\theta^*)$.

Because the optimizer of (2), say θ^0, has to satisfy the following equation,

$$\mathbb{E}\left[-\gamma G(\theta')e^{-\gamma\left(x+G(\theta^0)\right)}\right] = 0 \text{ for all } \theta' \in \Theta, \tag{13}$$

it holds

$$\theta^* = \gamma\theta^0.$$

Therefore,

$$\frac{dQ}{dP}(\omega) = \frac{e^{-\gamma G(\theta^0)(\omega)}}{\int_{\omega\in\Omega} e^{-\gamma G(\theta^0)(\omega)}dP(\omega)}.$$

Q.E.D.

Remark 3.1. Although the equation (13) is an expanded result of Section 3.1 of Frittelli (2000)[10], where the discrete model is discussed, we can directly deduce it in more general case. If $\theta \in \Theta$, it holds $\kappa\theta \in \Theta$ for any $\kappa > 0$. Let $\hat{\theta}$ be the optimizer of the problem (1). It holds for any $\vartheta \in \Theta$ and $\kappa > 0$,

$$\mathbb{E}\left[-e^{-\gamma(x+pq+\int_0^T \hat{\theta}_t^\top dS_t - Bq)}\right] \geq \mathbb{E}\left[-e^{-\gamma(x+pq+\int_0^T (\hat{\theta}_t^\top + \kappa\vartheta_t^\top) dS_t - Bq)}\right]$$

$$= \mathbb{E}\left[-e^{-\gamma\int_0^T \kappa\vartheta_t^\top dS_t} e^{-\gamma(x+pq+\int_0^T \hat{\theta}_t^\top dS_t - Bq)}\right]$$

$$= \mathbb{E}\left[\left(1 - \gamma\kappa\int_0^T \vartheta_t^\top dS_t + \gamma^2\kappa^2\int_0^T \frac{1}{2}\vartheta_t^\top\vartheta_t d\langle S\rangle_t\right)\left(-e^{-\gamma(x+pq+\int_0^T \hat{\theta}_t^\top dS_t - Bq)}\right)\right]$$

$$\leftrightarrow \mathbb{E}\left[\left(\int_0^T \vartheta_t dS_t\right)\left(e^{-\gamma(x+pq+\int_0^T \hat{\theta}_t^\top dS_t - Bq)}\right)\right] \leq \mathbb{E}\left[\left(\frac{1}{2}\gamma\kappa\int_0^T \vartheta_t^\top\vartheta_t d\langle S\rangle_t\right)\left(e^{-\gamma(x+pq+\int_0^T \hat{\theta}_t^\top dS_t -}\right.\right.$$

Similarly,

$$\mathbb{E}\left[-e^{-\gamma(x+pq+\int_0^T \hat{\theta}_t^\top dS_t - Bq)}\right] \geq \mathbb{E}\left[-e^{-\gamma(x+pq+\int_0^T (\hat{\theta}_t^\top - \kappa\vartheta_t^\top) dS_t - Bq)}\right]$$

$$= \mathbb{E}\left[\left(1 + \gamma\kappa\int_0^T \vartheta_t^\top dS_t + \frac{1}{2}\gamma^2\kappa^2\int_0^T \vartheta_t^\top\vartheta_t d\langle S\rangle_t\right)\left(-e^{-\gamma(x+pq+\int_0^T \hat{\theta}_t^\top dS_t - Bq)}\right)\right]$$

$$\leftrightarrow \mathbb{E}\left[\left(-\int_0^T \vartheta_t^\top dS_t\right)\left(e^{-\gamma(x+pq+\int_0^T \hat{\theta}_t^\top dS_t - Bq)}\right)\right] \leq \mathbb{E}\left[\left(\frac{1}{2}\gamma\kappa\int_0^T \vartheta_t^\top\vartheta_t d\langle S\rangle_t\right)\left(e^{-\gamma(x+pq+\int_0^T \hat{\theta}_t^\top d}\right.\right.$$

Therefore, for some $M \in \mathbb{R}$, it holds

$$\mathbb{E}\left[\left|\int_0^T \vartheta_t^\top dS_t\right|\left(e^{-\gamma(x+pq+\int_0^T \hat{\theta}_t^\top dS_t - Bq)}\right)\right] \leq \kappa M.$$

Since the above inequality holds for any κ, martingale property is proved. (13) is applicable for $q = 0$.

□

Corollary 3.1. *The upper bound of utility indifference sell price is given as follows,*

$$p^s(B;q) \leq \frac{1}{\gamma q}\ln \mathbb{E}^{Q_0}\left[e^{\gamma q B}\right].$$

The lower bound of utility indifference buy price is given as follows,

$$p^b(B;q) \geq -\frac{1}{\gamma q}\ln \mathbb{E}^{Q_0}\left[e^{-\gamma q B}\right].$$

Proof By definition of utility indifference price, it holds

$$\mathbb{E}\left[-e^{-\gamma(x+\int_0^T (\theta_t^0)^\top dS_t)}\right] = \mathbb{E}\left[-e^{-\gamma(x+p^s(B;q)q+\int_0^T \hat{\theta}_t^\top dS_t - Bq)}\right],$$

where θ^0 and $\hat{\theta}$ are optimizers of left-hand side and right hand side of (3), respectively.

From above equation,

$$\mathbb{E}\left[-e^{-\gamma(x+\int_0^T(\theta_t^0)^\top dS_t)}\right] \geq \mathbb{E}\left[-e^{-\gamma(x+p^s(B;q)q+\int_0^T(\theta_t^0)^\top dS_t-Bq)}\right]$$

$$\leftrightarrow e^{\gamma p^s(B;q)q} \leq \frac{\mathbb{E}\left[e^{-\gamma(x+\int_0^T(\theta_t^0)^\top dS_t-Bq)}\right]}{\mathbb{E}\left[e^{-\gamma(x+\int_0^T(\theta_t^0)^\top dS_t)}\right]} = \mathbb{E}^{Q_0}\left[e^{\gamma Bq}\right]$$

$$\leftrightarrow p^s(B;q) \leq \frac{1}{\gamma q}\ln \mathbb{E}^{Q_0}\left[e^{\gamma q B}\right].$$

Likewise, we can show the case of utility indifference buy price.

Q.E.D.

The following lemma shows the convergence of utility indifference price on the amount q. Similar convergence is shown by Proposition 3.2 of Becherer (2003) [1], where the convergence on the risk aversion γ is discussed.

Lemma 3.2. *On the utility indifference price based on the exponential utility $U(x) = -e^{-\gamma x}$, it holds that*

$$\lim_{q\downarrow 0} p^s(B;q) = \lim_{q\downarrow 0} p^b(B;q) = \mathbb{E}^{Q^0}[B].$$

Proof Let $\hat{Q}$ be the optimizer of the right hand side of (8); i.e.

$$p^s(B;q) = \sup_{Q\in\mathcal{M}}\left\{\mathbb{E}^Q[B] - \frac{1}{\gamma q}\left(H(Q|P) - H(Q^0|P)\right)\right\}.$$

Likewise the Lemma 3.1, the Radon-Nikodým derivative $\frac{d\hat{Q}}{dP}$ is given as follows,

$$\frac{d\hat{Q}}{dP}(\omega) = \frac{e^{-(-\gamma qB(\omega)+\langle\lambda,G(\Theta)\rangle(\omega))}}{\int_\Omega e^{-(-\gamma qB(\omega)+\langle\lambda,G(\Theta)\rangle(\omega))}dP(\omega)},$$

where the optimizer λ and $G(\Theta)$ are given for satisfying as follows,

$$\int_{\omega\in\Omega} G(\theta')(\omega)\frac{e^{-(-\gamma qB(\omega)+\langle\lambda,G(\Theta)\rangle(\omega))}}{\int_\Omega e^{-(-\gamma qB(\omega)+\langle\lambda,G(\Theta)\rangle(\omega))}dP(\omega)}dP(\omega) = 0, \text{ for all } \theta' \in \Theta.$$

For the right hand side of (3), the optimizer $\tilde{\theta} \in \Theta$ satisfies $\int_\Omega G(\theta')(\omega)e^{-\gamma(-qB(\omega)+G(\tilde{\theta})(\omega))}dP(\omega) = 0$, for all $\theta' \in \Theta$. From the uniqueness of λ, it holds that

$$\langle\lambda, G(\Theta)\rangle = \gamma G(\tilde{\theta}).$$

That is,

$$\frac{d\hat{Q}}{dP}(\omega) = \frac{e^{-\gamma(-qB(\omega)+G(\tilde{\theta})(\omega))}}{\int_\Omega e^{-\gamma(-qB(\omega)+G(\tilde{\theta})(\omega))}dP(\omega)}.$$

Next, we show that the convergence $\hat{Q} \to Q^0$, when $q \to 0$.

$$\begin{aligned}\lim_{q\downarrow 0}\sup_{\theta\in\Theta}\mathbb{E}\left[-e^{-\gamma\left(\int_0^T \theta_t^\top dS_t - Bq\right)}\right] &= \lim_{q\downarrow 0}\mathbb{E}\left[-e^{-\gamma\left(\int_0^T \tilde{\theta}_t^\top dS_t - Bq\right)}\right]\\ &\geq \lim_{q\downarrow 0}\mathbb{E}\left[-e^{-\gamma\left(\int_0^T (\theta_t^0)^\top dS_t - Bq\right)}\right]\\ &= \mathbb{E}\left[-e^{-\gamma\left(\int_0^T (\theta_t^0)^\top dS_t\right)}\right].\end{aligned}$$

On the other hand, using duality,

$$\begin{aligned}\lim_{q\downarrow 0}\sup_{\theta\in\Theta}\mathbb{E}\left[-e^{-\gamma\left(\int_0^T \theta_t^\top dS_t - Bq\right)}\right] &= \lim_{q\downarrow 0}\inf_{\eta\in\mathbb{R}_+}\left\{\frac{\eta}{\gamma}\ln\frac{\eta}{\gamma} - \frac{\eta}{\gamma} + \frac{\eta}{\gamma}\inf_{Q\in\mathcal{M}}\left\{H[Q|P] - \gamma q\mathbb{E}^Q[B]\right\}\right\}\\ &\leq \lim_{q\downarrow 0}\inf_{\eta\in\mathbb{R}_+}\left\{\frac{\eta}{\gamma}\ln\frac{\eta}{\gamma} - \frac{\eta}{\gamma} + \frac{\eta}{\gamma}\left(H[Q^0|P] - \gamma q\mathbb{E}^{Q^0}[B]\right)\right\}\end{aligned}$$

The solution of $\inf_{\eta\in\mathbb{R}_+}\left\{\frac{\eta}{\gamma}\ln\frac{\eta}{\gamma} - \frac{\eta}{\gamma} + \frac{\eta}{\gamma}\left(H[Q^0|P] - \gamma q\mathbb{E}^{Q^0}[B]\right)\right\}$ is given by $\gamma e^{-\gamma\left(\frac{1}{\gamma}H[Q^0|P] - q\mathbb{E}^{Q^0}[B]\right)}$. Therefore,

$$\begin{aligned}\lim_{q\downarrow 0}\sup_{\theta\in\Theta}\mathbb{E}\left[-e^{-\gamma\left(\int_0^T \theta_t^\top dS_t - Bq\right)}\right] &\leq \lim_{q\downarrow 0} -e^{-\gamma\left(\frac{1}{\gamma}H[Q^0|P] - q\mathbb{E}^{Q^0}[B]\right)}\\ &= -e^{-\gamma\left(\frac{1}{\gamma}H[Q^0|P]\right)} = \sup_{\theta\in\Theta}\mathbb{E}\left[-e^{-\gamma\int_0^T \theta_t^\top dS_t}\right].\end{aligned}$$

That is,

$$\lim_{q\downarrow 0}\mathbb{E}\left[-e^{-\gamma\left(\int_0^T \tilde{\theta}_t^\top dS_t - Bq\right)}\right] = \mathbb{E}\left[-e^{-\gamma\int_0^T (\theta_t^0)^\top dS_t}\right].$$

By the convexity of the exponential utility maximization problem, when $q \downarrow 0$,

$$-qB + G(\tilde{\theta}) \to G(\theta^0).$$

Therefore, when $q \downarrow 0$, the measure $\hat{Q}$ converges to MEMM, that is, $\hat{Q} \to Q^0$, which implies that $H(\tilde{Q}|P) \to H(Q^0|P)$.

Since Q^0 is MEMM, $H(Q|P) \geq H(Q^0|P)$ for any $Q \in \mathcal{M}$. So, $\frac{H(Q|P)-H(Q^0|P)}{q} \geq 0$, for $q > 0$. Assume that $\lim_{q\downarrow 0}\frac{H(Q|P)-H(Q^0|P)}{q} > 0$. Then, for any $Q \in \mathcal{M}$ and $q > 0$, some $\epsilon > 0$ exists, such that $\frac{H(Q|P)-H(Q^0|P)}{q} > \epsilon$. Thus, $H(Q|P) > H(Q^0|P) + q\epsilon$. This implies that, for $\hat{Q}$, $\lim_{q\downarrow 0} H(\hat{Q}|P) > H(Q^0|P) + \epsilon\lim_{q\downarrow 0} q$. That is, $\lim_{q\downarrow 0} H(\hat{Q}|P) > H(Q^0|P)$. However, $\lim_{q\downarrow 0} H(\hat{Q}|P) = H(Q^0|P)$. Therefore, $\lim_{q\downarrow 0}\frac{H(\hat{Q}|P)-H(Q^0|P)}{q} = 0$. The case of the utility indifference sell price is proved. By the same way, we can show the case of the utility indifference buy price.

Q.E.D.

Frittelli (2000b) [9] and Rouge (2000) [26] show that the utility indifference sell price is non-decreasing, when the risk-aversion increases. Since $p^b(B;q) = -p^s(-B;q)$ (Becherer (2003) [1]), it is easily shown that utility indifference buy price is non-increasing, when the risk-aversion increases. Following Lemma is similar with this.

Lemma 3.3. *For the increasing selling (buying) amount of random endowment B, utility indifference sell price is non-decreasing (utility indifference buy price is non-increasing).*

Proof It is sufficient to prove the case of utility indifference sell price, since we can similarly prove the case of utility indifference buy price.

The utility indifference sell price for a risk-aversion γ and an amount q is given by

$$\begin{aligned} p^s(B;q) &= \sup_{Q\in\mathcal{M}} \left\{ \mathbb{E}^Q[B] - \frac{1}{\gamma q}\left(H(Q|P) - H(Q^0|P)\right)\right\} \\ &= \mathbb{E}^{\hat{Q}}[B] - \frac{1}{\gamma q}\left(H(\hat{Q}|P) - H(Q^0|P)\right), \end{aligned}$$

where $\hat{Q}$ is the optimizer of $\sup_{Q\in\mathcal{M}}\left\{\mathbb{E}^Q[B] - H[Q|P]/\gamma q\right\}$. For $\epsilon > 0$, consider the utility indifference sell price $p^s(B;q+\epsilon)$. It is given by

$$\begin{aligned} p^s(B;q+\epsilon) = &\sup_{Q\in\mathcal{M}} \left\{ \mathbb{E}^Q[B] - \frac{1}{\gamma(q+\epsilon)}\left(H(Q|P) - H(Q^0|P)\right)\right\} \\ \geq &\mathbb{E}^{\hat{Q}}[B] - \frac{1}{\gamma(q+\epsilon)}\left(H(\hat{Q}|P) - H(Q^0|P)\right) \\ = &\mathbb{E}^{\hat{Q}}[B] - \frac{1}{\gamma q}\left(H(\hat{Q}|P) - H(Q^0|P)\right) + \frac{1}{\gamma q}\left(H(\hat{Q}|P) - H(Q^0|P)\right) \\ &- \frac{1}{\gamma(q+\epsilon)}\left(H(\hat{Q}|P) - H(Q^0|P)\right) \\ = &p^s(B;q) + \left(\frac{1}{\gamma q} - \frac{1}{\gamma(q+\epsilon)}\right)\left(H(\hat{Q}|P) - H(Q^0|P)\right) \geq p^s(B;q). \end{aligned}$$

Q.E.D.

Proof of Theorem 3.1 Assume that there are $I + J$ investors with risk-aversion $\{\gamma_i;\ i = 1,\cdots,I+J\}$. Let the allocation $q^s := \{q_i^s;\ i = 1,\cdots,I,$ for some $i, q_i^s > 0\}$ and $q^b := \{q_j^b;\ j = 1,\cdots,J,$ for some $j, q_j^b > 0\}$ be selling quantities and buying quantities, respectively.

Assume that some $p > \mathbb{E}^{Q^0}[B]$ and the allocation (q^s, q^b) is the equilibrium. From Lemma 3.2 and Lemma 3.3, the lower (upper) bound of the utility indifference sell price (utility indifference buy price) is $\mathbb{E}^{Q^0}[B]$. Therefore, any allocation (q^s, q^b) is not preferred for the buy side of the random endowment to not buying the random endowment. That is, the allocation (q^s, q^b) is impossible. Conversely, if $p < \mathbb{E}^{Q^0}[B]$, then any allocation (q^s, q^b) is not preferred for the sell side of the random endowment to not selling the random endowment. Therefore, the price $p^* = \mathbb{E}^{Q^0}[B]$ is only available price in which some allocation is available. Furthermore, the available allocation is given as $\{(q_i^s)^*;\ i = 1, \cdots, I, \text{for all } i, (q_i^s)^* = 0\}$, $\{(q_j^b)^*;\ j = 1, \cdots, J, \text{for all } j, (q_j^b)^* = 0\}$.

Q.E.D.

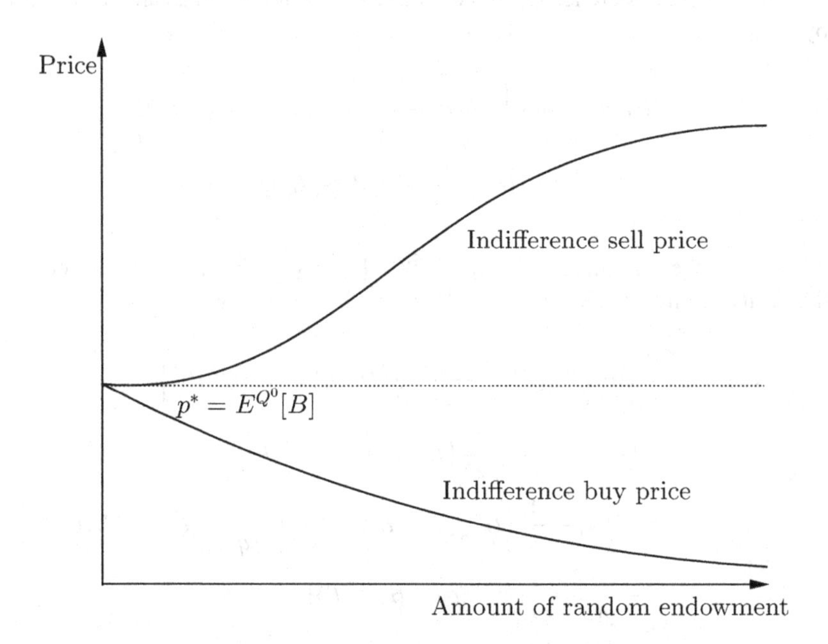

Figure 1. Utility indifference sell and buy price are non-decreasing and non-increasing on amount q of random endowment, respectively. The convergence to $\mathbb{E}^{Q^0}[B]$ implies the equilibrium price and equilibrium where any trade will not appear in the market of random endowment B.

Figure 1 shows the implication of Theorem 3.1. In fact, utility indifference sell price is non-decreasing on q and indifference buy price is non-increasing on q, and they converge to $\mathbb{E}^{Q^0}[B]$ when $q \to 0$. Indifference sell price implies supply curve of random endowment B and indifference buy price implies demand curve of random endowment B. The intuition, equilibrium is given by the intersection of demand curve and supply curve, tells us the equilibrium price is given by $\mathbb{E}^{Q^0}[B]$ and any trade of the random endowment will not appear in equilibrium.

4. Equilibrium with Model Uncertainty

In the previous section, we show that any random endowment B is not traded in equilibrium. However, this is counterintuitive. To overcome this, we contemplate the market incompleteness given in the previous section. Note that the market incompleteness discussed there is given by the existence of the random endowment B which might not necessarily have a perfect correlation with the underlying asset S. This is a very simple and straightforward setting for market incompleteness. However, there are various sources that generate market incompleteness; e.g. legal restrictions, the existence of untradeable assets, partial information, less liquidity, and so on. In this section, we consider the situation of partial information as the investors are not necessarily able to observe the true probability distribution P. This is considered as 'model risk' or 'model uncertainty'. Introducing this assumption, we consider utility indifference pricing and the equilibrium of random endowment.

For the introduction of model uncertainty, the Ellsberg paradox gives a good implication. As is well known, the Ellsberg paradox shows that people tend to prefer the situation where they can observe the true probability distribution to the situation where they cannot, even if the payoff is the same in each case. The disposition implied by the Ellsberg paradox, the so called uncertainty aversion, causes investors to be very cautious when they cannot observe the true probability distribution. The maxmin expected utility (MMEU) is one of the appropriate methods for depicting such a situation. The investor, who acts according to the MMEU, tries to maximize her (his) expected utility, while (s)he chooses subjective probability distribution which minimizes her (his) expected utility given their strong tendency towards caution.

4.1 Indifference pricing with model uncertainty

We consider the investor who don't necessarily observe the physical measure P and construct their strategy based on MMEU.

We define a subjective probability measure R which is equivalent to the physical probability measure P, and define the set $\mathcal{R}$ of subjective probability measures. Since the investor who cannot necessarily observe P, the martingale measure Q is given as the measure equivalent to the subjective probability measure R. For this assumption to make meaningful, we assume that $R \in \mathcal{R}$ is equivalent P, although investors cannot necessarily observe P. We define the set $\mathcal{Q}$ as the set of martingale measures; i.e.

$$\mathcal{X} := \{X \geq 0 : X \in L^1(Q) \text{ for all } Q \in \mathcal{Q} \text{ and } U(X) \in L^1(R) \text{ for all } R \in \mathcal{R}\}.$$

MMEU is the expected utility under the probability measure such that the expected utility will be the smallest under this probability; i.e. for the terminal wealth $X^{x,q}$, MMEU is given by,

$$\sup_{X^{x,q}-qB\in\mathcal{X}} \inf_{R\in\mathcal{R}} \mathbb{E}^R[U(X^{x,q}-qB)].$$

We have already known that for a given probability measure $R\in\mathcal{R}$, it holds that $\sup_{X^{x,q}-qB\in\mathcal{X}} \mathbb{E}^R\left[U(X^{x,q}-qB)\right] = -e^{-\inf_{Q\in\mathcal{Q}}\{\gamma(x+q(p-q\mathbb{E}^Q[B]))+H[Q|R]\}}$. Therefore, by the convexity,

$$\sup_{X^{x,q}-qB\in\mathcal{X}} \inf_{R\in\mathcal{R}} \mathbb{E}^R\left[U(X^{x,q}-qB)\right] = -e^{-\inf_{Q\in\mathcal{Q}}\inf_{R\in\mathcal{R}}\{\gamma(x+q(p-q\mathbb{E}^Q[B]))+H[Q|R]\}}.$$

We can proceed the calculation of superscript of the right hand side of the above equation as follows,

$$\begin{aligned}
&\inf_{Q\in\mathcal{Q}}\inf_{R\in\mathcal{R}}\left\{H[Q|R]+\gamma\left(x+q\left(p-\mathbb{E}^Q[B]\right)\right)\right\} \\
&\qquad = \inf_{Q\in\mathcal{Q}}\left\{\inf_{R\in\mathcal{R}} H[Q|R]+\gamma\left(x+q\left(p-\mathbb{E}^Q[B]\right)\right)\right\} \\
&\qquad = \inf_{Q\in\mathcal{Q}}\left\{\gamma\left(x+q\left(p-\mathbb{E}^Q[B]\right)\right)\right\}.
\end{aligned}$$

Similarly, $\sup_{X^{x,0}\in\mathcal{X}} \inf_{R\in\mathcal{R}} \mathbb{E}^P\left[U(X^{x,0})\right] = -e^{-\inf_{Q\in\mathcal{Q}}\inf_{R\in\mathcal{R}}\{H[Q|R]+\gamma x\}} = -e^{-\gamma x}$. Therefore, utility indifference price is given by,

$$x = \inf_{Q\in\mathcal{Q}}\left\{x+q\left(p-\mathbb{E}^Q[B]\right)\right\}.$$

Thus, utility indifference sell price is,

$$p^s(B;q) = \sup_{Q\in\mathcal{Q}} \mathbb{E}^Q[B],$$

and utility indifference buy price is,

$$p^b(B;q) = \inf_{Q\in\mathcal{Q}} \mathbb{E}^Q[B].$$

By definition, $\inf_{Q\in\mathcal{Q}} \mathbb{E}^Q[B] \le \mathbb{E}^{Q^0}[B] \le \sup_{Q\in\mathcal{Q}} \mathbb{E}^Q[B]$. We can depict the situation discussed above in Figure 2.

Note that, by (9)(10), utility indifference price is bounded by the price based on MMEU. Then, Figure 2 clearly implies that any trade will not appear in more extreme sense than Figure 1.

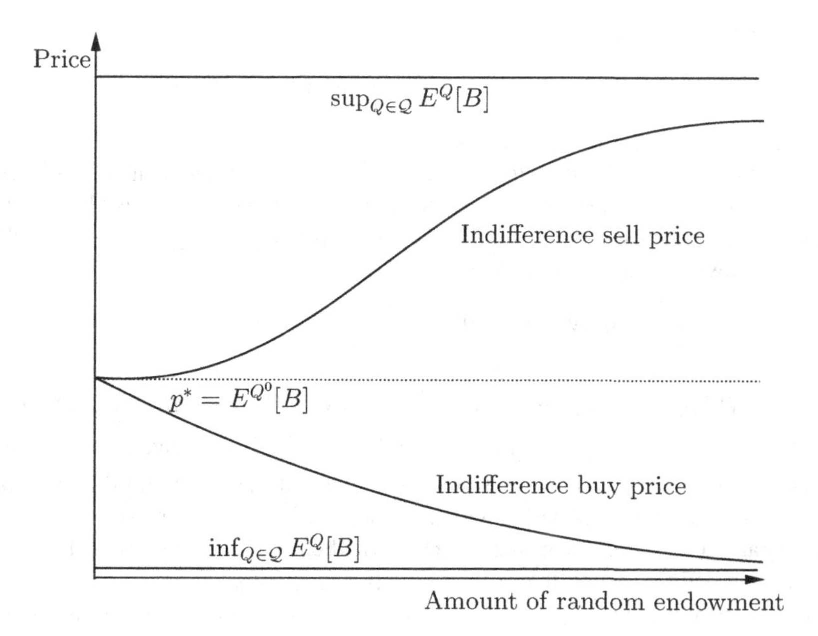

Figure 2. For utility indifference sell and buy price derived in the normal setting given in the previous section, the utility indifference prices with MMEU, $\sup_{Q\in\mathcal{Q}} \mathbb{E}^Q[B]$ and $\inf_{Q\in\mathcal{Q}} \mathbb{E}^Q[B]$, lie on their bounds. This shows that any equilibrium will not appear if all investor acts according to MMEU.

4.2 Equilibrium under model uncertainty and risk management

In the previous discussion, we show that trading of a random endowment does not appear in equilibrium even if MMEU is introduced. To determine the reason for this, we consider the following questions: Gilboa (2010)[11], "But why are your decision makers choosing the minimal EU of all those possible? Why are they so pessimistic?". These concerns clearly point out the high pessimistic nature of the investor with MMEU. In our study, we too agree that the investor with MMEU is extremely pessimistic. However, we pose a question: "But why doesn't the investor with MMEU construct a strategy with a risk management policy?" In this sense, the result shown in the previous subsection is natural: When the investor with MMEU does not have a risk management policy, the optimal strategy for the investor is to not participate in the market. Therefore, it is obvious that any trade of a random endowment will not appear in equilibrium.

Hereafter, we will show an appearance of trade of random endowment B by introducing a risk management policy under which the investor with MMEU construct the optimal strategy. Similarly to Gundel (2007)[15], let the risk management policy be described by a loss function $c(x)$ which is increasing function from

$\mathbb{R}$ to $[0, \infty)$. We define,

$$\mathcal{X}(\bar{x}) := \{X \in \mathcal{X} : \mathbb{E}^R[c(-X)] \leq \bar{x} \text{ for all } R \in \mathcal{R}\},$$

where $\bar{x}$ is given as the limit of loss.

Let us consider an investor who uses the random endowment B as a helpful hedging tool. Then, if (s)he can trade the random endowment, the risk limit can be loosen. In this sense, for the set $\mathcal{X}(\bar{x})$, by appropriately giving $x_2 > x_1$, we define the framework of utility indifference pricing,

$$\sup_{X^{x,0} \in \mathcal{X}(x_1)} \inf_{R \in \mathcal{R}} \mathbb{E}^R[U(X^{x,0})] = \sup_{X^{x,q} - qB \in \mathcal{X}(x_2)} \inf_{R \in \mathcal{R}} \mathbb{E}^R[U(X^{x,q} - qB)].$$

Let $\hat{X}^{x,q}$ be the optimizer of $\sup_{X^{x,q} \in \mathcal{X}} \inf_{R \in \mathcal{R}} \left\{\mathbb{E}^R[U(X^{x,q} - qB)]\right\}$ and $\tilde{X}^{x,q}$ be the optimizer of $\sup_{X^{x,q} - qB \in \mathcal{X}(x_2)} \inf_{R \in \mathcal{R}} \left\{\mathbb{E}^R[U(X^{x,q} - qB)]\right\}$. We have already known that $\hat{X}^{x,0} = x$. If the risk limit x_1 is loose such that the constraint doesn't work, the optimizer will be coincide with each other; i.e. $\hat{X}^{x,0} = \tilde{X}^{x,0}$. However, if the constraint x_1 is strict such that $c(-x) > x_1$, then $\tilde{X}^{x,0}$ will not coincide with x. Here, we define $\tilde{u}_0$ such that $-e^{-\gamma \tilde{u}_0} = \sup_{X^{x,0} \in \mathcal{X}(x_1)} \inf_{R \in \mathcal{R}} \mathbb{E}^R[U(X^{x,0})]$. Note that $\tilde{u}_0 \leq x$, since it holds $-e^{-\gamma \tilde{u}_0} \leq -e^{-\gamma x}$ by that the maximized expected utility with constraint is less than the maximized expected utility without constraint. If the risk limit works well such that $c(-x) > x_1$, then the above inequality strictly holds such that $\tilde{u}_0 < x$. In this subsection, we consider this case.

On the other hand, if the constraint x_2 is not strict such that it does not work, then the solution will be given by $\tilde{X}^{x,q} = \hat{X}^{x,q}$. Then, $\sup_{X^{x,q} \in \mathcal{X}(x_2)} \inf_{R \in \mathcal{R}} \mathbb{E}^R[U(X^{x,q} - qB)] = -e^{-\gamma\left(x + \left(pq - \sup_{Q \in \mathcal{Q}}\{q\mathbb{E}^Q[B]\}\right)\right)}$. Since the other case does not derive meaningful utility indifference price, we assume $x_2 > x_1$ satisfying above. Therefore, the framework of utility indifference pricing is given as follows,

$$\tilde{u}_0 = x + \left(pq - \sup_{Q \in \mathcal{Q}} \left\{q\mathbb{E}^Q[B]\right\}\right).$$

Utility indifference sell price is given by

$$\tilde{p}^s(B; q) = \sup_{Q \in \mathcal{Q}} \left\{\mathbb{E}^Q[B]\right\} + \frac{\tilde{u}_0 - x}{q} \leq \sup_{Q \in \mathcal{Q}} \left\{\mathbb{E}^Q[B]\right\},$$

and utility indifference buy price is,

$$\tilde{p}^b(B; q) = \inf_{Q \in \mathcal{Q}} \left\{\mathbb{E}^Q[B]\right\} - \frac{\tilde{u}_0 - x}{q} \geq \inf_{Q \in \mathcal{Q}} \left\{\mathbb{E}^Q[B]\right\}.$$

These pricing formula shows that $\tilde{p}^s(B; q)$ is increasing on q and $\tilde{p}^b(B; q)$ is decreasing on q. Therefore, for sufficiently small q, $\tilde{p}^s(B; q) < p^*$ and $\tilde{p}^b(B : q) > p^*$.

This implies that the existence of the intersection between the supply curve of the random endowment corresponding to utility indifference sell price and the demand curve corresponding to utility indifference buy price. We summarize the above discussion as follows,

Theorem 4.1. *Assume that the existence of investors with MMEU such that their loss functions satisfy* $c(-x) > x_1$ *and* $c\left(-\left(x + pq - \sup_{Q\in Q}\left\{q\mathbb{E}^Q[B]\right\}\right)\right) \leq x_2$, *for* $x_2 > x_1 > 0$. *Then, there exists trade of the random endowment in equilibrium.*

Figures 3, 4 and 5 show that the trade involving random endowment will appear in equilibrium, by introducing MMEU with a risk constraint, in three cases. In Figure 3, investors on both sell side and buy side have less information on the true probability distribution. Figure 4 shows the case where the investor on the buy side recognizes the true probability distribution, although the investor on the sell side has less information on this. Figure 5 depicts the opposite case, where the investor on the buy side does not have sufficient information on the true probability distribution, whereas the investor on the sell side has complete information on this. In any of these cases, the trade with positive amount of random endowment will appear in equilibrium.

5. Applying to Multi Assets Basis Model

In this section, we apply the above results to the basis model of Davis (2006) [5]. We consider the market which is constructed of tradable assets $S = \{S_t^i,\ t \in [0,\infty),\ i = 1,\cdots,d\}$ and untradeable assets $Y = \{Y_t,\ t \in [0,\infty)\}$. For simplicity, let S and Y be discounted price processes. On the given probability space, we assume stochastic processes for these assets as below; for $i = 1,\cdots,d+1$, let $W^i = \{W_t^i, 0 \leq t \leq \infty\}$ be $d+1$-dimensional Brownian Motion (Karatzas and Shreve (1998) [19]), that is, $\mathbb{E}[dW_t^i dW_t^j] = \begin{cases} dt & \text{if } i = j \\ 0 & \text{otherwise} \end{cases}$. For constant coefficients, SDE is described as follows,

$$\begin{pmatrix} dS_t/S_t \\ dY_t/Y_t \end{pmatrix} = \boldsymbol{\mu} dt + \Sigma d\mathbf{W}_t,$$

where $dS_t/S_t = (dS_t^1/S_t^1,\cdots,dS_t^d/S_t^d)^\top$, $\boldsymbol{\mu} = (\mu_1,\cdots,\mu_{d+1})^\top$, $d\mathbf{W}_t = (dW_t^1,\cdots,dW_t^{d+1})^\top$, $\Sigma = (\sigma_{i,j})_{i,j=1,\cdots,d+1}$.

We consider random endowment B as a contingent claim on the untradeable asset Y such that,

$$B := h(Y_T),$$

where $h(\cdot)$ is a European pay-off function.

We can find a local martingale measure for the process S. If Q is a probability measure equivalent to P on $\mathcal{F}_T$, then there exist adapted processes $g := \{g_t^i; t \in [0,T],\ i = 1,\cdots,d+1\} \in \mathcal{A}$, where g is square integrable on the time satisfying

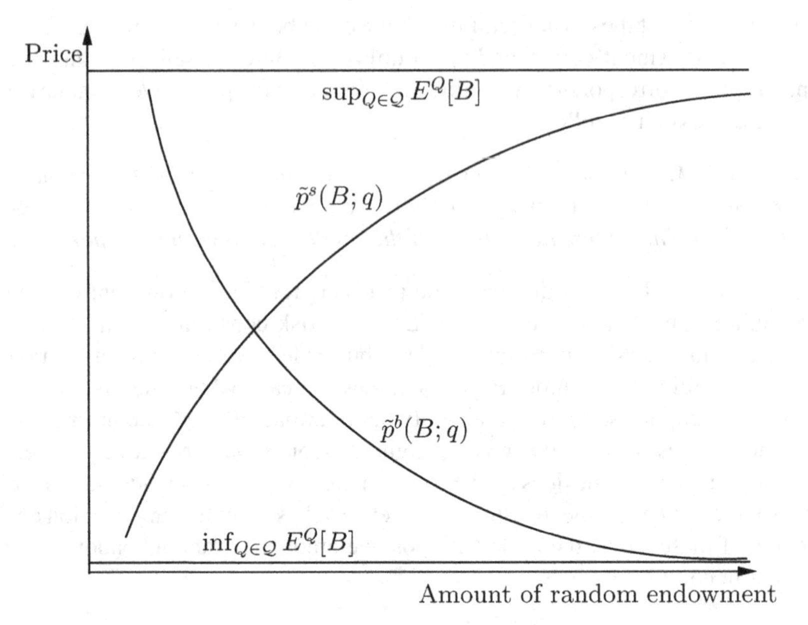

Figure 3. This figure depicts a case where investors on sell and buy side have less information on true probability distribution. Then, utility indifference sell and buy prices are derived under MMEU with risk constraint. The equilibrium shows the existence of trade of random endowment.

$\int_0^T g_t^i dt < \infty$, a.s. and $\mathcal{A}$ is the set of all adapted processes of square integrable on the time, and

$$\frac{dQ}{dP} = \exp\left(\sum_{i=1}^{d+1} \int_0^T g_t^i dW_t^i - \frac{1}{2}\sum_{i=1}^{d+1}\int_0^T (g_t^i)^2 dt\right).$$

For $\mathbf{g}_t := \left(g_t^1, \cdots, g_t^{d+1}\right)$, we can find that the transformation

$$d\tilde{\mathbf{W}}_t = d\mathbf{W}_t - \mathbf{g}_t dt,$$

where $\tilde{\mathbf{W}} = \{\tilde{W}_t^i; t \in [0,T], i = 1, \cdots, d+1\}$ is a Brownian motion under the measure Q. Then, Y is rewritten as follows,

$$\begin{aligned} dY_t/Y_t &= (\mu_{d+1} + \boldsymbol{\sigma}_{d+1}\mathbf{g}_t)dt + \boldsymbol{\sigma}_{d+1} d\tilde{\mathbf{W}}_t, \\ Y_0 &= y, \end{aligned}$$

where $\boldsymbol{\sigma}_{d+1} := (\sigma_{d+1,1}, \cdots, \sigma_{d+1,d+1})$ is $(d+1)$-th row vector of Σ, and y is constant.

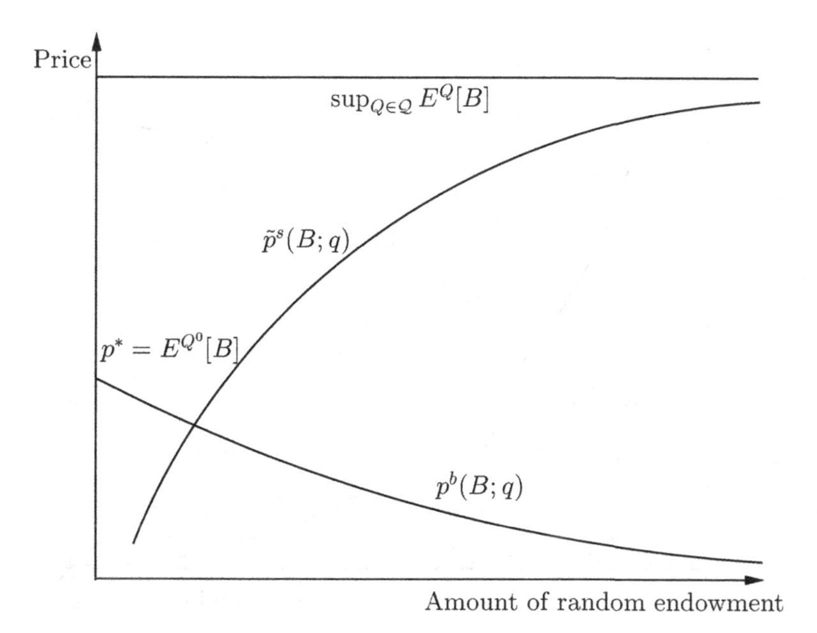

Figure 4. In this figure, only sell side investor adopt MMEU and risk constraint. Buy side investor have sufficient information on the true probability distribution. Then, her (his) utility indifference buy price is constructed under normal setting discussed in the previous section.

We define $\hat{\mathbf{g}} := -\Sigma^{-1}\boldsymbol{\mu}$. It is clear that $\hat{\mathbf{g}}$ is uniquely defined and we set $g_t^i = \hat{g}^i$ for $i = 1, \cdots, d$, where $\hat{g}^i$ is the i-the factor of $\hat{\mathbf{g}}$. Then S is local martingale under the measure Q.

To deduce the MEMM, we consider the problem $\inf_{Q\in\mathcal{M}} H[Q|P]$.

$$
\begin{aligned}
H[Q^0|P] &= \inf_{Q\in\mathcal{M}} H[Q|P] \\
&= \inf_{Q\in\mathcal{M}} \int \frac{dQ}{dP} \ln \frac{dQ}{dP} dP = \inf_{Q\in\mathcal{M}} \int \ln \frac{dQ}{dP} dQ \\
&= \inf_{g_t^{d+1}\in\mathcal{A}} \frac{1}{2} \sum_{i=1}^{d+1} \int_0^T (g_t^i)^2 dt = \inf_{g_t^{d+1}\in\mathcal{A}} \frac{1}{2} \left(\sum_{i=1}^{d} (\hat{g}^i)^2 T + \int_0^T (g_t^{d+1})^2 dt \right) \\
&= \frac{1}{2} \sum_{i=1}^{d} (\hat{g}^i)^2 T.
\end{aligned}
$$

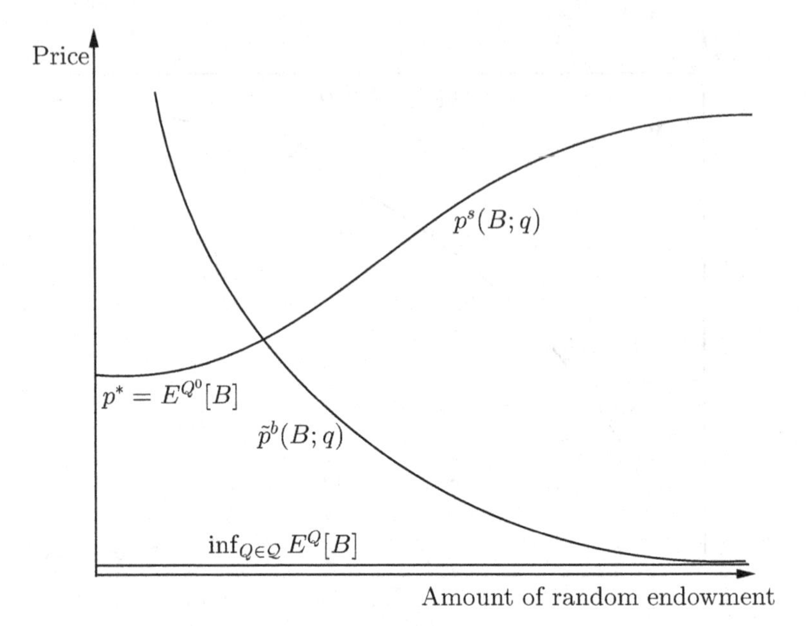

Figure 5. In this figure, only buy side investor adopt MMEU and risk constraint. Sell side investor have sufficient information on the true probability distribution. Then, her (his) utility indifference sell price is constructed under normal setting discussed in the previous section.

This implies that $\frac{dQ^0}{dP} = \exp\left(\sum_{i=1}^d \hat{g}^i W_T^i - \frac{1}{2}\sum_{i=1}^d (\hat{g}^i)^2 T\right)$. By Theorem 3.1, utility indifference price under MEMM is given as follows,

$$p^* = \mathbb{E}^{Q^0}[h(Y_T)] = \mathbb{E}^{Q^0}\left[h\left(ye^{\left(\mu_{d+1} - \frac{\sigma_{d+1}\sigma_{d+1}^\top}{2}\right)T + \sigma_{d+1}\tilde{\mathbf{W}}_T}\right)\right].$$

Next, we deduce the utility indifference sell price.

$$\begin{aligned}
p^s(B;q) &= \sup_{Q\in\mathcal{M}}\left\{\mathbb{E}^Q[B] - \frac{1}{\gamma q}(H(Q|P))\right\} + \frac{1}{\gamma q}H(Q^0|P) \\
&= \sup_{Q\in\mathcal{M}}\left\{\mathbb{E}^Q[h(Y_T)] - \frac{1}{\gamma q}\mathbb{E}^Q\left[\ln\left(\frac{dQ}{dP}\right)\right]\right\} + \frac{1}{\gamma q}\frac{1}{2}\sum_{i=1}^d (\hat{g}^i)^2 T \\
&= \sup_{Q\in\mathcal{M}}\mathbb{E}^Q\left\{h(Y_T) - \frac{1}{\gamma q}\left(\sum_{i=1}^{d+1}\int_0^T g_t^i d\tilde{W}_t^i + \frac{1}{2}\int_0^T (g_t^{d+1})^2 dt\right)\right\}.
\end{aligned}$$

We think a value function $V(t, y)$ as,

$$V(t,y) := \sup_{Q\in\mathcal{M}} \mathbb{E}^Q \left\{ h(Y_T) - \frac{1}{\gamma q}\left(\sum_{i=1}^{d+1} \int_t^T g_s^i d\tilde{W}_s^i + \frac{1}{2}\int_t^T (g_s^{d+1})^2 ds \right) \right\}$$
$$= \sup_{Q\in\mathcal{M}} \mathbb{E}^Q \left\{ -\frac{1}{\gamma q}\left(\sum_{i=1}^{d+1} g_t^i d\tilde{W}_t^i \right) - \frac{1}{2\gamma q}(g_t^{d+1})^2 dt + V(t+, y+) \right\}.$$

The HJB equation is deduced as follows,

(14)

$$V_t + \frac{1}{2}y^2 V_{yy}\boldsymbol{\sigma}_{d+1}\boldsymbol{\sigma}_{d+1}^\top + yV_y\left(\mu_{d+1} + \sum_{i=1}^{d} \sigma_{d+1,i}\hat{g}^i \right) + \sup_{g_t^k\in\mathcal{A}} \left\{ yV_y\sigma_{d+1,d+1}g_t^{d+1} - \frac{(g_t^{d+1})^2}{2\gamma q} \right\} = 0,$$

(15)

$$V(T, y) = h(y).$$

The optimal $\hat{g}_t^{d+1}$ is given by

$$\hat{g}_t^{d+1} = \gamma q \sigma_{d+1,d+1} y V_y$$

Then, (14) is rewritten as,

$$V_t + \frac{1}{2}y^2 V_{yy}\boldsymbol{\sigma}_{d+1}\boldsymbol{\sigma}_{d+1}^\top + \mu_{d+1}yV_y + \frac{1}{2}\gamma q\sigma_{d+1,d+1}^2 y^2 V_y^2 = 0, \tag{16}$$

where we used $\sum_{i=1}^{d} \sigma_{d+1,i}\hat{g}^i = 0$.

For $z := \ln y$ and $\tau = T - t$, we define a function $u(\tau, z)$ satisfying such that,

$$V(t,y) = \frac{\boldsymbol{\sigma}_{d+1}\boldsymbol{\sigma}_{d+1}^\top}{\gamma q \sigma_{d+1,d+1}^2}\left(\ln u(\tau, z) - \frac{1}{2}\boldsymbol{\sigma}_{d+1}\boldsymbol{\sigma}_{d+1}^\top \alpha^2 \tau + \alpha z \right),$$

where $\alpha := -(\mu_{d+1} - \frac{1}{2}\boldsymbol{\sigma}_{d+1}\boldsymbol{\sigma}_{d+1}^\top)/\boldsymbol{\sigma}_{d+1}\boldsymbol{\sigma}_{d+1}^\top$. Then, (15)(16) is rewritten as follows,

$$u_\tau = \frac{1}{2}\boldsymbol{\sigma}_{d+1}\boldsymbol{\sigma}_{d+1}^\top u_{zz},$$
$$\frac{\boldsymbol{\sigma}_{d+1}\boldsymbol{\sigma}_{d+1}^\top}{\gamma q \sigma_{d+1,d+1}^2}(\ln u(0, z) + \alpha z) = h(e^z).$$

Since it is well known heat equation, we can solve this PDE and derive an analytical form of $V(t, y)$ as follows,

$$V(t,y) = \frac{\boldsymbol{\sigma}_{d+1}\boldsymbol{\sigma}_{d+1}^\top}{\gamma q \sigma_{d+1,d+1}^2}\left(\ln \mathbb{E}^Q\left[e^{\frac{\gamma q\sigma_{d+1,d+1}^2}{\boldsymbol{\sigma}_{d+1}\boldsymbol{\sigma}_{d+1}^\top} h(\tilde{Y}_{T-t})} \right] \right),$$

where $\tilde{Y}_t := ye^{(\mu_{d+1}-\frac{1}{2}\sigma_{d+1}\sigma_{d+1}^\top)t+\sigma_{d+1}\tilde{\mathbf{W}}_t}$ is given by Y when $\mathbf{g} = \mathbf{0}$. Since $p^s(B;q) = V(0,y)$,

$$p^s(B;q) = \frac{\boldsymbol{\sigma}_{d+1}\boldsymbol{\sigma}_{d+1}^\top}{\gamma q \sigma_{d+1,d+1}^2}\left(\ln \mathbb{E}^Q\left[e^{\frac{\gamma q \sigma_{d+1,d+1}^2}{\sigma_{d+1}\sigma_{d+1}^\top}h(\tilde{Y}_T)}\right]\right).$$

Similarly, utility indifference buy price is given by

$$p^b(B;q) = -\frac{\boldsymbol{\sigma}_{d+1}\boldsymbol{\sigma}_{d+1}^\top}{\gamma q \sigma_{d+1,d+1}^2}\left(\ln \mathbb{E}^Q\left[e^{-\frac{\gamma q \sigma_{d+1,d+1}^2}{\sigma_{d+1}\sigma_{d+1}^\top}h(\tilde{Y}_T)}\right]\right).$$

Next, we consider indifference price with model uncertainty. First, consider $\sup_{Q\in\mathcal{Q}}\mathbb{E}^Q[B]$ and $\inf_{Q\in\mathcal{Q}}\mathbb{E}^Q[B]$. By the integrability of g, let $\hat{g}^{d+1} = \sup g_T^{d+1}$ and $\mathbf{g}^* := (\hat{g}^1,\cdots,\hat{g}^d,\hat{g}^{d+1})$. Then,

$$\sup_{Q\in\mathcal{M}}\mathbb{E}^Q[h(Y_T)] = \mathbb{E}[h(ye^{(\mu_{d+1}+\sigma_{d+1}\mathbf{g}^*)\mathbf{T}+\sigma_{\mathbf{d+1}}\tilde{\mathbf{W}}_\mathbf{T}}].$$

Similarly, we can derive

$$\inf_{Q\in\mathcal{M}}\mathbb{E}^Q[h(Y_T)] = \mathbb{E}[h(ye^{(\mu_{d+1}+\sigma_{d+1}\mathbf{g}_*)\mathbf{T}+\sigma_{\mathbf{d+1}}\tilde{\mathbf{W}}_\mathbf{T}}].$$

where $\tilde{g}^{d+1} := \inf g_T^{d+1}$ and $\mathbf{g}_* := (\hat{g}^1,\cdots,\hat{g}^d,\tilde{g}^{d+1})$. Therefore, for given $\tilde{u}_0$, utility indifference price with MMEU with risk management policy is given by,

$$\tilde{p}^s(B;q) = \mathbb{E}[h(ye^{(\mu_{d+1}+\sigma_{d+1}\mathbf{g}^*)\mathbf{T}+\sigma_{\mathbf{d+1}}\tilde{\mathbf{W}}_\mathbf{T}}] + \frac{\tilde{u}_0 - x}{q},$$

$$\tilde{p}^b(B;q) = \mathbb{E}[h(ye^{(\mu_{d+1}+\sigma_{d+1}\mathbf{g}_*)\mathbf{T}+\sigma_{\mathbf{d+1}}\tilde{\mathbf{W}}_\mathbf{T}}] - \frac{\tilde{u}_0 - x}{q}.$$

Finally, we show numerical examples. Let investor on sell side be with full information on true probability distribution, then the utility indifference price $p^s(B;q)$ is derived under the normal setting. On the other hand, let investor on buy side with less information on the true probability distribution, then the utility indifference price $\tilde{p}^b(B;q)$ is derived under MMEU and risk constraint.

For simplicity, we consider ATM Call option with 1 year maturity and give the parameter as $\mu = (0.05, 0.02, 0.01)$ and $\Sigma = \begin{pmatrix} 0.4 & 0.2 & 0.1 \\ 0.2 & 0.8 & 0.3 \\ 0.1 & 0.3 & 0.5 \end{pmatrix}$. Furthermore, we fix risk aversion of the investor on buy side be 0.05 and $u_0 = 10.1$.

Figure 6 and 7 show numerical examples, respectively.

Figure 6 depicts a case where utility indifference buy price $\tilde{p}^b(B;q)$ is derived under MMEU and risk constraint with given parameters and utility indifference sell price $p^s(B;q)$ is derived under normal setting. On the utility indifference sell price, we show three cases where risk aversions $\gamma = \{0.15, 0.1, 0.05\}$ of sell side investor are different from each other. It shows that the higher the risk aversion, the more amount of random endowment which is determined by the intersection of indifference sell and buy price is. This implies that the participation of less pessimistic investors makes market more active.

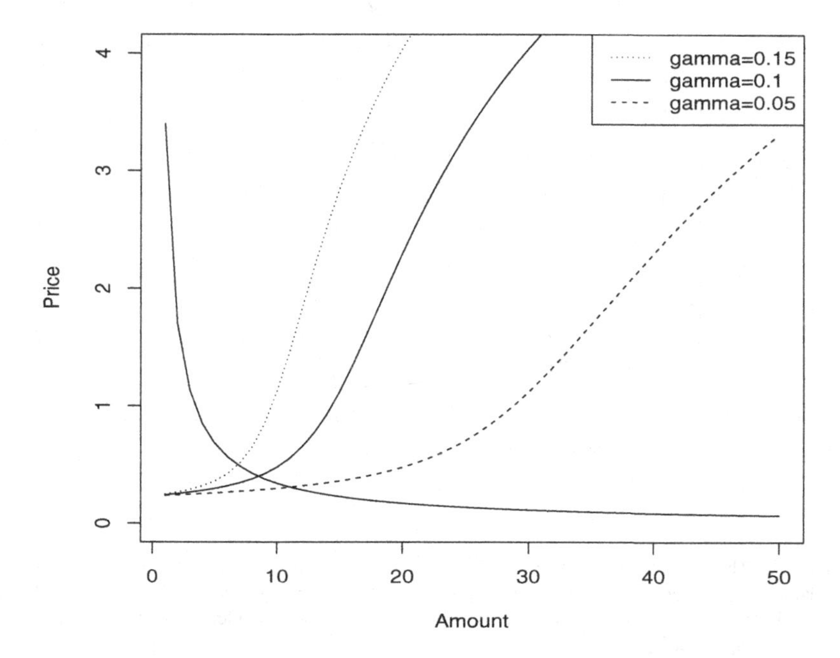

Figure 6. This figure depicts a case where utility indifference buy price is derived under MMEU with risk constraint and utility indifference sell price is derived under normal setting. On utility indifference sell price, we show three cases where risk aversions of sell side investor are different from each other. It shows that the higher the risk aversion, the more amount of random endowment which is defined by the intersection of indifference sell and buy price is.

Figure 7 shows a case where utility indifference buy price is derived under MMEU and risk constraint with given parameters and utility indifference sell price is derived under normal setting. However, in this figure, the risk aversion of sell side investor is fixed. On the other hand, utility indifference buy prices are depicted for three initial endowments $x = \{13.5, 14.5, 15.5\}$ of buy side investor. The equilibrium which is given by the intersection of indifference buy and sell price is getting larger, when the initial endowment is larger. This implies that traded amount of the random endowment is increased by the existence of wealthier investors.

Both of these two numerical examples is consistent with our intuition and we think these examples implies that our model depicts well the market reality.

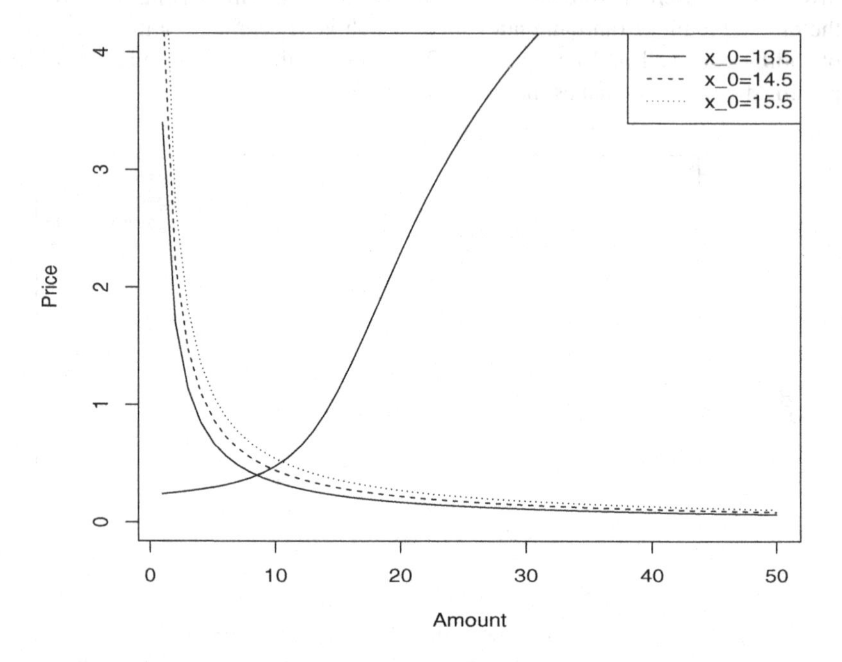

Figure 7. The parameters are same as the case of Figure 6, other than that three cases of the initial endowment of buy side investor are given and the risk aversion of sell side investor is fixed. The intersection between indifference sell and buy price moves from left to right, when the initial endowment is getting larger.

References

1. Becherer, D. (2003), "Rational hedging and valuation of integrated risks under constant absolute risk aversion," *Insuarance: Mathematics & Economics*, **33**, 1–28.
2. Biagini, F., Frittelli, M. and Grasselli, M. (2011), "Indifference price with general semimartingales," *Mathematical Finance*, **21**, 423–446.
3. Botelho, F.S. (2009), *Variational convex analysis*, Ph.D. dissertation, Virginia Polytechnic Institute and State University.
4. Cochrane, P.C. and Macgregor, T.H. (1978), "Fréchet differentiable functionals and support points for families of analytic functions," *Transactions of the American Mathematical Society*, **236**, 75–92.
5. Davis, M.H.A. (2006), "Optimal hedging with basis risk," in Y. Kabanov, R. Lipster, and J. Stoyanov, editors, *2nd bachelier colloquium on stochastic calculus and probability,9 - 15 January 2005, Metabief, FRANCE*, 169–187. Springer-Verlag.

6. Delbaen, F., Grandits, P., Rheinländer,Th., Samperi, D., Schweizer, M. and Stricker, Ch. (2002), "Exponential hedging and entropic penalties," *Mathematical Finance*, **12**, 99–123.
7. Favretti, M. (2005), "Isotropic submanifolds generated by the maximum entropy principle and Onsager reciprocity relations," *Journal of Functional Analysis*, **227**, 227–243.
8. Föllmer, H. and Schweizer, M. (1991), "Hedging of contingent claims under incomplete information," in M.H.A. Davis and R.J. Elliott, editors, *Applied Stochastic Analysis, Stochastics Monographs*, volume 5. Gordon and Breach, London.
9. Frittelli, M. (2000), "Introduction to a theory of value coherent with the no-arbitrage principle," *Finance and Stochastics*, **4**, 275–297.
10. Frittelli, M. (2000), "The minimal entropy martingale measure and the valuation problem in incomplete markets," *Mathematical Finance*, **10**, 39–52.
11. Gilboa, I. (2010), *Theory of decision under uncertainty*, Cambridge.
12. Goll, T. and Rüschendorf, L. (2001), "Minimax and minimal distance martingale measures and their relationship to portfolio optimization," *Finance and Stochastics*, **5**, 557–581.
13. Gombani, A., Jaschke, S. and Runggaldier, W. (2007), "Consistent price systems for subfiltrations," *ESAIM: P&S*, **11**, 35–39.
14. Grandits, P. (1999), "The p-optimal martingale measure and its claims under incomplete information," *Bernoulli*, **5**, 225–247.
15. Gundel, A. and Weber, S. (2007), "Robust utility maximization with limited downside risk in incomplete markets," *Stochastic Processes and their Applications*, **117**, 1663–1688.
16. Hugonnier, J. and Kramkov, D. (2004), "Optimal investment with random endowments in incomplete markets," *The Annals of Applied Probability*, **14**, 845–864.
17. Hugonnier, J., Kramkov, D. and Shachermayer, W. (2005), "On utility-based pricing of contingent claims in incomplete markets," *Mathematical Finance*, **15**, 203–212.
18. Ihara, S. (1993), *Information theory for continuous system*, World Scientific.
19. Karatzas, I. and Shreve, S.E. (1998), *Methods of Mathematical Finance*, Springer.
20. Kramkov, D. and Schachermayer, W. (1999), "The Asymptotic elasticity of utility functions and optimal investment in incomplete markets," *The Annals of Applied Probability*, **9**, 904–950.
21. Monoyios, M. (2004), "Performance of utility-based strategies for hedging basis risk," *Quantitative Finance*, **4**, 245–255.
22. Monoyios, M. (2007), "The minimal entropy measure and an Esscher transform in an incomplete market model," *Statistics and Probability Letters*, **77**, 1070–1076.
23. Monoyios, M. (2008), "Utility indifference pricing with market incompleteness," in Ehrhardt M., editor, *Nonlinear Models in Mathematical Finance: New Research Trends in Option Pricing*, 67–100, Nova Science Publishers.
24. Danilova, A., Monoyios, M. and Ng, A. (2010), "Optimal investment with inside information and parameter uncertainty," *Matematics and Financial Economics*, **3**, 13–38.
25. Mss-Colell, A., Whinston, M.D. and Green, J.R. (1995), *Microeconomic Theory*, Oxford University Press.
26. Rouge, R. and El Karoui, N. (2000), "Pricing via utility maximization and entropy," *Mathematical Finance*, **10**, 259–276.

27. Schachermayer, W. (2000), "Optimal investment in incomplete financial markets," in H. Geman, D. Madan, St.R. Pliska, and T. Vorst, editors, *Mathematical Finance: Bachelier Congress 2000*, 427–462, Springer-Verlag.
28. Schweizer, M. (1999), "A minimality property of the minimal martingale measure," *Statistics & probability letters*, **42**, 27–31.
29. Schweizer, M. (2010), "The Minimal Entropy Martingale Measure," in Cont, R., editor, *Encyclopedia of Quantitative Finance*, 1195–1200, Wiley.
30. Sircar, R. and Zariphopoulou, T. (2004), "Bounds and asymptotic approximations for utility prices when volatility is random," *SIAM Journal on Control and Optimization*, **43**, 1328 – 1353.
31. Zariphopoulou, T. (2001), "A solution approach to valuation with unhedgeable risks," *Finance and Stochastics*, **5**, 61–82.

Volume Imbalance and Market Making*

Álvaro Cartea[1], Ryan Donnelly[2], and Sebastian Jaimungal[3]

[1]Department of Mathematics, University College London,
25 Gordon St., London, England WC1E 6BT. Email: a.cartea@ucl.ac.uk
[2]Swiss Finance Institute, École Polytechnique Fédérale de Lausanne (EPFL),
Quartier UNIL-Dorigny, Extranef 218 CH-1015 Lausanne, Switzerland. Email:
ryan.donnelly@epfl.ch
[3]Department of Statistics & Mathematical Finance Program, University of Toronto,
100 St. George St., Toronto, Ont. Canada M5S 3G3. Email:
sebastian.jaimungal@utoronto.ca

Shortcomings of continuous and static microstructure models are noted with motivation provided by data from the NASDAQ. The influence of order imbalance on microstructure dynamics is incorporated in to a model which allows the agent to adjust their strategy based on an easily observable quantity. The predictive power of order imbalance allows the agent to decide when they should trade more agressively to take advantage of beneficial price movements, and when they should trade more conservatively to protect against adverse selection effects. High imbalance results in a stronger inclination to place limit buy orders with the opposite effect on limit sell orders.

Key words: Algorithmic trading, market making, stochastic control, volume order imbalance

1. Introduction

Many approaches to developing algorithmic trading strategies are based on market dynamics which are either too simplistic or driven by unobservable factors. Typical assumptions of this type are that the midprice of the asset evolves according to an arithmetic Brownian motion, that market orders arrive according to the increments of a Poisson process, and that limit orders can be placed at prices

*Send all correspondence to Ryan Donnelly, Swiss Finance Institute, École Polytechnique Fédérale de Lausanne (EPFL), Quartier UNIL-Dorigny, Extranef 218 CH-1015 Lausanne, Switzerland. Email:ryan.donnelly@epfl.ch

corresponding to any positive real number. In reality, the midprice is a pure jump process, market orders exhibit clustering, and limit orders can only be placed at integer multiples of the asset's tick size (usually one cent). Examples of these simplistic models can be seen in [1], [5], [6], and [2]. For models which incorporate some more realistic features, or attempt to take advantage of short term speculation, see [7] and [3]. In this work, a realistic model is proposed for the market dynamics which are influenced by observable quantities with motivation given by analysis of real data. The market maker (MM) optimizes their trading strategy by observing the state of the limit order book which has a strong effect on short term market behaviour, allowing the MM to take advantage of short term speculation.

2. Shortcomings of Continuous Model

This section will discuss a shortcoming of the models referenced in the introduction. The two main characteristics of the model which contribute to its shortcomings are its continuous structure and non-speculative nature. By continuous structure, this makes reference to the fact that limit orders can be posted at any real valued distance $\delta \geq 0$ from the midprice. In reality, this is not the case; agents may post limit orders at integer multiples of the "tick" size (often one cent). An agent acting according to such a model would have to compute their optimal trading strategy and then round it to an appropriate tick. The non-speculative nature of the model will be discussed in Section 3.

2.1 Tick activity

The side effects of a continuous limit order book will be illustrated here. In Figure 1, data for three different equities (Bed Bath & Beyond (BBBY), Microsoft (MSFT), and Teva Pharmaceutical Industries (TEVA)) is used to show how many market orders in a full month of trading are executed at the first available tick past the midprice, versus how many are executed beyond the first tick due to a large market order volume.

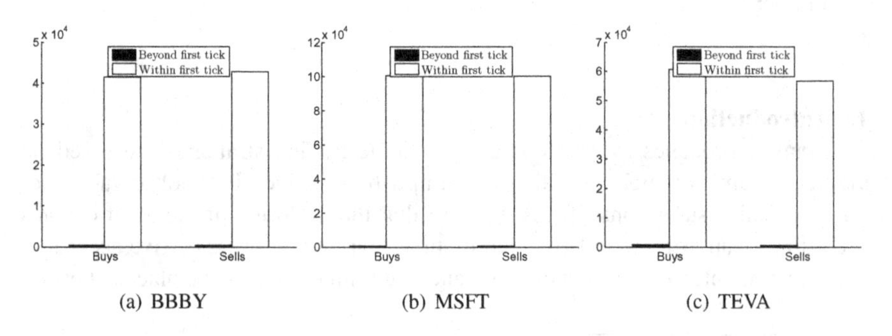

Figure 1. Number of market buy and sell orders that trade with limit orders at the first tick or beyond. Data is taken from a full month of trading in January 2011 on the NASDAQ exchange.

Clearly, a significant proportion of all trades occur at the first tick. With a continuous range of limit order prices as in the models referenced in the introduction, the likelihood of an order to be filled by a given market order can be finely adjusted by making small changes to the limit order price. But with discrete ticks that exhibit drastically different activity throughout the day, this fine adjustment is not possible. Instead, the agent's order will either be filled with some probability at the best tick, or essentially not filled if posted at the next tick. For this reason, we elect to use a different class of control processes than is often used in previous literature. In this setting, the market maker will only be able to post limit orders at the best bid and ask, and the control will simply be whether such an order is posted or not.

3. Imbalance as a Market Predictor

The previously referenced models are static in the sense that none of the dynamics are dependent on changing market conditions. A change in market conditions can have an effect on whether one specific trading strategy is beneficial or detrimental to the agent. Unfortunately, the agent is able to make no distinction between these cases if the model does not account for this behaviour. This section will introduce the quantity called volume order imbalance and show how it acts as an indicator of short term future market behaviour.

Volume order imbalance is defined as the proportion of limit order volume at the best bid with respect to the total volume at the best bid and ask. If the volume at the best bid (ask) is denoted V_t^b (V_t^a), then imbalance may be defined:

$$I_t = \frac{V_t^b}{V_t^a + V_t^b}.$$

Clearly, I_t is a stochastic process with $I_t \in [0, 1]$.

3.1 Market order activity

In Figure 1, the total number of market buy and sell orders that touched the first tick or beyond the first tick were counted. Here, a similar counting exercise will be performed, except instead of being concerned with the ticks that are involved with the market order, the likelihood of a buy or sell order will be shown to depend on the value of I_t. To do so, the interval $[0, 1]$ is divided into three subintervals: $[0, 0.3)$, $[0.3, 0.7]$, and $(0.7, 1]$ which will be referred to as the low, mid, and high regimes of imbalance respectively. The number of market buy and sell orders that occur with I_t lying in each of these intervals is counted and displayed in Figure 2. The three intervals are chosen symmetrically around 0.5, but otherwise arbitrarily, except to keep the total number of trades within each regime roughly the same over all three equities.

Clearly, acting under a model which does not take into account the state of the limit order book blinds the agent to a significant source of information, namely the likelihood of the next market order being a buy or a sell. This dependence on

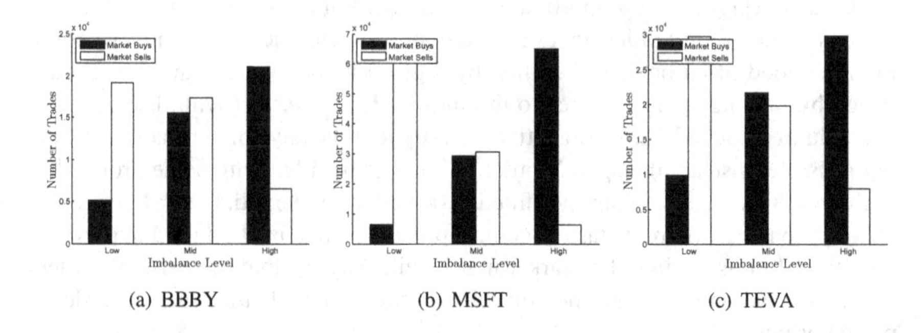

(a) BBBY (b) MSFT (c) TEVA

Figure 2. Number of market buy and sell orders that occur in each regime of imbalance. Data is taken from a full month of trading in January 2011.

market order intensity as a function of imbalance will be incorporated into the model in Section 5.

3.2 Midprice dynamics

In addition to the regime dependent intensity of market orders, the movement of the asset's midprice will also be shown to depend on the imbalance regime. Consider once again the three intervals $[0, 0.3)$, $[0.3, 0.7]$, and $(0.7, 1]$. For each of these intervals, the midprice will be observed $20ms$ after a market order, and the change with respect to the midprice immediately before the market order is computed.

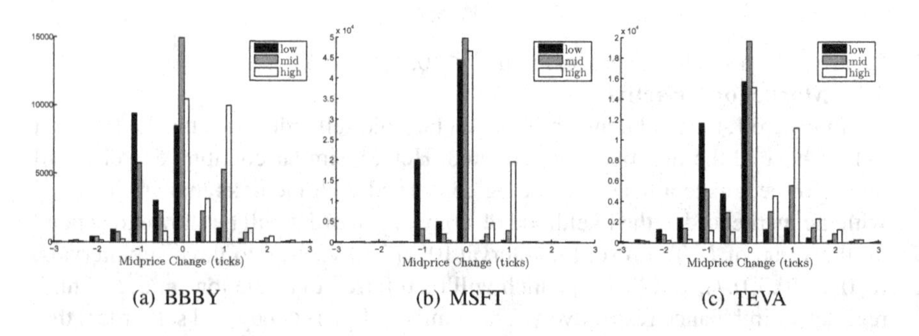

(a) BBBY (b) MSFT (c) TEVA

Figure 3. Distribution of midprice changes $20ms$ after a market order. Data is taken from a full month of trading (January 2011).

In Figure 3, similar behaviour is observed over all three equities. Specifically, when imbalance is low, the midprice generally decreases shortly after a market order. Similarly, it increases shortly after a market order when imbalance is high.

The changes are roughly symmetric when imbalance is in the mid regime. This is another significant observation of market behaviour which is ignored in many other models.

It will also be of importance to see how these midprice changes behave not only after market orders, but after a specific type of market order (buy or sell). This decomposition is displayed in Figure 4 for market buy orders only. Again it is clear that there is regime dependence on the distribution of these changes. When imbalance is high, the size of an upward movement after a market buy order is generally larger than in the other two regimes. In the low imbalance regime, the changes are generally still positive (as would be expected after a market buy order), but the magnitude of the changes are smaller than the other two regimes. Similar behaviour occurs for market sell orders, except changes are generally negative and larger for the low imbalance regime rather than the high imbalance regime. This is the second example of behaviour that is not captured by other models.

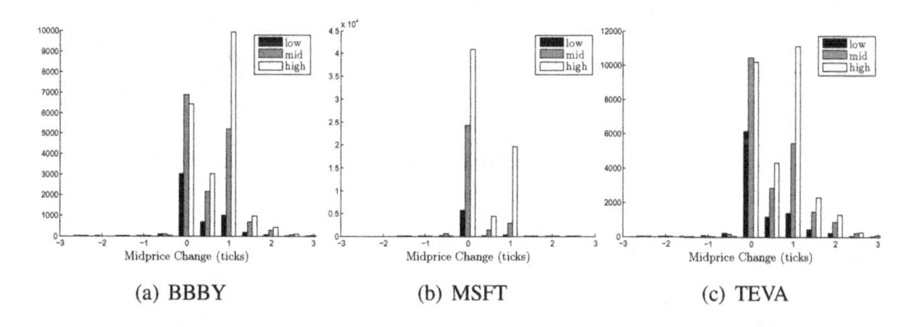

Figure 4. Distribution of midprice changes $20ms$ after a market buy order. Data is taken from a full month of trading (January 2011).

By ignoring the differences seen above in the short term dynamics of midprice movements and market order intensity, the agent is unable to protect themselves against adverse selection effects, and they are unable to take advantage of situations which would result in a benefit to their objective.

4. Modelling Imbalance

The nature of the limit order book makes the imbalance process itself a volatile pure jump process. Some of the volatility can be smoothed out by redefining the imbalance process to include limit order volumes at ticks beyond the first, or by defining it to be an average over a moving window. In Figure 5, the path of I_t is shown over a time period of one minute for each of the three equities considered above.

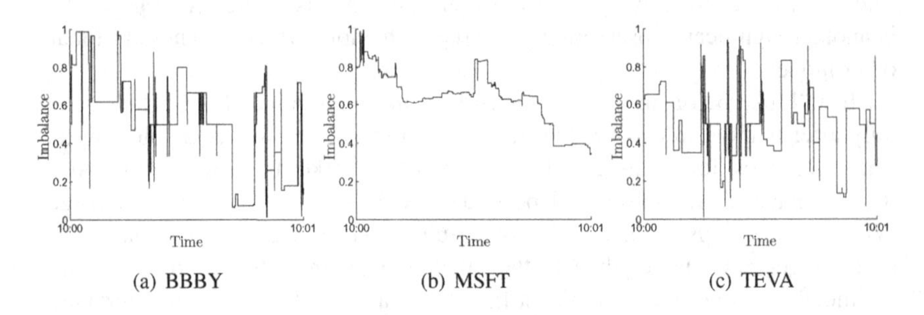

Figure 5. Imbalance process for one minute of data.

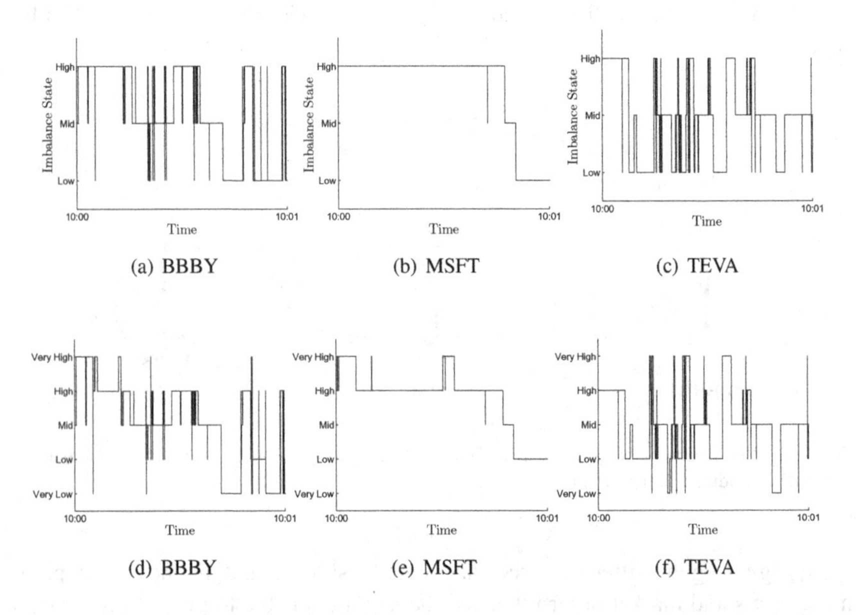

Figure 6. Imbalance regime processes for three states (top row) and five states (bottom row) over one minute.

At this point, a modelling decision must be made for the imbalance process. Rather than modelling the process of imbalance directly, the imbalance regime will be modelled using a continuous time Markov chain. This simplifying choice is motivated by the following observations:

- When imbalance is considered as a regime switching process with only three states, it has a significant amount of predictive power as seen in the previous section.

- The dependence of market order intensity and midprice changes on imbalance are very easy to calibrate using a regime switching process.

- Modelling the imbalance process directly would be significantly more complicated than modelling the regime switching process.

Up to this point, only three regime states of imbalance have been considered. This was done to simplify the demonstration of its predictive power, and also to demonstrate that even with a small number of regimes, its predictive power is significant. As one considers more regimes for imbalance, the resulting fine structure becomes more realistic, but the dependence of other market dynamics become more difficult to calibrate accurately. In Figure 6, the regime processes that correspond to the paths in Figure 5 are shown when the number of regime states is 3 and 5.

5. Market Model

In this section, dynamics for the market underliers are proposed which reflect the dependence on volume order imbalance. The factors to consider are i) imbalance itself, ii) the midprice of the asset, iii) the dynamics of market orders, and iv) the interaction between market orders and limit orders. Once these factors have specified dynamics, the agent's optimal control problem can be stated.

Order Imbalance: The imbalance process I_t is not modelled directly. Instead, a process which acts as an approximation to I_t is modelled as follows: divide the interval $[0, 1]$ into n subintervals, U_k, such that $\cup_{k=1}^{n} U_k = [0, 1]$ and $U_i \cap U_j = \emptyset$ for all $i \neq j$. Let Z_t be a continuous time Markov chain with generator matrix G that has elements $G_{i,j}$ (recall that $\mathbb{P}(Z_t = j | Z_s = i) = [\exp\{G(t-s)\}]_{i,j}$). The process Z_t is intended to serve as an approximation to I_t by the relation $Z_t = k$ when $I_t \in U_k$. Each subinterval will be referred to as an imbalance regime, and the set of regimes will be denoted $\mathcal{Z}$.

Market Orders: Let $\mu^{\pm}$ be doubly stochastic Poisson random measures with compensators $\nu_{\mathbb{P}}^{\pm}(dy, dt) = \lambda_t^{\pm} F_{Z_t}^{\pm}(y) dy dt$ where $\lambda_t^{\pm} = \lambda^{\pm}(Z_t)$. Each point in the support of $\mu^{\pm}$ represents a market buy order (+) or market sell order (−), and specifically, the total number of market buy orders is $M_t^+ = \int_0^t \int_{-\infty}^{\infty} \mu^+(dy, du)$, and of market sell orders is $M_t^- = \int_0^t \int_{-\infty}^{\infty} \mu^-(dy, du)$. With this formulation, market buy (sell) orders arrive according to a doubly stochastic Poisson process with intensity that depends directly on the imbalance regime Z_t through the function $\lambda^+(Z_t)$ ($\lambda^-(Z_t)$). This dependence is intended to reflect the information conveyed in Section 3.1.

Midprice: The midprice of the asset is defined as:

$$S_t = S_0 + \int_0^t \int_{-\infty}^{\infty} y\mu^+(dy, du) - \int_0^t \int_{-\infty}^{\infty} y\mu^-(dy, du) + \xi^+ J_t^+ - \xi^- J_t^-, \tag{1}$$

where $\xi^{\pm}$ are constants and $J_t^{\pm}$ are Poisson processes with intensities $\beta^{\pm}$. With this formulation, the y dimension of a point in the support of $\mu^{\pm}$ represents the distance that the midprice jumps after a market buy or sell order. This jump size has probability density function $F_{Z_t}^{\pm}(y)$, depending on the state at the time of the jump. The dependence of the jump distribution on Z_t reflects the information from Section 3.2. The terms $\xi^{\pm} J_t^{\pm}$ represent changes in the midprice that are not a result of market order executions. These can arise, for example, due to placements and cancellations of limit orders. Using constant values of $\xi^{\pm}$ may seem restrictive, but if the asset is very liquid then any change in the midprice due to limit order activity is very likely only a single value.

Consider the completed filtered probability space $(\Omega, \mathbb{F}, \mathcal{F} = \{\mathcal{F}_t\}_{0\leq t\leq T}, \mathbb{P})$ where $\mathcal{F}$ is generated by the midprice $(S_t)_{0\leq t\leq T}$, the number of market buy and sell orders that have arrived $(M_t^{\pm})_{0\leq t\leq T}$, and the imbalance regime process $(Z_t)_{0\leq t\leq T}$. The MM is given control processes $\gamma_t^{\pm} \in \{0, 1\}$ which are $\mathcal{F}_t$-predictable. The MM is restricted to posting their buy and sell limit orders at a fixed distance from the midprice, $\frac{\Delta}{2}$, which represent only posting at the best bid and ask. The control processes $\gamma_t^{\pm}$ represent whether the MM posts an order or not. If $\gamma_t^{+} = 1$, then the MM is posting a limit sell order at time t at a price of $S_t + \frac{\Delta}{2}$. If $\gamma_t^{+} = 0$, then the MM is not posting any limit sell order at time t. Similarly for the buy side. The MM is assumed to be filled by any market order that is executed as long as they have an active limit order at the time of the execution. It is not difficult to generalize to the case where the MM's limit orders are filled only by some proportion of all market orders, ρ. The restriction that the MM may only post at a fixed spread is motivated by the discussion in Section 2.1. This gives the following dynamics to the MM's inventory, $(q_t)_{0\leq t\leq T}$:

$$dq_t = \gamma_{t^-}^{-} dM_t^{-} - \gamma_{t^-}^{+} dM_t^{+}.$$

If a market buy order lifts the MM's limit sell order, then their wealth increases by $S_{t^-} + \frac{\Delta}{2}$. Similarly, a market sell order that hits the MM's bid will decrease the MM's wealth by $S_{t^-} - \frac{\Delta}{2}$. If the MM does not have a limit order posted on one side of the book, then activity on that side will not change their wealth. This results in the following dynamics of wealth:

$$dX_t = \gamma_{t^-}^{+}(S_{t^-} + \tfrac{\Delta}{2})dM_t^{+} - \gamma_{t^-}^{-}(S_{t^-} - \tfrac{\Delta}{2})dM_t^{-}.$$

The MM is risk-neutral but capital constrained, so they can not acquire large long or short inventory positions. The MM restricts their inventory such that $-\infty < \underline{q} \leq q_t \leq \overline{q} < \infty$. This is done by setting $\gamma_t^{+} = 0$ whenever $q_t = \underline{q}$ and $\gamma_t^{-} = 0$ whenever $q_t = \overline{q}$. This ensures that the inventory will never deviate beyond the finite bounds. The MM selects the strategy $(\gamma_t^{\pm})_{0\leq t\leq T}$ which maximizes the expected terminal wealth:

$$H(t, x, q, S, Z) = \sup_{(\gamma_s^{\pm})_{t\leq s\leq T} \in \mathcal{A}} \mathbb{E}_{t,x,q,S,Z}\Big[X_T + q_T\Big(S_T - \ell(q_T)\Big)\Big] \tag{2}$$

where T is the terminal time of the strategy, q_T final inventory, $\mathbb{E}_{t,x,q,S}[\cdot]$ denotes $\mathbb{P}$ expectation conditional on $X_{t^-} = x$, $q_{t^-} = q$ and $S_{t^-} = S$, and $\mathcal{A}$ denotes the set of admissible strategies which are $\mathcal{F}_t$-predictable processes such that inventories are bounded above by $\overline{q}$ and below by $\underline{q}$. Moreover, the function ℓ, with $\ell(0) = 0$ and $\ell(q)$ increasing in q, is a liquidation penalty which consists of fees and market impact costs when the MM sends a market order to unwind terminal inventory. For example $\ell(q) = \theta q$ represents a linear impact when liquidating q shares.

The approach to solving for the value function $H(t, x, q, S, Z)$ and the optimal posting strategies $\gamma_t^\pm$ proceeds from using a dynamic programming principle. The value function should satisfy a Hamilton-Jacobi-Bellman (HJB) equation (see [4] and [8]) of the form:

$$\partial_t H + \sup_{\gamma^\pm} \mathcal{L}H = 0, \tag{3}$$

subject to terminal conditions $H(T, x, q, S, Z) = x + q(S - \ell(q))$, and where the infinitesimal generator $\mathcal{L}$ is given by:

$$\begin{aligned}\mathcal{L}H(t, x, q, S, Z) =& \beta^+ \mathcal{D}_L^+ H + \beta^- \mathcal{D}_L^- H \\ &+ \lambda^+(Z) \int_{-\infty}^{\infty} \mathcal{D}_M^+ H \, F_Z^+(y) dy \, \mathbb{I}_{q \neq \underline{q}} + \lambda^-(Z) \int_{-\infty}^{\infty} \mathcal{D}_M^- H \, F_Z^-(y) dy \, \mathbb{I}_{q \neq \overline{q}} \\ &+ \sum_{k=1}^{n} G_{Z,k} \Big(H(t, x, q, S, k) - H(t, x, q, S, Z) \Big),\end{aligned}$$

where the operators $\mathcal{D}_L^\pm$ act as follows:

$$\mathcal{D}_L^\pm H = H(t, x, q, S \pm \xi^\pm, Z) - H(t, x, q, S, Z),$$

and the operators $\mathcal{D}_M^\pm$ act as follows:

$$\mathcal{D}_M^\pm H = H(t, x \pm \gamma^\pm (S \pm \tfrac{\Delta}{2}), q \mp \gamma^\pm, S \pm y, Z) - H(t, x, q, S, Z),$$

The first two terms in $\mathcal{L}$ represent the change in the value function due to changes in the midprice that are not the result of market orders. The second two terms represent the change in the value function due to market orders. These events happen with intensity $\lambda^\pm(Z)$, and the effect they have on the value function depends on both the strategy of the MM at that time, $\gamma^\pm$, and the magnitude of the midprice change after the market order, y. Since the infinitesimal generator represents the expected rate of change of the value function, an expectation must be taken over all possible jump sizes. This is what is represented by the integrals. The last term containing the summation represents the expected rate of change of the value function due to the imbalance regime switching to different states.

5.1 Feedback controls

The form of the infinitesimal generator and the terminal conditions allow for a convenient ansatz to be applied: $H(t,x,q,S,Z) = x + qS + h_{q,Z}(t)$. Substituting this into equation (3) gives a system of ODE's for the functions $h_{q,Z}(t)$:

$$\begin{aligned}
\partial_t h_{q,Z} + \beta^+\xi^+ q - \beta^-\xi^- q + \lambda^+(Z)\epsilon_Z^+ q - \lambda^-(Z)\epsilon_Z^- q \\
+\lambda^+(Z) \sup_{\gamma^+\in\{0,1\}} \left\{\gamma^+(\tfrac{\Delta}{2} - \epsilon_Z^+) + h_{q-\gamma^+,Z} - h_{q,Z}\right\}\mathbb{I}_{q\neq \underline{q}} \\
+\lambda^-(Z) \sup_{\gamma^-\in\{0,1\}} \left\{\gamma^-(\tfrac{\Delta}{2} - \epsilon_Z^-) + h_{q+\gamma^-,Z} - h_{q,Z}\right\}\mathbb{I}_{q\neq \overline{q}} \\
+ \sum_{k=1}^{n} G_{Z,k}\Big(h_{q,k} - h_{q,Z}\Big) = 0, \\
h_{q,Z}(T) = -q\ell(q),
\end{aligned} \tag{4}$$

where $\epsilon_Z^\pm = \int_{-\infty}^{\infty} yF_Z^\pm(y)dy$ represents the expected jump size of the midprice due to a market order. Due to discrete nature of the controls, there is little hope for finding an analytical solution to this equation, however the calculation of the feedback controls is quite simple:

Proposition 5.1. *The optimal feedback controls of the HJB equation (4) are given by:*

$$\gamma^{+*}(t,q,Z) = \begin{cases} 1, & \frac{\Delta}{2} - \epsilon_Z^+ + h_{q-1,Z}(t) - h_{q,Z}(t) > 0 \text{ and } q \neq \underline{q} \\ 0, & \frac{\Delta}{2} - \epsilon_Z^+ + h_{q-1,Z}(t) - h_{q,Z}(t) \leq 0 \text{ or } q = \underline{q} \end{cases},$$

$$\gamma^{-*}(t,q,Z) = \begin{cases} 1, & \frac{\Delta}{2} - \epsilon_Z^- + h_{q+1,Z}(t) - h_{q,Z}(t) > 0 \text{ and } q \neq \overline{q} \\ 0, & \frac{\Delta}{2} - \epsilon_Z^- + h_{q+1,Z}(t) - h_{q,Z}(t) \leq 0 \text{ or } q = \overline{q} \end{cases}.$$

Proof. This is clear from the binary nature of the controls and the form of the supremum terms in equation (4). □

In addition to the feedback controls, the existence of a unique classical solution is guaranteed:

Proposition 5.2. *Equation* (4) *has a unique classical solution.*

Proof. If we write equation (4) in the form:

$$\partial_t \boldsymbol{h} = F(\boldsymbol{h}),$$

we see that F is piecewise linear. Since there are a finite number of states q and Z, there are a finite number of domains in which the linear dependence of F on $\boldsymbol{h}$ is different. Therefore, F is globally Lipschitz, and so equation (4) has a unique classical solution. □

With the existence of a unique classical solution guaranteed, a verification theorem can be stated:

Theorem 5.1. *Let h be the solution to equation* (4) *and define* $\hat{H}(t,x,q,S,Z) = x + qS + h_{q,Z}(t)$*. Then* $\hat{H} = H$*, the value function as defined in* (2).

Proof. See appendix. □

6. Optimal Trading Strategy

With no analytical solution available, resorting to numerical solutions allows for the optimal strategy to be shown in Figure 7. The parameters corresponding to this optimal strategy are the following:

$$G = \begin{bmatrix} -1.00 & 0.75 & 0.25 \\ 1.00 & -2.00 & 1.00 \\ 0.25 & 0.75 & -1.00 \end{bmatrix} \quad \begin{array}{l} \overline{\lambda}^{+} = \begin{pmatrix} 0.25 & 0.50 & 1.00 \end{pmatrix} \quad \overline{\epsilon}^{+} = \begin{pmatrix} 0.002 & 0.003 & 0.005 \end{pmatrix} \\ \overline{\lambda}^{-} = \begin{pmatrix} 1.00 & 0.50 & 0.25 \end{pmatrix} \quad \overline{\epsilon}^{-} = \begin{pmatrix} 0.005 & 0.003 & 0.002 \end{pmatrix} \end{array}$$

$$\Delta = 0.01, \quad \beta^{\pm} = 1, \quad \xi^{\pm} = 0.005$$

This set of parameters has been chosen to reflect the impact of order imbalance on the dynamics of market orders and midprice movements. The states 1, 2, and 3 are intended to represent states of low, medium, and high imbalance respectively. The increasing (decreasing) values of λ^{+} and ϵ^{+} (λ^{-} and ϵ^{-}) are chosen to reflect the appropriate dependence of market order intensity and jump sizes on imbalance, as has been demonstrated in Sections 3.1 and 3.2.

It is clear from the visual representation of the trading strategy in Figure 7 that for any single state of imbalance, there is a boundary between the trade and no-trade region. For the sake of brevity, and so that the strategy with more imbalance states can easily be plotted in the future, all of the information in Figure 7 is condensed into Figure 8 by plotting only the boundary between the two regions for each state of imbalance.

There are two main features of the trade boundaries which can be easily explained from a financial perspective. The first is that all of the boundaries imply that at an inventory value of zero as time approaches maturity, trades are not made. This is in fact a consequence of the parameter values, namely that $\frac{\Delta}{2} - \theta - \epsilon^{\pm}(Z) < 0$ for each value of Z. This results in there being no benefit from executing a trade immediately before maturity when the agent is already at zero inventory, knowing that it will have to be liquidated for the same price as the initial transaction, and also suffering from a very small amount of inventory penalization. Indeed, in Figure 9, the trade boundaries are shown for the same set of parameters except with θ decreased to a value of 0.0005. Note now the willingness of the MM to make trades in some circumstances immediately before maturity, because the benefit of these trades now outweighs the terminal penalty that they cost.

The second feature is the ordering of the boundaries with respect to the imbalance state. The boundary for the limit sell orders is increasing with the state label

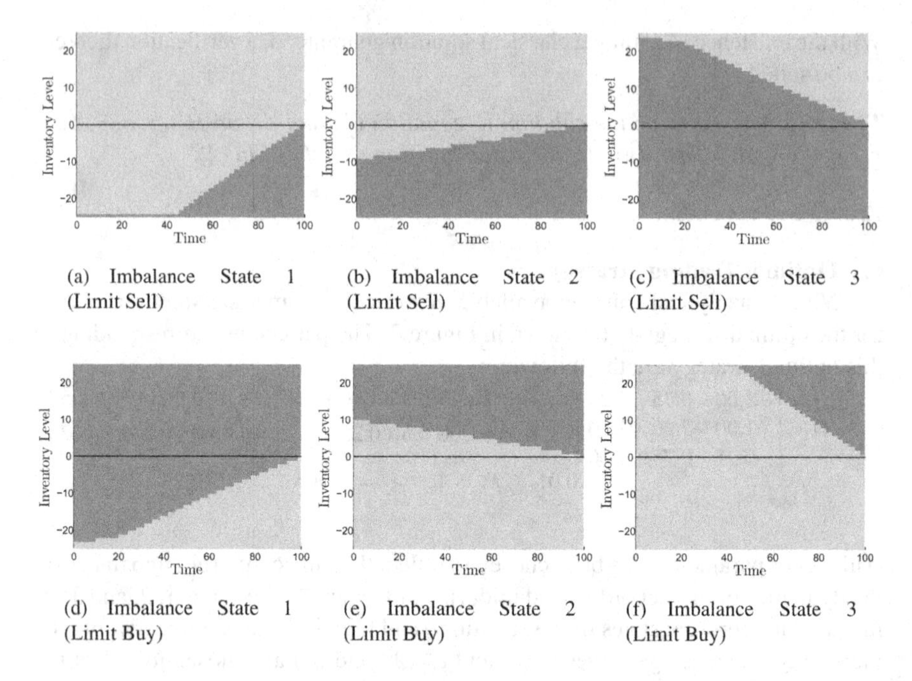

(a) Imbalance State 1 (Limit Sell)
(b) Imbalance State 2 (Limit Sell)
(c) Imbalance State 3 (Limit Sell)

(d) Imbalance State 1 (Limit Buy)
(e) Imbalance State 2 (Limit Buy)
(f) Imbalance State 3 (Limit Buy)

Figure 7. Optimal posting strategies for parameters listed above. The agent places the corresponding limit buy or sell order posting in the light shaded regions and does not place a posting in the dark shaded regions. Additional parameters are $T = 100$, $\ell(q) = -\theta q$, $\theta = 0.005$, $\overline{q} = -\underline{q} = 25$.

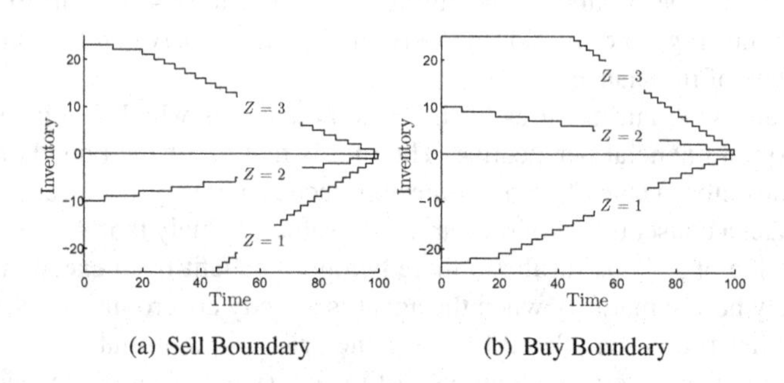

(a) Sell Boundary
(b) Buy Boundary

Figure 8. The market maker posts a sell (buy) limit order if their current inventory lies above (below) the corresponding imbalance state curve.

meaning that the MM is most willing to sell the asset when imbalance is low. This is a consequence of typical market behaviour when imbalance is low. Low

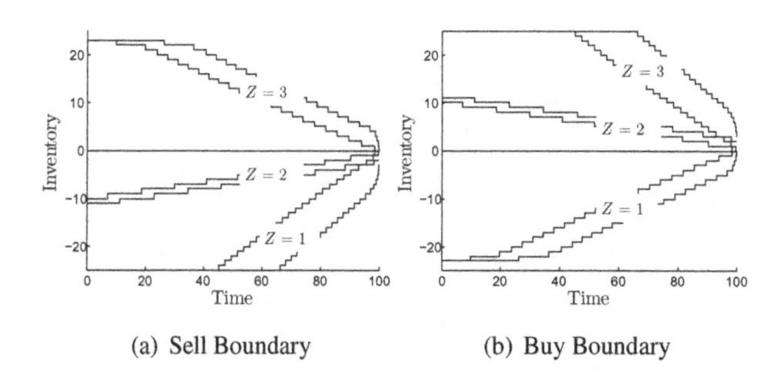

(a) Sell Boundary (b) Buy Boundary

Figure 9. The market maker posts a sell (buy) limit order if their current inventory lies above (below) the corresponding imbalance state curve. Dotted line represents original parameter set with $\theta = 0.005$. Solid line uses $\theta = 0.0005$.

imbalance makes market sell orders more likely, and it also makes the magnitude of a price change after a market sell order larger than after a market buy order. Together, these imply that the midprice of the asset is likely to decrease in the immediate future when imbalance is low, and so the MM should be more willing to sell shares in the low imbalance state compared to the other states. The MM will not always be willing to sell shares in the low imbalance state, however, due to both the possibility of having to pay a terminal penalty at maturity and due to the running inventory penalty.

As a final note, one additional case will be studied as a demonstration of incorporating additional imbalance regimes and observing the existence of no-trade regions. For this example, the parameter set is changed to the following:

$$G = \begin{bmatrix} -1.75 & 0.80 & 0.50 & 0.30 & 0.15 \\ 0.45 & -1.50 & 0.50 & 0.35 & 0.20 \\ 0.30 & 0.70 & -2.00 & 0.70 & 0.30 \\ 0.20 & 0.35 & 0.50 & -1.50 & 0.45 \\ 0.15 & 0.30 & 0.50 & 0.80 & -1.75 \end{bmatrix} \quad \begin{aligned} \overline{\lambda}^{+} &= \begin{pmatrix} 0.2 & 0.3 & 0.5 & 1.0 & 2.0 \end{pmatrix} \\ \overline{\lambda}^{-} &= \begin{pmatrix} 2.0 & 1.0 & 0.5 & 0.3 & 0.2 \end{pmatrix} \\ \overline{\epsilon}^{+} &= \begin{pmatrix} 0.001 & 0.002 & 0.003 & 0.006 & 0.011 \end{pmatrix} \\ \overline{\epsilon}^{-} &= \begin{pmatrix} 0.011 & 0.006 & 0.003 & 0.002 & 0.001 \end{pmatrix} \end{aligned}$$

$$\xi^{\pm} = 0.005, \quad \beta^{\pm} = 1, \quad \Delta = 0.01$$

This results in the optimal trade boundaries shown in Figure 10. In Figure 11, both the buy and sell trade boundaries are shown for each imbalance regime. This makes it clear that there are some regions in which the market maker posts both buy and sell limit orders, and some regions in which they post no limit orders. A no-trade region is indicated by the buy boundary (solid) being below the sell boundary (dashed).

7. Concluding remarks

In this paper, a microstructure model is proposed in which the dynamics are driven by an observable quantity, volume order imbalance. By observing the state

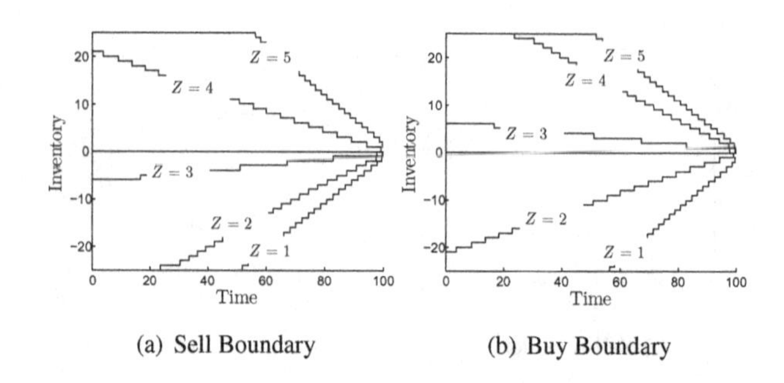

(a) Sell Boundary (b) Buy Boundary

Figure 10. The market maker posts a sell (buy) limit order if their current inventory lies above (below) the corresponding imbalance state curve. Additional parameters are $T = 100$, $\ell(q) = -\theta q, \theta = 0.005$, $\overline{q} = -\underline{q} = 25$.

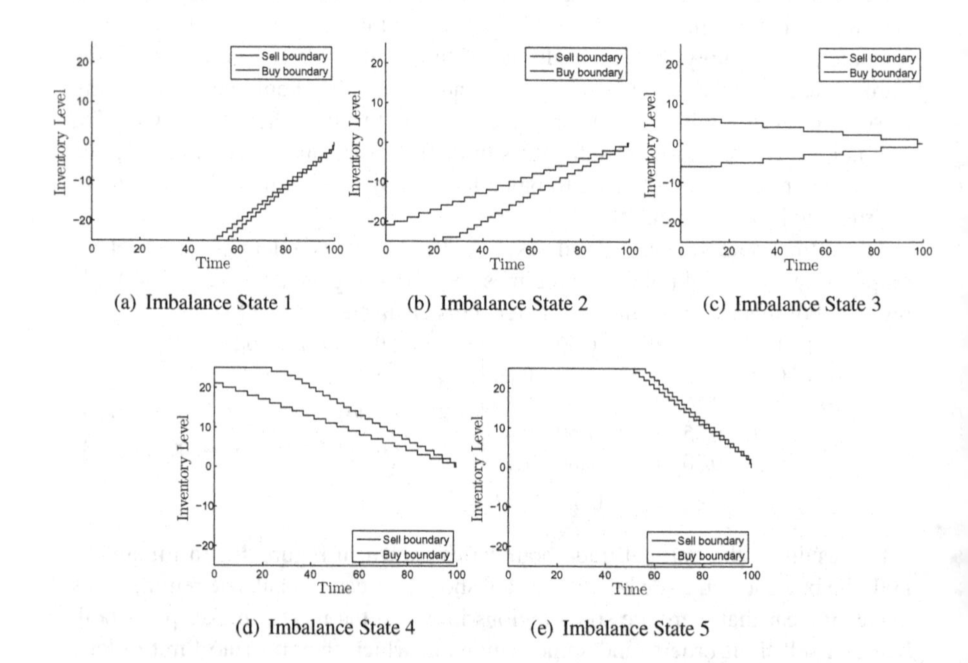

(a) Imbalance State 1 (b) Imbalance State 2 (c) Imbalance State 3

(d) Imbalance State 4 (e) Imbalance State 5

Figure 11. Buy (solid line) and sell (dashed line) trade boundaries for each imbalance state. No-trade regions are indicated by a dashed curve lying above a solid curve as in states 1 and 5.

of this quantity, an MM is able to modify their trading strategy in such a way that they take advantage of short term price changes and simultaneously protect themselves against adverse selection. We find that when imbalance is high, the MM is more willing to post limit buy orders and apprehensive about posting limit sell orders. The opposite behaviour becomes optimal when imbalance is low. The MM's value function is shown to be the unique classical solution to the corresponding HJB equation.

8. Appendix

Proof. [Proof of Theorem 5.1] Let h be the solution to equation (4) and define a candidate optimal value function $\hat{H}(t,x,q,S,Z) = x + qS + h_{q,Z}(t)$. From Ito's Lemma we have:

$$\begin{aligned}
\hat{H}(T, X^{\gamma^\pm}_{T^-}, q^{\gamma^\pm}_{T^-}, S_{T^-}, Z_{T^-}) =& \hat{H}(t,x,q,S,Z) + \int_t^T \partial_t h_{q^{\gamma^\pm}_u, Z_u}(u)\,du \\
&+ \int_t^T \xi^+ q^{\gamma^\pm}_u\, dJ^+_u - \int_t^T \xi^- q^{\gamma^\pm}_u\, dJ^-_u \\
&+ \int_t^T \int_{-\infty}^{\infty} q^{\gamma^\pm}_u\, y\mu^+(dy,du) - \int_t^T \int_{-\infty}^{\infty} q^{\gamma^\pm}_u\, y\mu^-(dy,du) \\
&+ \int_t^T \int_{-\infty}^{\infty} \gamma^+_u\Big(\frac{\Delta}{2} - y + h_{q^{\gamma^\pm}_u -1, Z_u}(t) - h_{q^{\gamma^\pm}_u, Z_u}(t)\Big)\mu^+(dy,du) \\
&+ \int_t^T \int_{-\infty}^{\infty} \gamma^-_u\Big(\frac{\Delta}{2} - y + h_{q^{\gamma^\pm}_u +1, Z_u}(t) - h_{q^{\gamma^\pm}_u, Z_u}(t)\Big)\mu^-(dy,du) \\
&+ \int_t^T \Big(h_{q^{\gamma^\pm}_u, Z_u}(t) - h_{q^{\gamma^\pm}_u, Z_{u^-}}(t)\Big)dZ_u\,.
\end{aligned}$$

Taking an expectation conditional on $\mathcal{F}_t$ on both sides, performing integrals with respect to y, and rearranging yields:

$$\begin{aligned}
\hat{H}(t,x,q,S,Z) =& \mathbb{E}\Big[\hat{H}(T, X^{\gamma^\pm}_{T^-}, q^{\gamma^\pm}_{T^-}, S_{T^-}, Z_{T^-})\Big|\mathcal{F}_t\Big] - \mathbb{E}\Big[\int_t^T \partial_t h_{q^{\gamma^\pm}_u, Z_u}(u)\,du \\
&+ \int_t^T \beta^+\xi^+ q^{\gamma^\pm}_u\,du - \int_t^T \beta^-\xi^- q^{\gamma^\pm}_u\,du + \int_t^T \lambda^+(Z_u) q^{\gamma^\pm}_u\, \epsilon^+_{Z_u}\,du - \int_t^T \lambda^-(Z_u) q^{\gamma^\pm}_u\, \epsilon^-_{Z_u}\,du \\
&+ \int_t^T \lambda^+(Z_u)\gamma^+_u\Big(\frac{\Delta}{2} - \epsilon^+_{Z_u} + h_{q^{\gamma^\pm}_u -1, Z_u}(t) - h_{q^{\gamma^\pm}_u, Z_u}(t)\Big)du \\
&+ \int_t^T \lambda^-(Z_u)\gamma^-_u\Big(\frac{\Delta}{2} - \epsilon^-_{Z_u} + h_{q^{\gamma^\pm}_u +1, Z_u}(t) - h_{q^{\gamma^\pm}_u, Z_u}(t)\Big)du \\
&+ \int_t^T \sum_{k=1}^{n} G_{Z_u,k}\Big(h_{q^{\gamma^\pm}_u, k}(t) - h_{q^{\gamma^\pm}_u, Z_{u^-}}(t)\Big)du\Big|\mathcal{F}_t\Big]\,.
\end{aligned}$$

(5)

Equation (4) then yields the inequality:

$$\begin{aligned}\hat{H}(t,x,q,S,Z) &\geq \mathbb{E}\left[\hat{H}(T,X^{\gamma^\pm}_{T^-},q^{\gamma^\pm}_{T^-},S_{T^-},Z_{T^-})\middle|\mathcal{F}_t\right]\\ &= \mathbb{E}\left[\hat{H}(T,X^{\gamma^\pm}_{T},q^{\gamma^\pm}_{T},S_{T},Z_{T})\middle|\mathcal{F}_t\right]\\ &= \mathbb{E}\left[X^{\gamma^\pm}_T + q^{\gamma^\pm}_T(S_T-\ell(q^{\gamma^\pm}_T))\middle|\mathcal{F}_t\right].\end{aligned}$$

Since this inequality holds for arbitrary controls $\gamma^\pm$, we have:

$$\begin{aligned}\hat{H}(t,x,q,S,Z) &\geq \sup_{(\gamma^\pm_s)_{t\leq s\leq T}\in\mathcal{A}} \mathbb{E}\left[X^{\gamma^\pm}_T + q^{\gamma^\pm}_T(S_T-\ell(q^{\gamma^\pm}_T))\middle|\mathcal{F}_t\right]\\ &= H(t,x,q,S,Z)\,. \end{aligned} \tag{6}$$

Now, if $\gamma^{\pm *}$ is selected to be of the form given in Proposition 5.1, then (5) implies:

$$\begin{aligned}\hat{H}(t,x,q,S,Z) &= \mathbb{E}\left[\hat{H}(T,X^{\gamma^{\pm*}}_{T^-},q^{\gamma^{\pm*}}_{T^-},S_{T^-},Z_{T^-})\middle|\mathcal{F}_t\right]\\ &= \mathbb{E}\left[\hat{H}(T,X^{\gamma^{\pm*}}_{T},q^{\gamma^{\pm*}}_{T},S_{T},Z_{T})\middle|\mathcal{F}_t\right]\\ &= \mathbb{E}\left[X^{\gamma^{\pm*}}_T + q^{\gamma^{\pm*}}_T(S_T-\ell(q^{\gamma^{\pm*}}_T))\middle|\mathcal{F}_t\right]\\ &\leq \sup_{(\gamma^\pm_s)_{t\leq s\leq T}\in\mathcal{A}} \mathbb{E}\left[X^{\gamma^\pm}_T + q^{\gamma^\pm}_T(S_T-\ell(q^{\gamma^\pm}_T))\middle|\mathcal{F}_t\right]\\ &= H(t,x,q,S,Z)\,. \end{aligned} \tag{7}$$

Combining (6) and (7) yields the result. □

References

1. Avellaneda, M. and Stoikov, S. (2008), "High-frequency trading in a limit order book," *Quantitative Finance*, **8(3)**, 217–224.
2. Cartea, Á. and Jaimungal, S. (2013), "Risk Metrics and Fine Tuning of High Frequency Trading Strategies," *Mathematical Finance*, doi:10.1111/mafi.12023.
3. Cartea, Á., Jaimungal, S., and Ricci, J. (2014), "Buy Low, Sell High: A High Frequency Trading Perspective," *SIAM Journal on Mathematical Finance*, **5(1)**, 415–444.
4. Fleming, W. H. and Soner, H. M. (2006), "Controlled Markov Processes and Viscosity Solutions," *Springer*.
5. Fodra, P. and Labadie, M. (2012), "High-frequency market-making with inventory constraints and directional bets," http://EconPapers.repec.org/RePEc:hal:wpaper:hal-00675925.
6. Guéant, O., Lehalle, C.A., and Fernandez-Tapia, J. (2012), "Dealing with the inventory risk: a solution to the market making problem," *Mathematics and Financial Economics*, **7(4)**, 477–507.

7. Guilbaud, F. and Pham, H. (2013), "Optimal High Frequency Trading with Limit and Market Orders," *Quantitative Finance*, **13(1)**, 79–94.
8. Pham, H. (2009), "Continuous-time Stochastic Control and Optimization with Financial Applications," *Springer-Verlag*.

Optimal Short-Covering with Regime Switching*

Tsz-Kin Chung

Graduate School of Social Sciences, Tokyo Metropolitan University, 1-1 Minami-Osawa, Hachiohji, Tokyo 192-0397, Japan. Email: btkchung@gmail.com

We formulate a short-selling strategy of a stock and seek the optimal timing of short covering in the presence of a random recall and a loan fee rate in an illiquid stock loan market. The aim is to study how the optimal trading strategy of the short-seller is influenced by the relevant features of the stock loan market. We consider a regime-switching stock price model that captures the transition in between the bull and the bear markets. The solution to the optimal stopping problem is obtained in closed-form based on the techniques in Guo and Zhang (2005). We provide the numerical example to illustrate of importance of a regime-dependent stopping rule for the short-seller's problem.

Key words: Short-selling, regime switching, optimal stopping, recall risk

1. Introduction

Short-selling is the selling of a financial security that the investor does not own and it provides an efficient means for investors to exploit the opportunity or hedge against downside risk when they anticipate the overpricing of a security and speculate its future decline in value. A typical situation involving short-selling transactions can be described as follows. Some institutional investors are long biased and hence only rebalance their portfolios on a quarterly or yearly basis. These institutional investors are usually mutual funds, pension funds or tracker funds. They place their stocks with the broker who acts as a custodian. The broker who holds the inventory of stocks has the discretion to lend out the stock in order to earn a loan fee income. On the other side is the short-seller (e.g., a hedge fund manager) who implements a short-selling position by borrowing the stock ("stock loan") from such a broker and selling it in the market. After that, the

*The author would like to thank Professor Keiichi Tanaka for guidance and helpful comments. The author is also grateful to the participants at the TMU Finance Workshop 2014 for their helpful comments.

short-seller's objective is to "buy low, sell high" such that the stock is first sold high and purchased later at a lower price. The buying back and returning of the stock to the broker is called *short covering*. At the end, the short-seller makes a profit from a price decline or a loss from a price rise of the stock.

There is a number of real-world complications involved in the implementation of a short-selling strategy. A unique feature in a stock loan contract is that there is no guaranteed maturity and it is effectively rolled over on a daily basis as documented in D'Avolio (2002). Hence, the broker is granted an option to recall the stock borrowing at any time. At a recall, the short-seller is then forced to cover the short position immediately regardless of a profit or loss. The risk of such an involuntary termination of a short-selling strategy is called the recall risk. Besides the capital profit/loss associated with the recall risk, the short-seller also has to take into account the running cost of the strategy. The broker charges the short-seller a loan fee, which is calculated as the loan fee rate times the stock price times the length of the period. As noted in D'Avolio (2002), the loan fee rate varies dramatically across different categories of stocks from 50 to 800 basis points. At the same time, the short-seller deposits the sale proceeds into a margin account, which generates interest income called the short interest rebate.

The optimal trading rule of a long position (i.e., optimal selling rule) has been formulated as an optimal stopping problem in Peskir and Shiryaev (2005) and extended in Guo and Zhang (2005), in which the investor initially holds a security and seeks the optimal timing to sell it in order to maximize the expected discounted payoff. In this paper, we formulate a short-selling strategy as an optimal stopping problem and seek the optimal timing of the short covering in the presence of a recall risk, loan fee and interest income. We assume a regime-switching stock price model in order to capture the transition in between the bull and the bear markets in the face of changing macroeconomic environment. This is important for the realistic modeling of a short-selling because the target price of short-covering should be sensitive to the market trend of the stock price. As a random recall with exponential distribution can be treated as a transition to an absorbing state, we are able to derive the value function and the optimal stopping rule by solving a system of differential equations along with the smooth-fit principle as in Guo and Zhang (2005). We calibrate the regime-switching model to the Nikkei 225 market and identify the bull-bear market transition based on the business cycles of the Japan's economy. Based on the calibrated parameters, our numerical results illustrate that it is important to derive a regime-dependent stopping rule for the short-seller's problem.

The paper is structured as follows. Section 2 presents the regime-switching stock price model and the corresponding short-seller's problem. Section 3 provides the detailed derivation of the closed-form solution to the short-seller's problem. Section 4 discusses the numerical example. Section 5 concludes.

2. Regime Switching Stock Price Model

2.1 Setup

We fix the filtered probability space $(\Omega, \mathcal{F}, (\mathcal{F}_t)_{t\geq 0}, \mathbb{P})$. The stock price is a regime-switching diffusion process $X = (X_t)_{t\geq 0}$ satisfying

$$dX_t = \mu_{J(t)} X_t dt + \sigma_{J(t)} X_t dW_t, \qquad X_0 = x > 0,$$

which is modulated by a continuous-time Markov chain $J = \{J(t)\}$ on a finite state space $E = \{1, 2\}$. Here, $\mu = \{\mu_1, \mu_2\}$ are the expected return and $\sigma = \{\sigma_1, \sigma_2\}$ with $\sigma_1 < \sigma_2$ are the discrete volatilities at each state. Furthermore, we take $\mu_1 \geq 0, \mu_2 < 0$ such that state 1 can be interpreted as a bull market (low volatility, upward trend) while state 2 as a bear market (high volatility, downward trend). The intensity matrix is

$$\mathbf{Q} = \begin{pmatrix} -\lambda_1 & \lambda_1 \\ \lambda_2 & -\lambda_2 \end{pmatrix},$$

with $\lambda_i > 0$ for $i = 1, 2$. Denote

$$\mathbf{M} = \begin{pmatrix} \mu_1 & 0 \\ 0 & \mu_2 \end{pmatrix}, \qquad \mathbf{\Theta}(z) = \begin{pmatrix} \theta_1(z) & 0 \\ 0 & \theta_2(z) \end{pmatrix}, \qquad \mathbf{\Lambda}(z) = \begin{pmatrix} l_1(z) & \lambda_1 \\ \lambda_2 & l_2(z) \end{pmatrix},$$

where

$$\theta_i(z) = \frac{1}{2}\sigma_i^2 z^2 + \left(\mu_i - \frac{1}{2}\sigma_i^2\right) z, \qquad l_i(z) = \theta_i(z) - r - \lambda_i,$$

in which $\theta_i(z)$ is the Levy exponent corresponding to the generator

$$\mathcal{L}_i = \frac{1}{2}\sigma_i^2 x^2 \frac{d^2}{dx^2} + \mu_i x \frac{d}{dx},$$

at each state $i = 1, 2$.

The random time of the broker's recall is governed by an independent exponential random variable $\tau_R \sim Exp(\lambda_0)$ with the parameter $\lambda_0 \geq 0$. We write $\mathcal{F}_t^{W,J} = \sigma(W_s, J(s); 0 \leq s \leq t)$, $\mathbb{F}^{W,J} = \left(\mathcal{F}_t^{W,J}\right)_{t\geq 0}$ and $\mathbb{F} = (\mathcal{F}_t)_{t\geq 0}$, and assume that $\mathcal{F}_t^{W,J} \vee \sigma(1_{\{\tau_R > t\}}) \subset \mathcal{F}_t$. We denote the expectation

$$\mathbb{E}_{x,i}[\cdot] = \mathbb{E}[\cdot \mid J(0) = i, X(0) = x],$$

under $\mathbb{P}$.

There is a stock loan market to borrow/lend the stock against the loan fee, although the liquidity may be limited in the sense that the borrowing is available only with specific brokers and the stock availability is not always guaranteed. The contract may be automatically renewed instantaneously, although there is a chance that the broker will not be able to find stock to replace. Hence, we assume that the lender does not renew the contract with probability λdt over the next instant period dt (random recall). Once the loan contract is terminated, the short-seller has to cover the short position immediately by buying stock at the market price.

The loan fee is charged instantaneously based on the current stock price, i.e., the borrower makes the loan fee payment $\delta X_t dt$ over a small time interval dt, where δ is the constant loan fee rate. The short-seller deposits the initial proceeds K from selling the stock into a margin account that pays interest continuously at a constant rate q. As a result, the net cash outflow is given by $(\delta X_t - qK)\,dt$ over a time interval dt, which can be positive or negative depending on the levels of the stock price, the loan fee rate and the interest rate. The net cash flow $\delta X_t - qK$ is sometimes referred to as the effective loan fee and can be interpreted as the net running cost of the short-selling strategy.

In the following, we assume the investor's own discount rate r is constant, while the margin account interest rate and the loan fee rate are regime dependent as $\mathbf{q} = (q_1, q_2)^\top$ and $\boldsymbol{\delta} = (\delta_1, \delta_2)^\top$. The short-seller is supposed to have already undertaken the short position when the stock price was equal to K and to hold the position until she buys back at the market price either at her own discretion or following a recall by the broker. When the current state is observable as i, the short-seller's problem is to optimize the expected net profit

$$(1)\quad v_i(x) = \sup_{\tau_E \in \mathcal{A}} \mathbb{E}_{x,i}\left[e^{-r(\tau_E \wedge \tau_R)}\left(K - X_{\tau_E \wedge \tau_R}\right) - \int_0^{\tau_E \wedge \tau_R} e^{-rs}\left(\delta_{J(s)} X_s - q_{J(s)} K\right) ds\right],$$

for $i = 1, 2$. Here, $\mathcal{A}$ is the set of all $\mathbb{F}^{W,J}$-stopping times taking values in $[0, \infty]$. We assume that the subjective discount rate r is sufficiently high, compared with the growth rate of the stock price: $r > \max(\max(\mu_1, \mu_2), 0)$.

2.2 Auxiliary problem and lower bounds

Let us denote

$$\alpha_i \delta_i = \mathbf{e}_i^\top (r\mathbf{I} - \mathbf{M} - \mathbf{Q})^{-1} \boldsymbol{\delta}, \qquad \beta_i q_i = \mathbf{e}_i^\top (r\mathbf{I} - \mathbf{Q})^{-1} \mathbf{q},$$

for $i = 1, 2$. The following lemma will be useful for the calculation of the value function.

Lemma 2.1. *Suppose that* $r > \max(\max(\mu_1, \mu_2), 0)$. *Then, for any stopping time* $t > 0$ *it holds that*

$$\mathbb{E}\left[\int_t^\infty e^{-ru} \delta_{J(u)} X_u du \,\middle|\, \mathcal{F}_t\right] = e^{-rt} \alpha_{J(t)} \delta_{J(t)} X_t, \quad \mathbb{E}\left[\int_t^\infty e^{-ru} q_{J(u)} du \,\middle|\, \mathcal{F}_t\right] = e^{-rt} \beta_{J(t)} q_{J(t)},$$

and

$$\mathbb{E}_{x,i}\left[\int_0^t e^{-ru} \delta_{J(u)} X_u du\right] = \alpha_i \delta_i x - \mathbb{E}_{x,i}\left[e^{-rt} \alpha_{J(t)} \delta_{J(t)} X_t\right],$$

$$\mathbb{E}_{x,i}\left[\int_0^t e^{-ru} q_{J(u)} K du\right] = \beta_i q_i K - \mathbb{E}_{x,i}\left[e^{-rt} \beta_{J(t)} q_{J(t)} K\right].$$

Proof. *See Appendix A.* □

By Lemma 2.1, the short-seller's problem can be re-written as

$$v_i(x) = u_i(x) + (1 - \eta_i)\, K - (1 - \rho_i)\, x, \tag{2}$$

with

$$\eta_i = 1 - \beta_i q_i, \qquad \rho_i = 1 - \alpha_i \delta_i,$$

for $i = 1, 2$. The original problem is reduced to the auxiliary optimal stopping problem

$$u_i(x) = \sup_{\tau_E \in \mathcal{A}} \mathbb{E}_{x,i} \left[e^{-r(\tau_E \wedge \tau_R)} \left(\eta_{J(\tau_E \wedge \tau_R)} K - \rho_{J(\tau_E \wedge \tau_R)} X_{\tau_E \wedge \tau_R} \right) \right], \tag{3}$$

in which the gain function is regime dependent as

$$g_i(x) = g(x, i) = \begin{cases} \eta_1 K - \rho_1 x, & i = 1, \\ \eta_2 K - \rho_2 x, & i = 2. \end{cases} \tag{4}$$

The coefficients $\boldsymbol{\eta} = (\eta_1, \eta_2)^\top$ and $\boldsymbol{\rho} = (\rho_1, \rho_2)^\top$ summarize the expected running reward and cost when the interest rate and loan fee rate are regime dependent.

The auxiliary optimal stopping problem has two lower bounds on each state i. The first one is $g_i(x)$ which corresponds to $\tau_E = 0$. The second one corresponds to $\tau_E = \infty$ as given below.

Proposition 2.1. *The lower bound of the value function $u_i(x)$ corresponding to $\tau_E = \infty$ is*

$$\psi_i(x) = a_i x + b_i, \quad i = 1, 2, \tag{5}$$

where

$$\mathbf{a} = \lambda_0 \left(\mathbf{M} + \mathbf{Q} - (r + \lambda_0)\, \mathbf{I} \right)^{-1} \boldsymbol{\rho}, \quad \mathbf{b} = \lambda_0 K \left((r + \lambda_0)\, \mathbf{I} - \mathbf{Q} \right)^{-1} \boldsymbol{\eta},$$

with $\mathbf{a} = (a_1, a_2)^\top$ and $\mathbf{b} = (b_1, b_2)^\top$.

Proof. *See Appendix B.* □

The two lower bounds g_i and ψ_i determine the type of stopping rule of the optimal stopping problem at each state. When there is no regime switching, Chung and Tanaka (2015) explicitly characterize the four possible types of (i) put-type problem, (ii) call-type problem, (iii) immediate stop and (iv) wait forever, depending on the model parameters. For simplicity, we shall focus only on the put-type problem in which the stopping rule is of down-and-out type.[1] This is consistent with the general views that a short-selling strategy is to speculate a decline in the stock price. To this end, we can assume that $\mathbf{q}$ and $\boldsymbol{\delta}$ are sufficiently small as

$$\frac{q_i}{r} < 1 \quad \text{and} \quad \frac{\delta_i}{r - \mu_i} < 1 \tag{6}$$

in order to ensure $\eta_i > 0$ and $\rho_i > 0$ for $i = 1, 2$ (see Appendix C for details). In this case, we see that both the two lower bounds at each regime are decreasing with the stock price x, hence it is natural to conjecture a down-and-out stopping rule.

3. Solution

3.1 Value function and optimal threshold

Since $\{X_t, J_t\}$ is a joint Markov process, we can conjecture the threshold-type stopping rule for the optimal stopping problem (3). Following Jobert and Rogers (2006), Guo and Zhang (2005) and Tanaka (2012), the candidate of optimal stopping time is

$$\tau_E = \min_{i=1,2} \tau_i, \qquad \tau_i = \inf\{t > 0; X_t \leq x_i, J(t) = i\}, \qquad i = 1, 2, \tag{7}$$

that is, the threshold is regime dependent. Let us assume the order

$$x_2 < x_1,$$

indicating that the threshold under state 2 (bear market) is lower than that under state 1 (bull market). The rationale is that the short-seller should be more aggressive and sets a lower target price for short-covering under a bear market than that of a bull market. Otherwise, we can rearrange the index and proceed similarly.

Case 1: when $x \in (0, x_2)$

The short-seller stops immediately for all regimes. We have

$$u_1(x) = \eta_1 K - \rho_1 x, \qquad u_2(x) = \eta_2 K - \rho_2 x.$$

Case 2: when $x \in [x_2, x_1)$

The short-seller stops immediately for state 1 and continues for state 2. We have

$$u_1(x) = \eta_1 K - \rho_1 x,$$

and u_2 solves the differential equation

$$(\mathcal{L}_2 - r)\, u_2(x) + \lambda_2 (u_1(x) - u_2(x)) + \lambda_0 (g_2(x) - u_2(x)) = 0.$$

---

[1] In general, there can be a switching in between the put-type and call-type problems and one needs to be more careful about the conjecture of the stopping rule.

The first term corresponds to the change in the value function u_2 without regime switching, the second term is the change when there is a regime switch from state 2 to state 1, and the third term is the change when there is a broker's random recall in state 2.

It is easy to see that the solution is

$$u_2(x) = C_1 x^{\gamma_1} + C_2 x^{\gamma_2} + \phi(x),$$

in which $\gamma_1 > 1$ and $\gamma_2 < 0$ are the roots of [2]

$$(r + \lambda_0 + \lambda_2) - \frac{1}{2}\sigma_2^2 z^2 - \left(\mu_2 - \frac{1}{2}\sigma_2^2\right) z = G_2(z) = 0,$$

and ϕ is the special solution of the form

$$\phi(x) = \frac{\lambda_2 \eta_1 + \lambda_0 \eta_2}{r + \lambda_2 + \lambda_0} K - \frac{\lambda_2 p_1 + \lambda_0 \rho_2}{r - \mu_2 + \lambda_2 + \lambda_0} x.$$

The coefficients C_1 and C_2 are to be determined by the smooth-fit principle.

Case 3: when $x \in [x_1, \infty)$

The short-seller continues for both the two regimes. We have to solve the system of differential equations

$$\begin{cases} (\mathcal{L}_1 - r) u_1(x) + \lambda_1 (u_2(x) - u_1(x)) + \lambda_0 (g_1(x) - u_1(x)) = 0 \\ (\mathcal{L}_2 - r) u_2(x) + \lambda_2 (u_1(x) - u_2(x)) + \lambda_0 (g_2(x) - u_2(x)) = 0 \end{cases}.$$

Define $\mathcal{A}_i = \mathcal{L}_i - (r + \lambda_0 + \lambda_i)$ and turn it into a matrix equation as

$$\begin{pmatrix} \mathcal{A}_1 & \lambda_1 \\ \lambda_2 & \mathcal{A}_2 \end{pmatrix} \begin{pmatrix} u_1(x) \\ u_2(x) \end{pmatrix} = -\lambda_0 \begin{pmatrix} \eta_1 K - \rho_1 x \\ \eta_2 K - \rho_2 x \end{pmatrix},$$

we are ready to apply the solution method as in Guo and Zhang (2005). Let us focus on the homogenous equation

$$\begin{pmatrix} \mathcal{A}_1 & \lambda_1 \\ \lambda_2 & \mathcal{A}_2 \end{pmatrix} \begin{pmatrix} f_1(x) \\ f_2(x) \end{pmatrix} = \begin{pmatrix} 0 \\ 0 \end{pmatrix}. \tag{8}$$

We can conjecture the solution to be the linear combination of the form

$$f_1(x) = \sum_{k=1}^{4} A_k x^{\beta_k}, \qquad f_2(x) = \sum_{k=1}^{4} B_k x^{\beta_k}.$$

[2]It can be checked that: $G_2(-\infty) < 0$, $G_2(0) > 0$, $G_2(1) > 0$ and $G_2(\infty) < 0$.

As shown in Appendix D, this leads to the characteristic equation

$$G_1(\beta)G_2(\beta) = \lambda_1\lambda_2, \tag{9}$$

with

$$G_1(\beta) = (r + \lambda_0 + \lambda_1) - \frac{1}{2}\sigma_1^2\beta^2 - \left(\mu_1 - \frac{1}{2}\sigma_1^2\right)\beta,$$
$$G_2(\beta) = (r + \lambda_0 + \lambda_2) - \frac{1}{2}\sigma_2^2\beta^2 - \left(\mu_2 - \frac{1}{2}\sigma_2^2\right)\beta,$$

which has the four distinct roots $\beta_4 > \beta_3 > 0 > \beta_2 > \beta_1$. Moreover, the coefficients A_k and B_k are related as

$$B_k = l_k A_k = \frac{G_1(\beta_k)}{\lambda_1}A_k = \frac{\lambda_2}{G_2(\beta_k)}A_k, \qquad k = 1, 2, 3, 4.$$

To ensure $f_1(x)$ and $f_2(x)$ are bounded as $x \to \infty$, we have to discard the positive roots β_3 and β_4. The special solution is

$$\psi_i(x) = a_i x + b_i, \qquad i = 1, 2,$$

which is just the lower bound function as obtained in Proposition 2.1. As a result, the value functions for $x \in [x_1, \infty)$ are given by

$$u_1(x) = A_1 x^{\beta_1} + A_2 x^{\beta_2} + \psi_1(x),$$
$$u_2(x) = l_1 A_1 x^{\beta_1} + l_2 A_2 x^{\beta_2} + \psi_2(x),$$

in which the coefficients A_1 and A_2 are to be determined.

3.2 Smooth-fit

The solution for the 3 regions $(0, x_2)$, $[x_2, x_1)$ and $[x_1, \infty)$ is summarized as

$$u_1(x) = \begin{cases} \eta_1 K - \rho_1 x, & \text{if} \quad x \in (0, x_1), \\ A_1 x^{\beta_1} + A_2 x^{\beta_2} + \psi_1(x), & \text{if} \quad x \in [x_1, \infty). \end{cases} \tag{10}$$

$$u_2(x) = \begin{cases} \eta_2 K - \rho_2 x, & \text{if} \quad x \in (0, x_2), \\ C_1 x^{\gamma_1} + C_2 x^{\gamma_2} + \phi(x), & \text{if} \quad x \in [x_2, x_1), \\ l_1 A_1 x^{\beta_1} + l_2 A_2 x^{\beta_2} + \psi_2(x), & \text{if} \quad x \in [x_1, \infty). \end{cases} \tag{11}$$

Due to the exponential termination of broker's recall, we see that the special solutions appear in the value functions under all the continuation regions. The 6 unknown coefficients $A_1, A_2, C_1, C_2, x_1, x_2$ can be obtained by the smooth-fit principle as follows:

1. Matching of $u_2(x)$ at $x = x_1$:

$$l_1 A_1 x_1^{\beta_1} + l_2 A_2 x_1^{\beta_2} + \psi_2(x_1) = C_1 x_1^{\gamma_1} + C_2 x_1^{\gamma_2} + \phi(x_1),$$
$$\beta_1 l_1 A_1 x_1^{\beta_1} + \beta_2 l_2 A_2 x_1^{\beta_2} + x_1 \psi_2'(x_1) = \gamma_1 C_1 x_1^{\gamma_1} + \gamma_2 C_2 x_1^{\gamma_2} + x_1 \phi'(x_1).$$

2. Matching of $u_1(x)$ at $x = x_1$:

$$A_1 x_1^{\beta_1} + A_2 x_1^{\beta_2} + \psi_1(x_1) = \eta_1 K - \rho_1 x_1,$$
$$\beta_1 A_1 x_1^{\beta_1} + \beta_2 A_2 x_1^{\beta_2} + x_1 \psi_1'(x_1) = -\rho_1 x_1.$$

3. Matching of $u_2(x)$ at $x = x_2$:

$$C_1 x_2^{\gamma_1} + C_2 x_2^{\gamma_2} + \phi(x_2) = \eta_2 K - \rho_2 x_2,$$
$$\gamma_1 C_1 x_2^{\gamma_1} + \gamma_2 C_2 x_2^{\gamma_2} + x_2 \phi'(x_2) = -\rho_2 x_2.$$

We can obtain a equation solving the thresholds (x_1, x_2) as

$$\begin{pmatrix} x_1^{-\gamma_1} & 0 \\ 0 & x_1^{-\gamma_2} \end{pmatrix} F_1(x_1) = \begin{pmatrix} x_2^{-\gamma_1} & 0 \\ 0 & x_2^{-\gamma_2} \end{pmatrix} F_2(x_2), \tag{12}$$

where $F_1(x_1)$ and $F_2(x_2)$ are given by

$$\begin{aligned} F_1(x_1) &= \begin{pmatrix} 1 & 1 \\ \gamma_1 & \gamma_2 \end{pmatrix}^{-1} \left[\begin{pmatrix} l_1 & l_2 \\ \beta_1 l_1 & \beta_2 l_2 \end{pmatrix} \begin{pmatrix} 1 & 1 \\ \beta_1 & \beta_2 \end{pmatrix}^{-1} \begin{pmatrix} \eta_1 K - \rho_1 x_1 - \psi_1(x_1) \\ -\rho_1 x_1 - x_1 \psi_1'(x_1) \end{pmatrix} \right. \\ &\quad \left. - \begin{pmatrix} \phi(x_1) - \psi_2(x_1) \\ x_1 \phi'(x_1) - x_1 \psi_2'(x_1) \end{pmatrix} \right], \\ F_2(x_2) &= \begin{pmatrix} 1 & 1 \\ \gamma_1 & \gamma_2 \end{pmatrix}^{-1} \begin{pmatrix} \eta_2 K - \rho_2 x_2 - \phi(x_2) \\ -\rho_2 x_2 - x_2 \phi'(x_2) \end{pmatrix}. \end{aligned}$$

Once (x_1, x_2) are obtained, we can solve (A_1, A_2) and (C_1, C_2) as

$$\begin{pmatrix} A_1 \\ A_2 \end{pmatrix} = \begin{pmatrix} x_1^{\beta_1} & x_1^{\beta_2} \\ \beta_1 x_1^{\beta_1} & \beta_2 x_1^{\beta_2} \end{pmatrix}^{-1} \begin{pmatrix} \eta_1 K - \rho_1 x_1 - \psi_1(x_1) \\ -\rho_1 x_1 - x_1 \psi_1'(x_1) \end{pmatrix}, \tag{13}$$

$$\begin{pmatrix} C_1 \\ C_2 \end{pmatrix} = \begin{pmatrix} x_2^{\gamma_1} & x_2^{\gamma_2} \\ \gamma_1 x_2^{\gamma_1} & \gamma_2 x_2^{\gamma_2} \end{pmatrix}^{-1} \begin{pmatrix} \eta_2 K - \rho_2 x_2 - \phi(x_2) \\ -\rho_2 x_2 - x_2 \phi'(x_2) \end{pmatrix}. \tag{14}$$

The optimality of the value functions and thresholds can be verified by following the procedure in Guo and Zhang (2005). Lastly, it is worth to note that the derivation herein works for a two-state regime switching model only. For multiple regimes with $n > 2$, we refer the readers to Tanaka (2012) for the application of linear algebra techniques on a general regime switching model.

For illustration purpose, let us also report the value function and the optimal threshold for a single regime case. Take $\mu = \mu_i$, $\sigma = \sigma_i$, $q = q_i$ and $\delta = \delta_i$, such that

$$g(x) = \eta K - \rho x, \quad \eta = 1 - \frac{q}{r}, \quad \rho = 1 - \frac{\delta}{r - \mu},$$

the value function u for the auxiliary problem can be expressed as

$$u(x) = \begin{cases} \left(g(b_p) - \phi(b_p)\right)\left(\dfrac{x}{b_p}\right)^m + \phi(x), & x \in \left(b_p, \infty\right), \\ g(x), & x \in \left(0, b_p\right], \end{cases}$$

where

$$\phi(x) = \lambda\left[\frac{1}{r + \lambda_0}\eta K - \frac{1}{r + \lambda_0 - \mu}\rho x\right],$$

and the optimal threshold is

$$b_p = \frac{m}{m-1} aK, \qquad a = \frac{\beta}{\beta + \lambda}\frac{\beta + \lambda - \mu}{\beta - \mu}\frac{\eta}{\rho}.$$

Here, $m < 0$ is the negative root to the quadratic equation $G_1(z) = 0$ after setting $\lambda_1 = \lambda_2 = 0$. The value function v corresponding to the short-seller's problem is given by $v(x) = u(x) + (1 - \eta)K - (1 - \rho)x$.

4. Numerical Examples

4.1 Parameter calibration

We obtain the daily closing prices of the Nikkei 225 index since early 1990s and identify the bull-bear market transition based on the business cycles as announced by the Economic and Social Research Institute (ESRI) in Japan.[3] The Nikkei 225 market is an ideal venue to test our short-selling strategy because it has declined by more than 70% since the burst of the financial bubble in early 1990s. Figure 1 reports the Nikkei 225 index from March 1991 to November 2012 which covers several business cycles of the Japan's economy. A recession (boom) starts at the peak (trough) of a business cycle and ends at the trough (peak) as measured by the economic activity. As can be seen, the stock market declines coincide with the ESRI resession periods, suggesting that it is appropriate to determine the stock market trend based on the ESRI business cycles. To be specific, we take a boom period and a recession period to be a bull market and a bear market respectively. As such, we estimate that the average annualized log-returns during a boom period and a recession period are about 3% and -20% respectively over the sample period. Furthermore, we compute the historical volatility of Nikkei 225 using the

---

[3] See http://www.esri.cao.go.jp/jp/stat/di/140530hiduke.html (in Japanese). We are grateful to the anonymous referee for the suggestion.

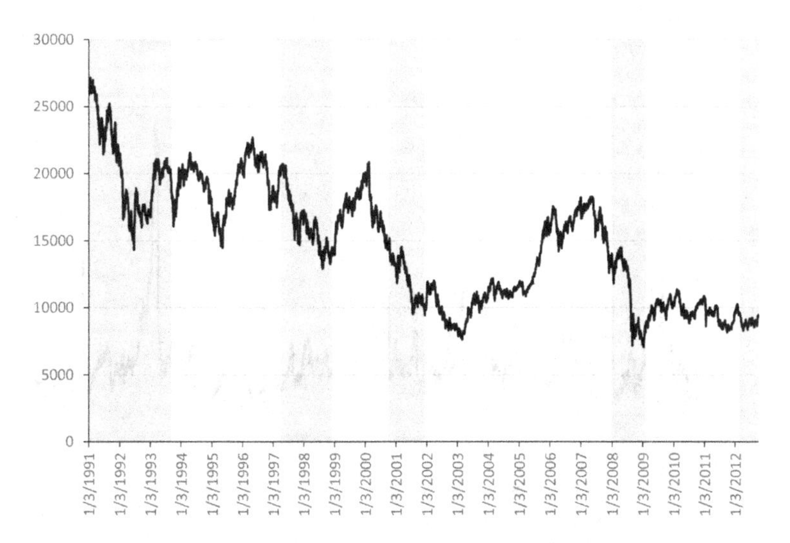

Figure 1. : The Nikkei 225 index since early 1990s. The shaded area is the recession period as announced by the Economic and Social Research Institute (ESRI) in Japan.

exponentially weighted moving average (EWMA) method. As shown in Figure 2, the stock market volatility is usually higher during a recession period (around 30% to 40%) than that of a boom period (around 15% to 25%). This demonstrates that a regime-switching model provides a better description to the stock market dynamics than the standard Black-Scholes model with constant parameters.

μ_1	σ_1	λ_1	μ_2	σ_2	λ_2
0.05	0.20	0.25	-0.10	0.40	0.75

Table 1: Model parameters for the two-state regime-switching model.

Based on the analysis, we calibrate the regime-switching model as in Table 1 which is explained as follows. We take state 1 to be a bull market (low volatility, upward trend) and state 2 to be a bear market (high volatility, downward trend).

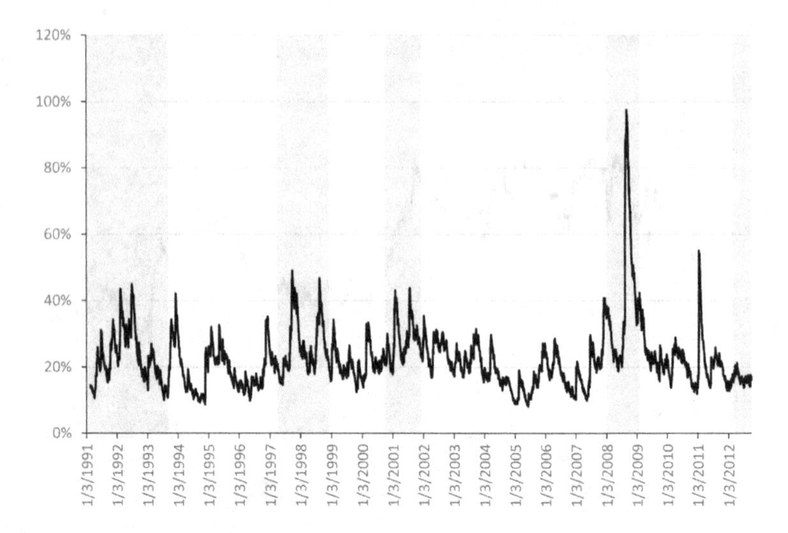

Figure 2. : The historical volatility of Nikkei 225 based on the EWMA method with a decay factor of 0.06.

Conditional on the regime state i, the log-return is normally distributed as

$$\ln(X_{t+\Delta t}) - \ln(X_t) \sim N\left((\mu_i - \frac{1}{2}\sigma_i^2)\Delta t,\ \sigma_i^2 \Delta t\right),$$

where Δt is the time interval. Plugging in the values in Table 1, we get $\mu_1 - \sigma_1^2/2 = 0.03$ and $\mu_2 - \sigma_2^2/2 = -0.18$, which match with the empirical data. To calibrate the transition intensity, we note that on average a boom period lasts for 36 months while a recession period lasts for 16 months as indicated by the ESRI business cycles. Because the expected time to be spent on the state i is $1/\lambda_i$, the transition intensities can be chosen as

$$\lambda_1 = 0.25 \approx (1/36) \times 12, \quad \text{and} \quad \lambda_2 = 0.75 \approx (1/16) \times 12.$$

We set the recall intensity to be $\lambda_0 = 0.05$ such that a recall is a rare event as noted in D'Avolio (2002). Since the Japanese interest rates are close to zero during the sample period, we take the rebate interest rates to be $q_1 = q_2 = 0.01$ and set $\delta_1 = \delta_2 = 0.02$ for the loan fee rates. The short-seller's subjective discount rate is taken to be $r = 0.10$, while the initial stock price and entry price are normalized to be $x = K = 100$.

λ_0	x_1^*	x_2^*	$v_1(100)$	$v_2(100)$
0	77.68	38.04	6.25	16.08
0.02	84.37	38.51	2.74	13.12
0.05	92.30	38.91	0.59	10.87
0.1	102.25	39.30	0.00	9.16
0.2	116.06	39.77	0.00	7.71

Table 2: Impact of recall risk on the optimal thresholds and value functions.

Figure 3 illustrates the value-matching and smooth-fit of the value functions v_1 and v_2 at the respective optimal thresholds $(x_1^*, x_2^*) = (92.30, 38.91)$, obtained by solving equation (12). The optimal threshold for short-covering in state 1 is much higher than that in state 2, suggesting that the short-seller should set a higher target price and take profit earlier under a bull market while being more aggressive in a bear market. The large difference in the optimal thresholds (x_1^*, x_2^*) highlights the importance to incorporate regime-switching into the short-seller's problem and derive the corresponding regime-dependent optimal stopping rule.

In contrast to a standard real option problem, the value functions can become negative for large x due to the broker's recall and running cost.[4] This is illustrated more clearly by plotting the lower bounds of the value functions, $h_i(x) = \psi_i(x) + (1 - \eta_i) K - (1 - \rho_i) x$ for $i = 1, 2$, which represent the expected present values of the trading strategy when the short-seller chooses to wait forever ($\tau_E = \infty$). We find that the intersection of the lower bound $h_1(x)$ and the payoff $K - x$, is quite close to x_1^* in state 1 but this is not the case in state 2. Moreover, the difference $v_2(x) - h_2(x)$ is also larger than $v_1(x) - h_1(x)$. This suggests that the optionality to stop at a finite time is more valuable in state 2 than that in state 1. Intuitively, this can be explained by the fact that a short-selling strategy is more opportunistic when the stock price tumbles in a bear market. As the value function in state 2 is quite different from that in state 1, a financial implication is that a short-seller should book the net present value of a trade (P&L) according to the current market regime.

4.2 The impact of recall risk

Table 2 reports the impact of recall risk on the optimal thresholds as the intensity λ_0 varies. It can be seen that the impact of recall risk is higher in state 1 than that in state 2. In a bull market, the stock price tends to go up and the broker's recall is more likely to lead to a trading loss. As a result, the short-seller has to be more careful about the timing of taking profit (short covering) and she needs to modify the target price accordingly as the recall risk increases. Furthermore, we

[4] For a detailed comparative analysis for the short-seller's problem, see Chung and Tanaka (2015).

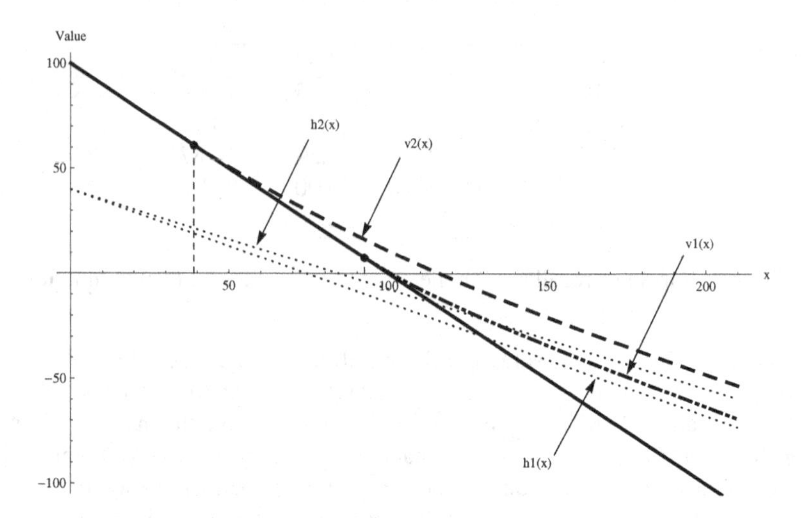

Figure 3. : Value functions and optimal thresholds based on Table 1.

note that the corresponding optimal threshold x_1^* may go over $K = 100$, indicating that the short-seller should not enter the trade in a bull market when the recall risk and stock price x are high.

4.3 The impact of transition intensity

We vary the transition intensities $\{\lambda_i, i = 1, 2\}$ as 0.25, 0.50, 0.75 and 1.00 to study the effect of regime-switchings to the optimal thresholds. From Table 3, we find that the optimal threshold under state 1 is very sensitive to the intensities λ_1 and λ_2. As the intensity λ_1 increases, there is a higher probability to switch from state 1 to state 2. This encourages the short-seller in state 1 to wait further for a transition to state 2 which is more opportunistic. When the intensity λ_2 increases, state 2 becomes relatively short-lived. The short-seller is more time constrained and has less chance to make a good profit during a bear market. Consequently, the trade in state 1 also become less opportunistic and the short-seller should cover the position earlier.

In contrast, the optimal threshold under state 2 is quite robust to the variation in transition intensities. This can be explained as follows. When the current state is 2 and $x_2^* << x_1^*$, a regime switching $(2 \to 1)$ indicates a jump from the continuation region under state 2 to the stopping region under state 1. This effectively leads to a forced termination of the strategy and such a switching is similar to a random recall. When the market is trending downward, such a forced termination is likely to lead to an early profit taking which is not too harmful to the short-seller. This is consistent with the analysis in Chung and Tanaka (2015) which explicitly show that the optimal threshold is less sensitive to a forced termination in a down market.

	λ_2			
λ_1	0.25	0.50	0.75	1.00
0.25	(77.26, 35.63)	(85.54, 37.56)	(92.30, 38.91)	(97.92, 39.91)
0.50	(63.58, 35.18)	(69.49, 37.03)	(74.73, 38.44)	(79.40, 39.52)
0.75	(57.35, 34.86)	(61.93, 36.57)	(66.13, 37.97)	(69.98, 39.09)
1.00	(53.66, 34.61)	(57.41, 36.20)	(60.90, 37.54)	(64.17, 38.67)

Table 3: Impact of transition intensity on the optimal thresholds (x_1^*, x_2^*).

5. Concluding Remarks

In this paper, we study the optimal stopping problem related to a short-selling strategy in a financial market. We consider a two-state regime-switching stock price model and derive the optimal stopping rule to the short-seller's problem. When the random recall is an independent exponential variable, we are able to obtain the closed-form solution by extending the results (i.e., maturity randomization) in Guo and Zhang (2005) and Tanaka (2012). Although we focus on the put-type problem in this paper, it would be interesting to evaluate a short-selling strategy when there is a switch in between the put-type and call-type problems. Moreover, the intensity of broker's recall can also depend on the regime. These are left for future research.

References

1. Asmussen, S. (2003) Applied Probabilities and Queues. Springer.
2. Chung, T.K. and Tanaka, K. (2015) Optimal Timing for Short Covering of an Illiquid Security. Journal of the Operations Research Society of Japan (58), 165–183.
3. G. D'Avolio (2002) The Market of Borrowing Stock. Journal of Financial Economics (26), 271-306.
4. Guo, X. and Zhang, Q. (2005) Optimal Selling Rules in a Regime Switching Model. IEEE Transactions on Automatic Control (50), 1450–1455.
5. Jobert, A. and Rogers, C. (2006) Option Pricing with Markov-modulated Dynamics. SIAM Journal on Control and Optimization (44), 2063-2078.
6. G. Peskir and A. Shiryaev (2005) Optimal Stopping and Free-Boundary Problems, Birkhäuser Verlag.
7. Tanaka, K. (2012) Irreversible Investment with Regime Switching: Revisit with Linear Algebra. Working paper, Tokyo Metropolitan University.

A. Proof of Lemma 2.1

Note that

$$\mathbf{\Lambda}(1) = -(r\mathbf{I} - \mathbf{M} - \mathbf{Q}), \qquad \mathbf{\Lambda}(0) = -(r\mathbf{I} - \mathbf{Q}),$$

which will be useful in the derivation. For $T \geq t$ and a real number ξ, Proposition 2.2 in Asmussen (2003) gives

$$\begin{aligned}\mathbb{E}\left[e^{-rT}\delta_{J(T)}X(T)^{\xi}\mathbf{1}_{\{J(T)=j\}}\middle| J(t)=i, X(t)\right] &= e^{-rT}X(t)^{\xi}\left\{\exp\left[\left(\mathbf{Q}+\boldsymbol{\Theta}(\xi)\right)(T-t)\right]\right\}_{ij}\\ &= e^{-rt}X(t)^{\xi}\left\{\exp\left[\boldsymbol{\Lambda}(\xi)(T-t)\right]\right\}_{ij},\end{aligned}$$

in which the second equality is due to $\mathbf{Q}+\boldsymbol{\Theta}(z) = \boldsymbol{\Lambda}(z) + r\mathbf{I}$. Applying this result, we have

$$\begin{aligned}\mathbb{E}\left[e^{-rT}\delta_{J(T)}X(T)^{\xi}\middle| J(t)=i, X(t)\right] &= \sum_{j\in E}\mathbb{E}\left[e^{-rT}\delta_{J(T)}X(T)^{\xi}\mathbf{1}_{\{J(T)=j\}}\middle| J(t)=i, X(t)\right]\\ &= \sum_{j\in E}\mathbb{E}\left[e^{-rT}X(T)^{\xi}\mathbf{1}_{\{J(T)=j\}}\middle| J(t)=i, X(t)\right]\delta_j\\ &= e^{-rt}X(t)^{\xi}\sum_{j\in E}\left\{\exp\left[\boldsymbol{\Lambda}(\xi)(T-t)\right]\right\}_{ij}\delta_j\\ &= e^{-rt}X(t)^{\xi}\mathbf{e}_i^{\top}\exp\left[\boldsymbol{\Lambda}(\xi)(T-t)\right]\boldsymbol{\delta}.\end{aligned}$$

Hence, we can compute

$$\begin{aligned}\mathbb{E}\left[\int_t^{\infty} e^{-rt}\delta_{J(u)}X_u du\middle| J(t)=i, X(t)\right] &= \int_t^{\infty}\mathbb{E}\left[e^{-rt}\delta_{J(u)}X_u\middle| J(t)=i, X(t)\right]du\\ &= e^{-rt}X(t)\mathbf{e}_i^{\top}\int_t^{\infty}\exp\left[\boldsymbol{\Lambda}(1)(u-t)\right]\boldsymbol{\delta}du\\ &= e^{-rt}X(t)\mathbf{e}_i^{\top}\left\{\int_t^{\infty}\exp\left[-\left(r\mathbf{I}-\mathbf{M}-\mathbf{Q}\right)(u-t)\right]du\right.\\ &= e^{-rt}X(t)\mathbf{e}_i^{\top}\left(r\mathbf{I}-\mathbf{M}-\mathbf{Q}\right)^{-1}\boldsymbol{\delta}\\ &= e^{-rt}\alpha_i\delta_i X_t,\end{aligned}$$

in which we denote $\alpha_i\delta_i = \mathbf{e}_i^{\top}\left(r\mathbf{I}-\mathbf{M}-\mathbf{Q}\right)^{-1}\boldsymbol{\delta}$. Similarly, we have

$$\begin{aligned}\mathbb{E}_{x,i}\left[\int_t^{\infty} e^{-rt}q_{J(u)}du\middle| J(t)=i, X(t)\right] &= e^{-rt}\mathbf{e}_i^{\top}\int_t^{\infty}\exp\left[\boldsymbol{\Lambda}(0)(u-t)\right]\mathbf{q}du\\ &= e^{-rt}\mathbf{e}_i^{\top}\left\{\int_t^{\infty}\exp\left[-\left(r\mathbf{I}-\mathbf{Q}\right)(u-t)\right]du\right\}\mathbf{q}\\ &= e^{-rt}\mathbf{e}_i^{\top}\left(r\mathbf{I}-\mathbf{Q}\right)^{-1}\mathbf{q}\\ &= e^{-rt}\beta_i q_i,\end{aligned}$$

in which we denote $\beta_i q_i = \mathbf{e}_i^{\top}\left(r\mathbf{I}-\mathbf{Q}\right)^{-1}\mathbf{q}$.

By the strong Markov property of (X, J), we have

$$\mathbb{E}_{x,i}\left[\int_0^\infty e^{-ru}\delta_{J(u)}X_u du\right] = \alpha_i\delta_i x,$$

$$\mathbb{E}_{x,i}\left[\int_t^\infty e^{-ru}\delta_{J(u)}X_u du\right] = \mathbb{E}_{x,i}\left[\mathbb{E}\left[\int_t^\infty e^{-ru}\delta_{J(u)}X_u du\middle|\mathcal{F}_t\right]\right] = \mathbb{E}_{x,i}\left[e^{-rt}\alpha_{J(t)}\delta_{J(t)}X_t\right],$$

such that

$$\mathbb{E}_{x,i}\left[\int_0^t e^{-ru}\delta_{J(u)}X_u du\right] = \alpha_i\delta_i x - \mathbb{E}_{x,i}\left[e^{-rt}\alpha_{J(t)}\delta_{J(t)}X_t\right].$$

For the second identity, we have

$$\mathbb{E}_{x,i}\left[\int_0^\infty e^{-ru}q_{J(u)}K du\right] = \beta_i q_i K,$$

$$\mathbb{E}_{x,i}\left[\int_t^\infty e^{-ru}q_{J(u)}K du\right] = \mathbb{E}_{x,i}\left[\mathbb{E}\left[\int_t^\infty e^{-ru}q_{J(u)}K du\middle|\mathcal{F}_t\right]\right] = \mathbb{E}_{x,i}\left[e^{-rt}\beta_{J(t)}q_{J(t)}K\right],$$

such that

$$\mathbb{E}_{x,i}\left[\int_0^t e^{-ru}q_{J(u)}K du\right] = \beta_i q_i K - \mathbb{E}_{x,i}\left[e^{-rt}\beta_{J(t)}q_{J(t)}K\right].$$

B. Proof of Proposition 2.1

The lower bound of u_i corresponding to $\tau_E = \infty$ is

$$\begin{aligned}\psi_i(x) &= \mathbb{E}_{x,i}\left[e^{-r\tau_R}g\left(X_{\tau_R}\right)\right]\\ &= \int_0^\infty \lambda_0 e^{-(r+\lambda_0)u}\mathbb{E}_{x,i}\left[\eta_{J(u)}K\right]du - \int_0^\infty \lambda_0 e^{-(r+\lambda_0)u}\mathbb{E}_{x,i}\left[\rho_{J(u)}X_u\right]du,\end{aligned}$$

since τ_R is exponentially distributed. By Proposition 2.2 in Asmussen (2003), we have

$$\begin{aligned}\int_0^\infty \lambda_0 e^{-(r+\lambda_0)u}\mathbb{E}_{x,i}\left[\eta_{J(u)}K\right]du &= \int_0^\infty \lambda_0 K\mathbf{e}_i^\top \exp\left[-\left(\left(r+\lambda_0\right)\mathbf{I}-\mathbf{Q}\right)u\right]\boldsymbol{\eta}du\\ &= \lambda_0 K\mathbf{e}_i^\top\left(\left(r+\lambda_0\right)\mathbf{I}-\mathbf{Q}\right)^{-1}\boldsymbol{\eta},\end{aligned}$$

and

$$\begin{aligned}\int_0^\infty \lambda_0 e^{-(r+\lambda_0)u}\mathbb{E}_{x,i}\left[\rho_{J(u)}X_u\right]du &= \int_0^\infty \lambda_0 x\mathbf{e}_i^\top \exp\left[-\left(\left(r+\lambda_0\right)\mathbf{I}-\mathbf{M}-\mathbf{Q}\right)u\right]\boldsymbol{\rho}du\\ &= -\lambda_0 x\mathbf{e}_i^\top\left(\left(r+\lambda_0\right)\mathbf{I}-\mathbf{M}-\mathbf{Q}\right)^{-1}\boldsymbol{\rho}.\end{aligned}$$

Hence, we can express

$$\psi_i(x) = a_i x + b_i, \quad i = 1, 2,$$

where

$$\mathbf{a} = \lambda_0 \left(\mathbf{M} + \mathbf{Q} - (r + \lambda_0)\, \mathbf{I}\right)^{-1} \rho, \quad \mathbf{b} = \lambda_0 K \left((r + \lambda_0)\, \mathbf{I} - \mathbf{Q}\right)^{-1} \eta,$$

with $\mathbf{a} = (a_1, a_2)^\top$ and $\mathbf{b} = (b_1, b_2)^\top$.

C. Condition (6)

We can compute the coefficients $\alpha_i\delta_i$ and $\beta_i q_i$ explicitly and show that the condition (6) implies a put-type problem with $\rho_i > 0$ and $\eta_i > 0$, for $i = 1, 2$.

1. $\alpha_i\delta_i = \mathbf{e}_i^\top (r\mathbf{I} - \mathbf{M} - \mathbf{Q})^{-1} \boldsymbol{\delta}$. Note that

$$(r\mathbf{I} - \mathbf{M} - \mathbf{Q})^{-1} = \frac{1}{\det(r\mathbf{I} - \mathbf{M} - \mathbf{Q})} \begin{pmatrix} r + \lambda_2 - \mu_2 & \lambda_1 \\ \lambda_2 & r + \lambda_1 - \mu_1 \end{pmatrix},$$

where $\det(r\mathbf{I} - \mathbf{M} - \mathbf{Q}) = (r + \lambda_1 - \mu_1)(r + \lambda_2 - \mu_2) - \lambda_1\lambda_2$. We have

$$\alpha_1\delta_1 = \frac{(r + \lambda_2 - \mu_2)\,\delta_1 + \lambda_1\delta_2}{(r + \lambda_1 - \mu_1)(r + \lambda_2 - \mu_2) - \lambda_1\lambda_2},$$
$$\alpha_2\delta_2 = \frac{(r + \lambda_1 - \mu_1)\,\delta_2 + \lambda_2\delta_1}{(r + \lambda_1 - \mu_1)(r + \lambda_2 - \mu_2) - \lambda_1\lambda_2}.$$

Under the condition $\delta_i/(r - \mu_i) < 1$ for $i = 1, 2$ from (6), we see that

$$\begin{aligned} 0 < (r + \lambda_2 - \mu_2)\,\delta_1 + \lambda_1\delta_2 &< (r + \lambda_2 - \mu_2)(r - \mu_1) + \lambda_1 (r - \mu_2) \\ &= (r + \lambda_1 - \mu_1)(r + \lambda_2 - \mu_2) - \lambda_1\lambda_2, \end{aligned}$$

with $r > \max(\max(\mu_1, \mu_2), 0)$, $\lambda_i > 0$ and $\delta_i > 0$. By rearranging the terms, we can show that $\alpha_1\delta_1 < 1$ and $\rho_1 = 1 - \alpha_1\delta_1 > 0$. Similarly, we can show that $\alpha_2\delta_2 < 1$ and $\rho_2 = 1 - \alpha_2\delta_2 > 0$.

2. $\beta_i q_i = \mathbf{e}_i^\top (r\mathbf{I} - \mathbf{Q})^{-1} \mathbf{q}$. Note that

$$(r\mathbf{I} - \mathbf{Q})^{-1} = \frac{1}{\det(r\mathbf{I} - \mathbf{Q})} \begin{pmatrix} r + \lambda_2 & \lambda_1 \\ \lambda_2 & r + \lambda_1 \end{pmatrix},$$

where $\det(r\mathbf{I} - \mathbf{Q}) = (r + \lambda_1)(r + \lambda_2) - \lambda_1\lambda_2$. We have

$$\beta_1 q_1 = \frac{(r + \lambda_2)\, q_1 + \lambda_1 q_2}{(r + \lambda_1)(r + \lambda_2) - \lambda_1\lambda_2},$$
$$\beta_2 q_2 = \frac{(r + \lambda_1)\, q_2 + \lambda_2 q_1}{(r + \lambda_1)(r + \lambda_2) - \lambda_1\lambda_2}.$$

Under the condition $q_i/r < 1$ for $i = 1, 2$ from (6), we see that

$$\begin{aligned} 0 < (r + \lambda_2)\, q_1 + \lambda_1 q_2 &< (r + \lambda_2)\, r + \lambda_1 r \\ &= (r + \lambda_1)(r + \lambda_2) - \lambda_1\lambda_2, \end{aligned}$$

with $r > 0$, $\lambda_i > 0$ and $q_i > 0$. By rearranging the terms, we can show that $\beta_1 q_1 < 1$ and $\eta_1 = 1 - \beta_1 q_1 > 0$. Similarly, we can show that $\beta_2 q_2 < 1$ and $\eta_2 = \beta_2 q_2 > 0$.

D. Derivation of (9)

Re-write the homogenous equation (8) as

$$\begin{cases} \mathcal{A}_1 f_1(x) + \lambda_1 f_2(x) = 0 \\ \mathcal{A}_2 f_2(x) + \lambda_2 f_1(x) = 0 \end{cases}.$$

We can solve $f_2(x)$ as a function of $f_1(x)$ from the first equation and substitute it into the second equation, which gives

$$\mathcal{A}_2 \mathcal{A}_1 f_1(x) = \lambda_1 \lambda_2 f_1(x).$$

Similarly, we can solve $f_1(x)$ as a function of $f_2(x)$ and obtain

$$\mathcal{A}_1 \mathcal{A}_2 f_2(x) = \lambda_1 \lambda_2 f_2(x).$$

Take the solution form

$$f_1(x) = \sum_{k=1}^{4} A_k x^{\beta_k}, \quad f_2(x) = \sum_{k=1}^{4} B_k x^{\beta_k},$$

we have

$$\begin{cases} G_1(\beta_k) G_2(\beta_k) A_k x^{\beta_k} = \lambda_1 \lambda_2 A_k x^{\beta_k} \\ G_1(\beta_k) G_2(\beta_k) B_k x^{\beta_k} = \lambda_1 \lambda_2 B_k x^{\beta_k} \end{cases},$$

which holds for $k = 1, 2, 3, 4$. Hence, we need to solve the characteristic equation

$$G_1(\beta) G_2(\beta) = \lambda_1 \lambda_2,$$

for the roots β_k with $k = 1, 2, 3, 4$. It can be shown that the four roots are distinct by checking the quartic equation

$$p(\beta) = G_1(\beta) G_2(\beta) - \lambda_1 \lambda_2,$$

which has the properties: $p(-\infty) > 0$, $p\left(\beta_1^{(1)}\right) < 0$, $p(0) > 0$, $p\left(\beta_1^{(2)}\right) < 0$, $p(\infty) > 0$, where $\beta_1^{(2)} > 1$ and $\beta_1^{(1)} < 0$ are the two distinct roots of $G_1(\beta) = 0$. Furthermore, by a direct substitution we see that the coefficients A_k and B_k are related as

$$B_k = l_k A_k = \frac{G_1(\beta_k)}{\lambda_1} A_k = \frac{\lambda_2}{G_2(\beta_k)} A_k,$$

for $k = 1, 2, 3, 4$.

Effects of Reversibility on Investment Timing and Quantity Under Asymmetric Information

Xue Cui[a] and Takashi Shibata[a]

[a] *Graduate School of Social Sciences, Tokyo Metropolitan University, 1-1 Minami-Ohsawa, Hachiohji, Tokyo 192-0397, Japan.*
xcuiaa@gmail.com *tshibata@tmu.ac.jp*

The paper examines how changes in reversibility of investment affect a firm's investment timing and quantity strategies, in the presence of manager's private information. We find that even under asymmetric information, higher reversibility of investment decreases the investment trigger. More importantly, the quantity under asymmetric information is no longer independent of the degree of reversibility of investment, but increases with it.

Key words: investment timing; quantity; reversibility; asymmetric information

1. Introduction

The purpose of this paper is to examine the effects of reversibility on investment timing and quantity strategies under asymmetric information. The reversibility of investment enables the firm to sell capital when the marginal profitability of capital is low.[1] Thus, under the assumption of investment reversibility, the firm owns an abandonment option that allows the firm to shut down the operation of a facility after the investment in that facility. On the other hand, in most modern corporations, the firm owner delegates the management to the manager, taking advantage of the manager's expertise. In this situation, asymmetric information is likely to exist because the manager owns private information. In this paper, we investigate, in the presence of manager's private information, how changes in reversibility of investment affect the firm's investment timing (trigger) and quantity strategies.

A number of studies have analyzed investment timing decision strategies since the seminal work of McDonald and Siegel (1986). On the one hand, Bar-Ilan and

[1] See Abel and Eberly (1999).

Strange (1999) provide a framework to model both investment timing and quantity (intensity). Wong (2010) extends the model by Bar-Ilan and Strange (1999) to incorporate the reversibility of investment. The main result is that an increase in reversibility of investment decreases the investment trigger but has no impact on the quantity. On the other hand, Grenadier and Wang (2005) consider the investment timing decision problem in the presence of asymmetric information between owners and managers.[2] They show that asymmetric information leads to an increase in the investment trigger.

In this paper, we consider the interactions between Grenadier and Wang (2005) and Wong (2010). Thus, our model extends Wong (2010) by incorporating asymmetric information. Alternatively, our model extends Grenadier and Wang (2005) by incorporating a quantity decision and an abandonment option.

We find that, even under asymmetric information, higher reversibility of investment induces a smaller investment trigger. This result is the same as that under full (symmetric) information. More importantly, under asymmetric information, higher reversibility of investment induces a larger quantity. This result is different from that under full information. We explain the result following the mechanism by Wong (2010). An increase in reversibility of investment creates two opposing effects on quantity. First, higher reversibility induces a smaller investment trigger, thereby decreasing the marginal return on investment. This effect induces the firm to reduce the quantity. Second, higher reversibility enhances the value of the abandonment option and thus decreases the marginal cost of investment. This effect induces the firm to increase the quantity. Under asymmetric information, since the investment trigger is increased due to the distortion, the effect of increased reversibility on decreasing the marginal return is somewhat mitigated and the effect on decreasing the marginal cost becomes dominant. As a result, the firm increases the quantity in response to a higher reversibility of investment.

The remainder of the paper is organized as follows. Section 2 describes the framework of model and solves the problem using backward induction. Section 3 provides the model solution. Section 4 presents related empirical findings. Section 5 concludes the paper.

2. The Model

In this section, we begin with a description of the model setup. We then solve the problem backward. In Subsection 2.2, we derive the firm value at investment. In Subsection 2.3, we derive the firm value before investment. Before discussing the asymmetric information model, we provide the solution to the full information model as a benchmark.

---

[2]See Shibata (2009) and Shibata and Nishihara (2011) for the extension of Grenadier and Wang (2005) model.

2.1 Model setup

Consider a risk-neutral firm that is endowed with an option to invest in a production facility at any time. We assume that the firm owner delegates the investment decision to a manager. Throughout our analysis, we assume that both the owner and the manager are risk-neutral and aim to maximize their expected payoffs.

The investment is undertaken by incurring an investment cost $I(q) := F + C(q) > 0$, where $F > 0$ is the fixed cost and $C(q) > 0$ is the variable cost of producing $q \geq 1$ units of commodities. We assume that the fixed cost, F, takes one of two possible values, F_1 or F_2, with $F_2 > F_1 > 0$. We denote $\triangle F := F_2 - F_1$. Here, F_1 represents the *lower-fixed cost* expenditure and F_2 represents the *higher-fixed cost* expenditure. The probability of drawing $F = F_1$, $P(F_1) = p > 0$, is an exogenous variable. Further, the variable cost $C(q)$ is assumed to be a strictly increasing and convex function of q, that is, $C'(q) > 0$ and $C''(q) > 0$ for all $q \geq 1$ with $C(1) > 0$.

Once the investment is made, the facility starts to produce q units of a single commodity per unit of time. The firm sells the commodity in a perfectly competitive market at a per-unit price, X_t, at time t. The commodity price is a stochastic process that evolves over time according to the following geometric Brownian motion:

$$dX_t = \mu X_t dt + \sigma X_t dZ_t, \quad X_0 = x > 0, \tag{1}$$

where Z_t is a standard Brownian motion. $\mu > 0$ and $\sigma > 0$ are the constant expected growth rate and standard deviation of the commodity price per unit of time, respectively. We assume that the initial price x is too small to make an immediate investment optimal. Let r be the constant interest rate. For convergence, we assume $r > \mu$. [3]

After the investment option has been exercised and the facility is in operation, the firm is allowed to shut down the operation of the facility if the commodity price turns out to be unfavorable. The abandonment decision, once made, is irreversible. At the time of abandonment, the firm gets a value of $sI(q)$ with $s \in [0, 1]$. Importantly, the size of s gauges the degree of reversibility of investment. $s = 0$ ($s = 1$) means that the investment is completely irreversible (reversible).

The investment decision is made by the manager who has private information about the fixed cost F. Specifically, the commodity price X_t is publicly observable from the market. Meanwhile, given q, the cost of producing q units of commodities $C(q)$ is also known to both the owner and the manager. Immediately after making a contract with the owner, the manager observes whether F is equal to F_1 or F_2. However, the owner cannot observe the true value of F.[4] In this situation, the

[3] The assumption $r > \mu$ is needed to ensure a finite firm value.

[4] In the asymmetric information structure, it is quite common to assume that a portion of investment

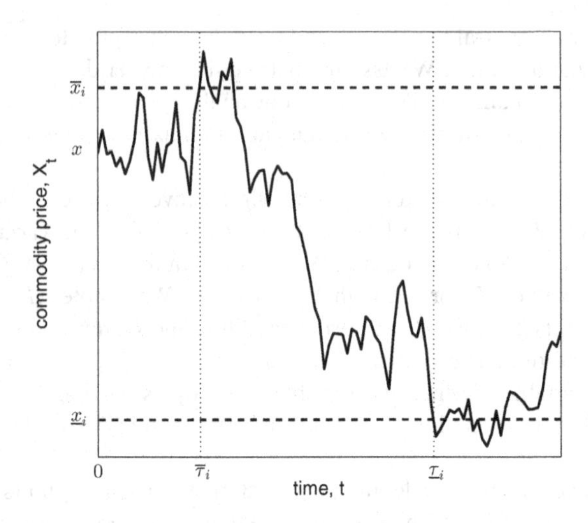

Figure 1. Investment and abandonment strategies

manager could attempt to divert the value of ΔF to himself by reporting F_2 when he truly observes F_1. Consequently, the owner suffers losses from the diversion. Thus, to prevent the diversion, the owner must induce the manager to reveal the private information truthfully by providing incentives.

Figure 1 depicts the scenario for the investment and abandonment strategies. Recall that the initial price $X_0 = x > 0$ is small. Let $\overline{x}_i$ and $\underline{x}_i$ individually denote the triggers for investment and abandonment, and q_i denote the quantity for $F = F_i$ ($i \in \{1, 2\}$). Once X_t increases and arrives at $\overline{x}_i$, the firm exercises the investment and decides q_i endogenously at the time of investment. After investment, if X_t declines and reaches $\underline{x}_i$, the firm exercises the abandonment and gets a value of $sI(q_i)$. Correspondingly, $\overline{\tau}_i = \inf\{t \geq 0; X_t = \overline{x}_i\}$ and $\underline{\tau}_i = \inf\{t \geq \overline{\tau}_i; X_t = \underline{x}_i\}$ represent the investment timing and the abandonment timing, respectively. Let $\mathbb{E}^x(\cdot)$ and $\mathbb{E}^{\overline{x}_i}(\cdot)$ be the conditional expectation on x and on $\overline{x}_i$, respectively. Using standard arguments (e.g., Dixit and Pindyck, 1994), we have

$$\text{(2)} \qquad \mathbb{E}^x[e^{-r\overline{\tau}_i}] = \left(\frac{x}{\overline{x}_i}\right)^{\beta}, \quad \mathbb{E}^{\overline{x}_i}[e^{-r(\underline{\tau}_i - \overline{\tau}_i)}] = \left(\frac{\overline{x}_i}{\underline{x}_i}\right)^{\gamma}, \quad i \in \{1, 2\},$$

value is privately observed by one party (here, the manager) and not observed by the other party (here, the owner). An excellent overview of situations with asymmetric information can be found in Laffont and Martimort (2002).

where

$$\beta = \frac{1}{2} - \frac{\mu}{\sigma^2} + \sqrt{(\frac{1}{2} - \frac{\mu}{\sigma^2})^2 + \frac{2r}{\sigma^2}} > 1, \tag{3}$$

and

$$\gamma = \frac{1}{2} - \frac{\mu}{\sigma^2} - \sqrt{(\frac{1}{2} - \frac{\mu}{\sigma^2})^2 + \frac{2r}{\sigma^2}} < 0. \tag{4}$$

The two terms of (2) provide the present value operators of one dollar received at the investment timing and the abandonment timing, respectively.

2.2 Value at investment

In this subsection, we derive the firm value at the time of investment, while the abandonment option is still alive. Given q_i and $\overline{x}_i$, the firm value at investment, $V(q_i, \overline{x}_i)$, is formulated as

$$V(q_i, \overline{x}_i) = \sup_{\underline{\tau}_i \geq \overline{\tau}_i} \mathbb{E}^{\overline{x}_i} \left[\int_{\overline{\tau}_i}^{\underline{\tau}_i} e^{-r(t-\overline{\tau}_i)} q_i X_t dt + e^{-r(\underline{\tau}_i - \overline{\tau}_i)} s(F_i + C(q_i)) \right] \tag{5}$$

$$= \max_{\underline{x}_i \leq \overline{x}_i} \frac{q_i \overline{x}_i}{r - \mu} + \left(s(F_i + C(q_i)) - \frac{q_i \underline{x}_i}{r - \mu} \right) \left(\frac{\overline{x}_i}{\underline{x}_i} \right)^{\gamma}. \tag{6}$$

The optimal abandonment trigger is solved to be

$$\underline{x}_i(q_i) = \frac{\gamma}{\gamma - 1} \frac{s(F_i + C(q_i))}{q_i} (r - \mu) \geq 0. \tag{7}$$

The inspection of (7) reveals that the optimal abandonment trigger depends on the quantity.

Substituting (7) into (6) yields

$$V(q_i, \overline{x}_i) = \frac{q_i \overline{x}_i}{r - \mu} + AO_i(q_i, \overline{x}_i), \tag{8}$$

where

$$AO_i(q_i, \overline{x}_i) = \left(\frac{q_i \overline{x}_i}{-\gamma(r - \mu)} \right)^{\gamma} \left(\frac{1 - \gamma}{s(F_i + C(q_i))} \right)^{\gamma - 1} \geq 0. \tag{9}$$

Note that the value of the abandonment option $AO_i(q_i, \overline{x}_i) = 0$ when $s = 0$.

2.3 Value before investment

We derive the firm value at $t = 0$, before the investment is undertaken. Subsection 2.3.1 considers the full information model as a benchmark. Subsection 2.3.2 discusses the asymmetric information model.

2.3.1 Full information benchmark

In this subsection, we consider the investment timing problem under full information. Suppose that the owner observes the true value of F. This implies that there is no delegation of the investment decision because the manager has no informational advantage. The contract $\mathcal{M}^*$ in the full information problem is modeled as

$$\mathcal{M}^* = (q_i^*, \overline{x}_i^*, \underline{x}_i^*), \quad i \in \{1, 2\}.$$

The superscript "*" refers to the *full (symmetric) information* problem.

The owner's optimization problem is

$$\max_{q_1, q_2, \overline{x}_1, \overline{x}_2} pH(x, q_1, \overline{x}_1; F_1) + (1 - p)H(x, q_2, \overline{x}_2; F_2), \tag{10}$$

where $x < \overline{x}_i$ for any i ($i \in \{1, 2\}$) and

$$H(x, q_i, \overline{x}_i; F_i) := \left(\frac{q_i \overline{x}_i}{r - \mu} + AO_i(q_i, \overline{x}_i) - F_i - C(q_i)\right)\left(\frac{x}{\overline{x}_i}\right)^{\beta}. \tag{11}$$

Then, as shown in Wong (2010), we have the following result.

Lemma 1 *In the full information model, for any i ($i \in \{1, 2\}$), q_i^* and $\overline{x}_i^*$ are determined by solving the following equations:*

$$\frac{q_i^* \overline{x}_i^*}{r - \mu} + \frac{\beta - \gamma}{\beta - 1} AO_i(q_i^*, \overline{x}_i^*) = \frac{\beta}{\beta - 1}(F_i + C(q_i^*)), \tag{12}$$

and

$$\frac{\overline{x}_i^*}{r - \mu} + \frac{\gamma}{q_i^*} AO_i(q_i^*, \overline{x}_i^*) = \left(1 - \frac{1 - \gamma}{F_i + C(q_i^*)} AO_i(q_i^*, \overline{x}_i^*)\right) C'(q_i^*). \tag{13}$$

In addition, we have $\underline{x}_i^* = \underline{x}_i(q_i^*)$.

(12) and (13) are obtained from the first-order conditions with respect to $\overline{x}_i$ and q_i, respectively. Intuitively, (12) implies that the firm optimally exercises the investment when the value of facility, plus the augmented value of the abandonment option, equals an option value multiple times the investment cost. In addition, holding the investment cost fixed, the left-hand side of (13) is the marginal return on investment, while the right-hand side is the marginal cost of investment. Then, (13) implies that the optimal quantity is the one that equates the marginal return to the marginal cost.

2.3.2 Asymmetric information model

In this subsection, we formulate the investment timing problem under asymmetric information. Suppose that the owner delegates the investment decision to the manager who has private information on F. At time 0, the owner designs a contract that commits the owner to give an incentive to the manager at the time of investment. No renegotiation is allowed afterwards. Here, we assume that the owner provides a bonus incentive w_i to motivate the manager to reveal the private information. Thus, the contract $\mathcal{M}^{**}$ under asymmetric information is modeled as

$$\mathcal{M}^{**} = (q_i^{**}, \overline{x}_i^{**}, \underline{x}_i^{**}, w_i^{**}), \quad i \in \{1, 2\}.$$

The superscript "**" refers to the *asymmetric information* problem.

Then, the owner's optimization problem is

$$\max_{q_i, \overline{x}_i, w_i} \sum_{i \in \{1,2\}} P(F_i) \left(\frac{q_i \overline{x}_i}{r - \mu} + AO_i(q_i, \overline{x}_i) - F_i - C(q_i) - w_i \right) \left(\frac{x}{\overline{x}_i} \right)^{\beta}, \tag{14}$$

subject to

$$\left(\frac{x}{\overline{x}_1} \right)^{\beta} w_1 \geq \left(\frac{x}{\overline{x}_2} \right)^{\beta} (w_2 + \triangle F), \tag{15}$$

$$\left(\frac{x}{\overline{x}_2} \right)^{\beta} w_2 \geq \left(\frac{x}{\overline{x}_1} \right)^{\beta} (w_1 - \triangle F), \tag{16}$$

$$w_1 \geq 0, \tag{17}$$

$$w_2 \geq 0, \tag{18}$$

$$\sum_{i \in \{1,2\}} P(F_i) \left(\frac{x}{\overline{x}_i} \right)^{\beta} w_i \geq 0. \tag{19}$$

Here, the objective function (14) is the owner's *ex-ante* option value. (15) and (16) are the *ex-post* incentive compatibility constraints for the manager when observing F_1 and F_2, respectively. Taking (15) as an example, for the manager who observes F_1, the manager's payoff is $(x/\overline{x}_1)^{\beta} w_1$ if he tells the truth and it is $(x/\overline{x}_2)^{\beta} (w_2 + \triangle F)$ if he instead reports F_2. When (15) is satisfied, the manager has no incentive to tell a lie. Similarly, (16) follows. (17) and (18) are the *ex-post* limited-liability constraints to ensure that the manager makes an agreement with the owner. (19) is the manager's *ex-ante* participation constraint. Note that for an exogenous quantity (e.g., $q = 1$) and a completely irreversible investment ($s = 0$), our model corresponds to Grenadier and Wang (2005).

3. Model Solution

In this section, we provide the solution to the problem under asymmetric information that is described in the previous section.

The constraints (15) - (19) can be simplified in three steps. First, (17) and (18) imply (19). Second, unlike a manager who observes F_1, a manager who observes F_2 has no incentive to tell a lie. This is because the manager who observes F_2 suffers a loss from the untruthful report. Thus, (16) is satisfied automatically, and $w_2^{**} = 0$ at the optimum. Finally, if (15) holds as a strict inequality, by decreasing w_1, the owner's value is increased. Thus, (15) is binding and we have $w_1^{**} = (\overline{x}_1^{**}/\overline{x}_2^{**})^{\beta}\Delta F$ at the optimum.

As a result, the owner's optimization problem is simplified as follows:

$$\max_{q_1, q_2, \overline{x}_1, \overline{x}_2} pH(x, q_1, \overline{x}_1; F_1) + (1-p)H(x, q_2, \overline{x}_2; F_2 + \phi\Delta F), \tag{20}$$

where $\phi = p/(1-p)$ and $x < \overline{x}_i$ for any i ($i \in \{1, 2\}$). (20) implies that under asymmetric information, the owner's value is reduced by the term $\phi\Delta F > 0$, compared with (10) under full information. This term is interpreted as an *inefficiency cost* in the presence of asymmetric information. Then, we have the following result.

Proposition 2 *In the asymmetric information model, for $F = F_1$, the solutions are*

$$\left(q_1^{**}, \overline{x}_1^{**}, \underline{x}_1^{**}, w_1^{**}\right) = \left(q_1^{*}, \overline{x}_1^{*}, \underline{x}_1^{*}, \left(\frac{\overline{x}_1^{*}}{\overline{x}_2^{**}}\right)^{\beta} \Delta F\right). \tag{21}$$

For $F = F_2$, q_2^{**} and $\overline{x}_2^{**}$ are determined by solving the following equations:

$$\frac{q_2^{**}\overline{x}_2^{**}}{r-\mu} + \frac{\beta-\gamma}{\beta-1}AO_2(q_2^{**}, \overline{x}_2^{**}) = \frac{\beta}{\beta-1}\left(F_2 + C(q_2^{**}) + \phi\Delta F\right), \tag{22}$$

and

$$\frac{\overline{x}_2^{**}}{r-\mu} + \frac{\gamma}{q_2^{**}}AO_2(q_2^{**}, \overline{x}_2^{**}) = \left(1 - \frac{1-\gamma}{F_2 + C(q_2^{**})}AO_2(q_2^{**}, \overline{x}_2^{**})\right)C'(q_2^{**}). \tag{23}$$

In addition, $\underline{x}_2^{**} = \underline{x}_2(q_2^{**})$ and $w_2^{**} = 0$.

Proposition 2 implies two properties. First, for $F = F_1$, we have $q_1^{**} = q_1^{*}$, $\overline{x}_1^{**} = \overline{x}_1^{*}$; for $F = F_2$, we have $q_2^{**} \neq q_2^{*}$, $\overline{x}_2^{**} \neq \overline{x}_2^{*}$. It is less costly for the owner to distort $(q_2^{**}, \overline{x}_2^{**})$ away from $(q_2^{*}, \overline{x}_2^{*})$ than to distort $(q_1^{**}, \overline{x}_1^{**})$ away from $(q_1^{*}, \overline{x}_1^{*})$. Second, we have $w_1^{**} \in (0, \Delta F)$ and $w_2^{**} = 0$. Here, $\Delta F > 0$ can be regarded as the informational rent for the manager who observes F_1. Thus, under asymmetric information, the owner gives the manager who observes F_1 a fraction of the informational rent as the bonus incentive to induce him to reveal the private information.

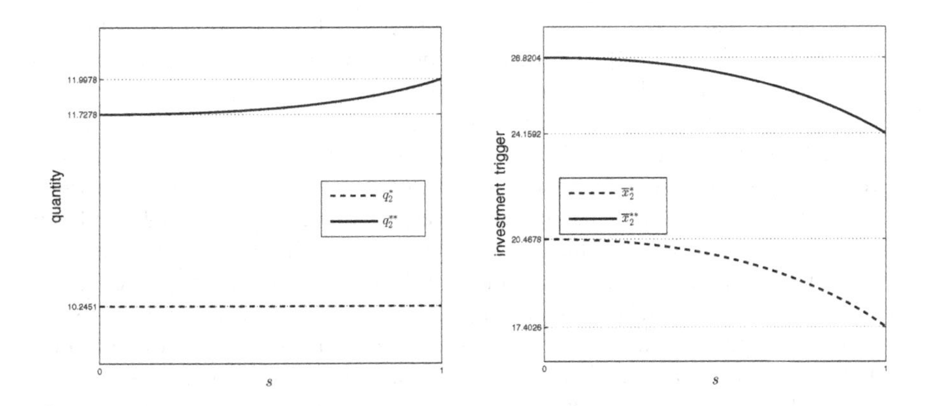

Figure 2. Effects of reversibility on quantity and investment trigger

To see more properties of the solutions, we consider numerical examples. In order to do so, the variable cost is assumed to be

$$C(q_i) = q_i^3. \tag{24}$$

Suppose that the basic parameters are $r = 0.09$, $\mu = 0.025$, $\sigma = 0.3$, $F_1 = 100$, $F_2 = 200$, $p = 0.5$ and $x = 10$.

The left panel of Figure 2 shows the optimal quantities with s, while the right panel shows the optimal investment triggers with s. We observe that $q_2^{**} > q_2^*$ and $\overline{x}_2^{**} > \overline{x}_2^*$ for all $s \in [0, 1]$. First, the ordering of $\overline{x}_2^{**} > \overline{x}_2^*$ implies that the presence of asymmetric information leads to an increase in the investment trigger, thereby delaying the exercise of investment.[5] We obtain the same result as Grenadier and Wang (2005) even for an endogenous quantity and a reversible investment. Second, the ordering of $q_2^{**} > q_2^*$ implies that the quantity under asymmetric information is larger than that under full information. Because the firm suffers losses under asymmetric information due to the delayed investment, the firm undertakes a larger quantity to compensate for losses. That is, there are trade-offs between efficiencies in investment timing and quantity strategies. This is similar to the result in Shibata and Nishihara (2011) that there are trade-offs between efficiencies in investment timing and management effort.

Furthermore, we see that $\overline{x}_2^{**}$ is decreasing with s. This means that even under asymmetric information, higher reversibility of investment induces a smaller

[5]The investment trigger and the investment timing are positively related. See Sarkar (2000) and Wong (2007).

investment trigger. This result is the same as that under full information. Interestingly, unlike that q_2^* is constant, q_2^{**} is increasing with s. This implies that under asymmetric information, higher reversibility of investment induces a larger quantity. This is a new result compared to that under full information.

We give an intuitive explanation for the new result. Recall that $\overline{x}_2^{**}$ and q_2^{**} are obtained by solving (22) and (23) simultaneously. First, by (22), holding the investment cost fixed, with higher reversibility of investment and thus a larger value of the abandonment option, the investment trigger is decreased. Second, similar to (13), the left-hand side and the right-hand side of (23) individually represent the marginal return on investment and the marginal cost of investment under asymmetric information. The optimal quantity is the one that equates the marginal return to the marginal cost. Then, the reversibility of investment shows impact on the quantity through the effects on the marginal return and the marginal cost, respectively. On the one hand, higher reversibility decreases the investment trigger, thereby decreasing the marginal return. This effect induces the firm to reduce the quantity. On the other hand, by (9), the decreased investment trigger, with the increased reversibility, enhances the value of the abandonment option and thus decreases the marginal cost. This effect induces the firm to increase the quantity. Under full information, Wong (2010) shows that an increase in reversibility decreases the marginal return and the marginal cost by the same amount, the firm has no incentive to alter the quantity. However, under asymmetric information, because the investment trigger is increased due to the distortion, which is captured by the term $\phi\Delta F > 0$ in (22), the increase of investment trigger somewhat mitigates the effect of reversibility on decreasing the marginal return. As a result, under asymmetric information, an increase in reversibility decreases the marginal cost by a larger amount than that on the marginal return, the firm is induced to increase the quantity in response to a higher reversibility of investment.

4. Related Empirical Findings

In this section, we present the related empirical findings that correspond to our three results. First, asymmetric information increases the investment trigger, thereby delaying the investment. Second, asymmetric information induces a larger investment quantity. Third, an increase in reversibility decreases the investment trigger and thus increases the likelihood of investment even under asymmetric information.

First, Weiss (1994) finds that firms with more uncertainty delay the adoption of new technology. It is well agreed that the "information effect" is an important factor in uncertainty. In the example of Weiss (1994), to adopt new technology, innovative and advanced facilities are required to replace the old facilities. The timing of the adoption of new technology can be regarded as the timing of the investment in the innovative facilities. As the innovative facilities are lack of pioneers' experiments, the information about the innovative facilities revealed to the market is very limited. In this situation, "information asymmetry" is likely to exist as one type of uncertainty, and it affects decisions regarding the adoption of new technology. Thus, our first result corresponds to the empirical finding of Weiss (1994).

Second, Biddle et al. (2009) find that lower quality of financial reporting is associated with larger investment. We recognize that lower reporting quality corresponds to increased asymmetric information. Thus, our second result coincides with the empirical finding of Biddle et al. (2009).

Third, Folta et al. (2006) examine the interactive effects of uncertainty and reversibility on the likelihood of investment. They show that, with higher reversibility, the effect of uncertainty on decreasing the likelihood of investment is reduced. Since uncertainty includes the presence of asymmetric information, our third result is consistent with the finding of Folta et al. (2006).

5. Conclusion

This paper investigates the effects of reversibility on investment timing and quantity strategies in the presence of manager's private information. We find that even under asymmetric information, higher reversibility decreases the investment trigger. More importantly, the quantity under asymmetric information is no longer independent of the degree of reversibility of investment, but increases with it. Under asymmetric information, since the investment trigger is increased due to the distortion, the effect of increased reversibility on decreasing the marginal cost dominates the effect on decreasing the marginal return. As a result, the firm increases the quantity in response to a higher reversibility of investment.

There are some potential results to be developed, such as the effects of asymmetric information on the duration of a facility's production, as well as the effects of reversibility on the value of the owner and the manager. As this is a preliminary work, we leave these questions for further study.

Acknowledgments

We acknowledge the financial support of the Asian Human Resources Fund of the Tokyo Metropolitan Government and JSPS KAKENHI (26242028 and 26285071). We also express our appreciation to the organization of the TMU Finance Workshop 2014.

References

1. Abel, A.B., Eberly, J.C., 1999. The effects of irreversibility and uncertainty on capital accumulation. Journal of Monetary Economics 44, 339-377.
2. Bar-Ilan, A., Strange, W.C., 1999. The timing and intensity of investment. Journal of Macroeconomics 21, 57-77.
3. Biddle, G.C., Hilary, G., Verdi, R.S., 2009. How does financial reporting quality relate to investment efficiency? Journal of Accounting and Economics 48, 112-131.

4. Dixit, A.K., Pindyck, R.S., 1994. Investment Under Uncertainty. Princeton University Press, Princeton, NJ.
5. Folta, T.B., Johnson, D.R., O'Brien, J., 2006. Uncertainty, irreversibility, and the likelihood of entry: an empirical assessment of the option to defer. Journal of Economic Behavior and Organization 61, 432-452.
6. Grenadier, S.R., Wang, N., 2005. Investment timing, agency, and information. Journal of Financial Economics 75, 493-533.
7. Laffont, J.J., Martimort, D., 2002. The Theory of Incentives: The Principal-Agent Model. Princeton University Press, Princeton, NJ.
8. McDonald, R.L., Siegel, D.R., 1986. The value of waiting to invest. Quarterly Journal of Economics 101, 707-728.
9. Sarkar, S., 2000. On the investment-uncertainty relationship in a real options model. Journal of Economic Dynamics and Control 24, 219-225.
10. Shibata, T., 2009. Investment timing, asymmetric information, and audit structure: a real options framework. Journal of Economic Dynamics and Control 33, 903-921.
11. Shibata, T., Nishihara, M., 2011. Interactions between investment timing and management effort under asymmetric information: costs and benefits of privatized firms. European Journal of Operational Research 215, 688-696.
12. Weiss, A.M., 1994. The effects of expectations on technology adoption: some empirical evidence. Journal of Industrial Economics 42, 341-360.
13. Wong, K.P., 2007. The effect of uncertainty on investment timing in a real options model. Journal of Economic Dynamics and Control 31, 2152-2167.
14. Wong, K.P., 2010. The effects of irreversibility on the timing and intensity of lumpy investment. Economic Modelling 27, 97-102.

Quadratic Gaussian Joint Pricing Model for Stocks and Bonds: Theory and Empirical Analysis*

Kentaro Kikuchi

Faculty of Economics, Shiga University, 1-1-1 Banba,
Hikone-shi, Shiga, 522-8522, Japan.
kentaro-kikuchi@biwako.shiga-u.ac.jp

This study proposes a joint pricing model for stocks and bonds in a no-arbitrage framework. A stock price representation is obtained in a manner consistent with the quadratic Gaussian term structure model, in which the short rate is the quadratic form of the state variables. In this study, specifying the dividend as a function using the quadratic form of the state variables leads to a stock price representation that is exponential-quadratic in the state variables. We prove that the coefficients determining the stock price have to satisfy some matrix equations, including an algebraic Riccati equation. Moreover, we specify the sufficient condition in which the matrix equations do have a unique solution. In our empirical analysis using Japanese data, we obtain estimates with a good fit to the actual data. Furthermore, we estimate the risk premiums for stocks and bonds and analyze how the BOJ's unconventional monetary policy has affected these risk premiums.

Key words: risk premium, quadratic Gaussian term structure model, unscented Kalman filter, algebraic Riccati equation, controllability, portfolio rebalance

1. Introduction

Risk premiums are basic inputs for investors' asset allocations. Needless to say, since investors make decisions on the asset allocations of several financial assets such as stocks and bonds, they need to simultaneously estimate their risk premiums. Moreover, risk premiums are essential for not only investors but also

*Send all correspondence to Kentaro Kikuchi, Faculty of Economics, Shiga University, 1-1-1 Banba, Hikone-shi, Shiga, 522-8522, Japan. kentaro-kikuchi@biwako.shiga-u.ac.jp

central bankers. Particularly, central banks have greater difficulty ignoring how risk premiums have evolved since the financial crisis that began in August 2007. After the collapse of Lehman Brothers, some central banks in developed countries conducted unconventional monetary easing measures. For example, the U.S. Federal Reserve (Fed) began its quantitative easing in December 2008 and ended this policy in October 2014. The Bank of Japan (BOJ) started its comprehensive monetary easing (CME) in October 2010. When this policy ended, the BOJ began the qualitative and quantitative easing (QQE) in April 2013. Furthermore, the BOJ decided to expand the scale of its QQE in November 2014. These policies consisted of large-scale purchases of longer-term government bonds by the Fed and the BOJ in their attempts to lower investors' risk premiums through portfolio rebalancing. Therefore, it is important for many investors and central bankers to simultaneously estimate the risk premiums of several financial assets.

To this end, we need a unified framework that jointly handles the prices of these assets. One potential candidate is the no-arbitrage pricing framework. Almost all the earlier studies that price multiple assets within the no-arbitrage framework target stocks and bonds. Although few studies exist on this subject, [3], [12], [5], [11], and [2] developed a joint-pricing model for stocks and bonds using the no-arbitrage framework. Except for [2], the rest are based on the Gaussian affine framework. In their models, both stock and bond prices are represented using an affine function of the Gaussian state variables. They sustain theoretical consistency in the sense that there is no arbitrage opportunity in financial markets. However, these models do not ensure the positivity of nominal interest rates. A main problem is that this situation may lead to inaccurate estimation results in the ongoing low interest rate environment.

Regarding previous studies that focused only on pricing bonds, some models that guarantee the positivity of interest rates have been proposed. For example, the Cox–Ingersoll–Ross (CIR) model proposed in [4] would be the most popular among positive interest rate models. In addition, the potential approach in [14] and the shadow-rate approach in [8] ensure positive interest rates. Furthermore, the quadratic Gaussian term structure model (QGTM) studied by [1] and [10] is one of these types of studies. The QGTM has an advantage over the CIR model, which is a similar and popular short rate model: while the interest rate in the CIR model may take a negative value in time discretization even though it ensures positivity in a continuous time setting, the interest rate in the QGTM always takes a positive value, even in a discrete time setting. Moreover, the QGTM with multivariate factors enables us to represent more flexible correlations among the variables. In contrast, the correlations of the multivariate CIR model are obliged to have some restrictions in obtaining well-defined bond prices.

In this study, we aim to incorporate the QGTM into the dividend discounted cash flow model of stocks in a manner consistent with the no-arbitrage condition[1]. We assume that the stock dividend is paid to stockholders on a continuous basis and that the dividend yield depends on the state variables. Furthermore, setting the ex-dividend stock price as the exponential quadratic form of the state variables leads to the necessary condition for the existence of the stock price. Then, we provide the sufficient condition of the well-defined stock price. The joint pricing model for stocks and bonds proposed herein allows us to provide more accurate estimation of market prices and their risk premiums than the affine Gaussian framework used in previous studies because our model ensures the positivity of the nominal interest rates. In particular, our model could enable us to elaborate an empirical analysis for financial markets under a low interest rate environment.

The remainder of this paper is organized as follows. In section 2, we present the theoretical basis of the study, which consists of the setup and the bond and stock pricing elements of our model. In section 3, we explain the estimation methodology. Section 4 presents the estimation results. The conclusions are presented in section 5.

2. Theory

In this section, we explain the theoretical basis of the study. First, we prepare the setup of our model and define the state variable processes and the short rate. Second, to provide bond pricing, we review the QGTM studied by [1] and [10]. The QGTM serves as the basis for the interest rate models in our study. Next, we aim to provide the stock price representation. In the part of our study, after defining the finite maturity stock, we represent the price as the general form of the conditional expectation of the discounted cash flow. Then, we specify the amount of dividends paid continuously to the stockholders, which depends on the state variables. The Feynman–Kac theorem enables us to derive the necessary condition for the stock price representation to be satisfied. For the finite maturity stock price to become well defined, we discuss the sufficient condition for the unique existence of the stock price. Next, we define the infinite maturity stock and provide the price as the discounted cash flow representation in the conditional expectation form. Finally, we prove that imposing two transversality conditions for the dividend and the infinite maturity stock leads to equality of the infinite maturity stock price and the limit of the finite maturity stock as maturity approaches infinity. Hence, under certain proper conditions, we obtain the well-defined stock price.

2.1 Setup

In this study, we fix a probability space $(\Omega, \mathcal{F}, \mathbb{F}, \mathbf{P})$ that satisfies the usual condition. Let $\mathbf{P}$ denote the physical measure. In addition, we assume the market

---

[1] [2] works on a joint pricing model for stocks and bonds. They build a bond pricing model with the potential approach and they incorporate it into the dividend discounted cash flow pricing of stocks that is consistent with the no-arbitrage condition. Our work is particularly concerned with the estimation of risk premiums; on the other hand, their work focuses on estimates of stock and bond prices.

to be complete such that the risk-neutral measure $\mathbf{Q}$ uniquely exists.

Let us consider the state variable X_t following the Ornstein–Uhlenbeck process under the physical measure $\mathbf{P}$:

$$dX_t = K_X^{\mathbf{P}}(\theta^{\mathbf{P}} - X_t)dt + \Sigma_X dW_{t,1}^{\mathbf{P}}, \tag{1}$$

where $W_{t,1}^{\mathbf{P}}$ is an N-dimensional Brownian motion. We also assume that Σ_X is a diagonal matrix with positive diagonal elements.

In addition to X_t, we define another state variable Y_t:

$$dY_t = \mu^{\mathbf{P}}dt + K_Y^{\mathbf{P}}X_t dt + \Sigma_{Y,1}dW_{t,1}^{\mathbf{P}} + \Sigma_{Y,2}dW_{t,2}^{\mathbf{P}}, \tag{2}$$

where $W_{t,2}^{\mathbf{P}}$ is an M-dimensional Brownian motion with $cov(W_{t,1}^{\mathbf{P}}, W_{t,2}^{\mathbf{P}}) = 0_{N\times M}$. As equation (2) shows, Y_t is a nonstationary process.

In order to model the state variable process under the risk-neutral measure $\mathbf{Q}$, we define the market price of risk. This allows us to bridge the gap between $\mathbf{P}$ and $\mathbf{Q}$. We assume that the market price of risk exists for $W_{t,1}^{\mathbf{P}}$ and $W_{t,2}^{\mathbf{P}}$. Let $\Lambda_{t,1}$ be defined such that $dW_{t,1}^{\mathbf{Q}} = dW_{t,1}^{\mathbf{P}} + \Lambda_{t,1}dt$, where $W_{t,1}^{\mathbf{Q}}$ is an N-dimensional Brownian motion under the risk-neutral measure. We also assume that $dW_{t,2}^{\mathbf{Q}} = dW_{t,2}^{\mathbf{P}} + \Lambda_{t,2}dt$, where $W_{t,2}^{\mathbf{Q}}$ is an M-dimensional Brownian motion under the risk-neutral measure. Specifically, we model $\Lambda_{t,1}$ as $\Lambda_{t,1} = \lambda_1 + \Lambda_1 X_t$ and $\Lambda_{t,2}$ as $\Lambda_{t,2} = \lambda_2 + \Lambda_2 X_t$ according to [7]. This situation is considered the essentially affine setting. Describing X_t under $\mathbf{Q}$ as follows:

$$dX_t = K_X^{\mathbf{Q}}(\theta^{\mathbf{Q}} - X_t)dt + \Sigma_X dW_{t,1}^{\mathbf{Q}}, \tag{3}$$

we find the following relationships from equations (1) and (3)

$$K_X^{\mathbf{P}}\theta^{\mathbf{P}} = K_X^{\mathbf{Q}}\theta^{\mathbf{Q}} + \Sigma_X\lambda_1, \quad K_X^{\mathbf{P}} = K_X^{\mathbf{Q}} - \Sigma_X\Lambda_1. \tag{4}$$

Under $\mathbf{Q}$, the nonstationary state process, Y_t, is given by

$$dY_t = \mu^{\mathbf{Q}}dt + K_Y^{\mathbf{Q}}X_t dt + \Sigma_{Y,1}dW_{t,1}^{\mathbf{Q}} + \Sigma_{Y,2}dW_{t,2}^{\mathbf{Q}}. \tag{5}$$

Thus, the essentially affine setting, equations (2) and (5) lead to the following relationships:

$$\mu^{\mathbf{P}} = \mu^{\mathbf{Q}} + \Sigma_{Y,1}\lambda_1 + \Sigma_{Y,2}\lambda_2, \quad K_Y^{\mathbf{P}} = K_Y^{\mathbf{Q}} + \Sigma_{Y,1}\Lambda_1 + \Sigma_{Y,2}\Lambda_2. \tag{6}$$

Next, the risk-free short rate r_t is defined as the quadratic form of the state variable X_t in this study:

$$r_t = X_t'\Psi X_t, \tag{7}$$

where the superscript of X_t represents the transposition of X_t and Ψ is assumed to be positive definite. This setting ensures the positivity of the short rate r_t. In addition, note that we sometimes denote r_t by $r(t, X_t)$ to emphasize on that r_t depends on X_t.

2.2 Bond Pricing

Under the aforementioned setting, we derive the zero-coupon bond pricing formula. First, the zero-coupon bond price $P^{T-t}(t, X_t)$ at time t with maturity T is described as

$$P^{T-t}(t, X_t) = E_t^{\mathbf{Q}}\left[\exp\left(-\int_t^T r(u, X_u)du\right)\right], \tag{8}$$

where $E_t^{\mathbf{Q}}[\ ]$ is the conditional expectation operator under $\mathbf{Q}$ with respect to filtration $\mathcal{F}_t$, which is generated by $W_t^{\mathbf{Q}}$. We immediately find that the zero-coupon bond price always becomes less than one because the short rate takes a positive value from equation (7). Therefore, the zero-coupon yields also take positive values.

Applying the Feynman–Kac theorem to equation (8), we obtain the following partial differential equation for $P^\tau(t, X_t)$ $(\tau = T - t)$:

$$\begin{aligned} &\frac{\partial P^\tau(t, X_t)}{\partial t} + \kappa^{\mathbf{Q}}(t, X_t)' \frac{\partial P^\tau(t, X_t)}{\partial X_t} \\ &\qquad - r(t, X_t)P^\tau(t, X_t) + \frac{1}{2}\mathrm{Tr}\left(\Sigma_X \Sigma_X' \frac{\partial^2 P^\tau(t, X_t)}{\partial^2 X_t}\right) = 0, \\ &P^0(T, X_T) = 1, \end{aligned} \tag{9}$$

where $\kappa^{\mathbf{Q}}(t, X_t) = K_X^{\mathbf{Q}}(\theta^{\mathbf{Q}} - X_t)$.

An attempting at finding the solution to equation (9) takes the form given by

$$P^\tau(t, X_t) = \exp\left(X_t' A_\tau X_t + b_\tau' X_t + c_\tau\right). \tag{10}$$

Computing derivatives based on equation (10), we obtain:

$$\begin{aligned} \frac{\partial P^\tau(t,X)}{\partial t} &= \left(X' \frac{\mathrm{d}A_\tau}{\mathrm{d}t} X + \frac{\mathrm{d}b_\tau'}{\mathrm{d}t} X + \frac{\mathrm{d}c_\tau'}{\mathrm{d}t}\right) P^\tau(t, X), \\ \frac{\partial P^\tau(t,X)}{\partial X} &= \left((A_\tau' + A_\tau)X + b_\tau\right) P^\tau(t, X), \\ \frac{\partial^2 P^\tau(t,X)}{\partial^2 X} &= \{(A_\tau' + A_\tau)XX'(A_\tau' + A_\tau) + (A_\tau' + A_\tau)Xb_\tau' \\ &\quad + b_\tau X'(A_\tau' + A_\tau) + (A_\tau' + A_\tau + b_\tau b_\tau')\}P^\tau(t, X). \end{aligned} \tag{11}$$

We substitute equation (11) into equation (9) and obtain the differential equations from the conditions that the coefficient with respect to each degree of X_t

must become equal to zero, respectively:

$$\frac{dA_\tau}{d\tau} = K_X^{\mathbf{Q}\prime}(A_\tau + A_\tau') + \Psi - \tfrac{1}{2}(A_\tau' + A_\tau)\Sigma_X\Sigma_X'(A_\tau' + A_\tau),\ \ A_0 = 0_{N\times N}$$

$$(12)\quad \frac{db_\tau'}{d\tau} = -(K_X^{\mathbf{Q}}\theta^{\mathbf{Q}})'(A_\tau' + A_\tau) + b_\tau' K_X^{\mathbf{Q}} - b_\tau'\Sigma_X\Sigma_X'(A_\tau' + A_\tau),\ \ b_0 = 0_{N\times 1}$$

$$\frac{dc_\tau}{d\tau} = -(K_X^{\mathbf{Q}}\theta^{\mathbf{Q}})'b_\tau - \tfrac{1}{2}\mathrm{Tr}(\Sigma_X\Sigma_X'(A_\tau' + A_\tau) + b_\tau b_\tau'),\ \ c_0 = 0$$

In general, equation (12) does not have the closed-form solution. Therefore, to compute the solutions, we rest on a numerical method such as the Runge–Kutta method.

2.3 Stock Price Modeling

2.3.1 Finite Maturity Stock

In this subsection, we define the finite maturity stock and provide a sufficient condition for the unique existence of this stock price.

First, we consider a stock with the finite maturity. This stock continuously pays dividend $D(t, Z_t)$ per time to stockholders, where Z_t denotes $Z_t = (X_t', Y_t')'$. At maturity T, this stock pays the terminal dividend $\overline{D}(T, Z_T)$. For the period after T, this stock is assumed to generate no cash flows. This finite maturity stock is bought and sold for the price $S^T(t)$ at time t.

The cumulative discounted gain from time 0 to time t $(t \leq T)$, $g^T(t)$ is described as

$$(13)\quad g^T(t) = \int_0^t \exp\left(-\int_0^s r(u, X_u)du\right) D(s, Z_s)ds + \exp\left(-\int_0^t r(s, X_s)ds\right) S^T(t).$$

Under the $\mathbf{Q}$ measure, $g^T(t)$ must become a martingale. Thus, the relationship $g^T(t) = E_t^{\mathbf{Q}}[g^T(T)]$ holds. This relationship and equation (13) lead to the following equation for the stock price at t.

(14)

$$S^T(t) = E_t^{\mathbf{Q}}\left[\int_t^T \exp\left(-\int_t^s r(u, X_u)du\right) D(s, Z_s)ds + \exp\left(-\int_t^T r(s, X_s)ds\right)\overline{D}(T, Z_T)\right].$$

Assuming the transversality condition for the terminal dividend given by

$$(15)\quad \lim_{T\to\infty} E_t^{\mathbf{Q}}\left[\exp\left(-\int_t^T r(s, X_s)ds\right)\overline{D}(T, Z_T)\right] = 0,$$

from equation (14), we derive the limit of the finite maturity stock price as the maturity approaches infinity:

$$(16)\quad \lim_{T\to\infty} S^T(t) = \lim_{T\to\infty} E_t^{\mathbf{Q}}\left[\int_t^T \exp\left(-\int_t^s r(u, X_u)du\right) D(s, Z_s)ds\right].$$

Next, let us specify the dividend. Here, we model the dividend $D(t, Z_t)$ as

$$D(t, Z_t) = (\delta_0 + \delta_1' X_t + X_t' \Phi X_t) \exp(kt + d' X_t + X_t' E X_t + c' Y_t), \tag{17}$$

where Φ and E are assumed to be symmetric. Regarding the terminal dividend at maturity T, we model it by

$$\overline{D}(T, Z_T) = \exp(kT + d' X_T + X_T' E X_T + c' Y_T). \tag{18}$$

Equations (14), (17), and (18) and the Feynman–Kac theorem allow us to derive the partial differential equation for $S^T(t)$. Note that we sometimes denote $S^T(t)$ by $S^T(t, Z_t)$ to emphasize that the price depends on the state variable Z_t:

$$\begin{aligned}
&\tfrac{\partial S^T(t,Z)}{\partial t} + \tfrac{\partial S^T(t,Z)}{\partial X} \kappa^{\mathbf{Q}}(X) + \tfrac{\partial S^T(t,Z)}{\partial Y} \tilde{\kappa}^{\mathbf{Q}}(X) + \tfrac{1}{2} \mathrm{Tr}\left(\Sigma_X \Sigma_X' \tfrac{\partial^2 S^T(t,Z)}{\partial^2 X}\right) \\
&+ \tfrac{1}{2} \mathrm{Tr}\left(\Sigma_Y \Sigma_Y' \tfrac{\partial^2 S^T(t,Z)}{\partial^2 Y}\right) + \mathrm{Tr}\left(\Sigma_X \Sigma_Y' \tfrac{\partial^2 S^T(t,Z)}{\partial X \partial Y}\right) \\
&- r(t, X) S^T(t, Z) + D(t, Z) = 0, \\
&S^T(T, Z_T) = \overline{D}(T, Z_T) = \exp(kT + d' X_T + X_T' E X_T + c' Y_T),
\end{aligned} \tag{19}$$

where $\kappa^{\mathbf{Q}}(X) = K_X^{\mathbf{Q}}(\theta^{\mathbf{Q}} - X)$ and $\tilde{\kappa}^{\mathbf{Q}}(X) = \mu^{\mathbf{Q}} + K_Y^{\mathbf{Q}} X$.

A guess of the solution to equation (19) is $S^T(t, Z) = \exp(kt + d' X_t + X_t' E X_t + c' Y_t)$. Then, we substitute this form and equation (17) into equation (19). The coefficient with respect to each degree of X_t has to be a zero matrix, zero vector, or zero; hence,

$$\begin{aligned}
&-2 E K_X^{\mathbf{Q}} + 2 E \Sigma_X \Sigma_X' E + \Phi - \Psi = 0_{N \times N}, \\
&(-K_X^{\mathbf{Q}\prime} + 2 E \Sigma_X \Sigma_X') d + 2 E K_X^{\mathbf{Q}} \theta^{\mathbf{Q}} + 2 E \Sigma_X \Sigma_Y c + \delta_1 = 0_{N \times 1}, \\
&k + d' K_X^{\mathbf{Q}} \theta^{\mathbf{Q}} + \tfrac{1}{2} \mathrm{Tr}(\Sigma_X \Sigma_X' (2E + dd')) + \tfrac{1}{2} \mathrm{Tr}(\Sigma_Y \Sigma_Y' cc') \\
&+ c' \Sigma_Y \Sigma_X + c' \Sigma_Y \Sigma_X + \delta_0 + c' \mu^{\mathbf{Q}} = 0.
\end{aligned} \tag{20}$$

Note that c is a free parameter and E, d, and k are variables that need to be solved.

We take the transposition of the first equation of equation (20), add it to the first equation, and multiply it by one half to obtain the following equation:

$$E K_X^{\mathbf{Q}} + K_X^{\mathbf{Q}\prime} E - 2 E \Sigma_X \Sigma_X' E + \Psi - \Phi = 0_{N \times N}. \tag{21}$$

Equation (21) is called an algebraic Riccati equation. In general, this equation has no solution. However, a sufficient condition for the unique existence of a solution in an algebraic Riccati equation is known. The condition is that a matrix pair $(K_X^{\mathbf{Q}}, \Sigma_X)$ is controllable and $\Psi - \Phi$ is positive definite. The solution $\hat{E}$ to this equation becomes positive definite.

In terms of the existence of the solution to the second equation of equation (20), we must examine whether the coefficient of d, $-K_X^{\mathbf{Q}\prime} + 2\hat{E}\Sigma_X\Sigma_X'$, has an inverse matrix. This coefficient becomes equal to $(\Psi - \Phi + \hat{E}K_X^{\mathbf{Q}})\hat{E}^{-1}$ from the first equation. Hence, we have to only examine whether $\Psi - \Phi + \hat{E}K_X^{\mathbf{Q}}$ has an inverse matrix. The resulting matrix obtained by adding this matrix to the transposition of this matrix is positive definite because the matrix becomes equal to $\Psi - \Phi + 2\hat{E}\Sigma_X\Sigma_X'\hat{E}$, which is positive given the assumption of the positivity of the diagonal elements of the diagonal matrix Σ_X. In this way, since the matrix is symmetric and positive, it has an inverse matrix. Consequently, we find that the coefficient of d, $-K_X^{\mathbf{Q}\prime} + 2\hat{E}\Sigma_X\Sigma_X'$, has an inverse matrix, such that d in the second equation always has a solution. Once we find the solution of E and d, we can easily compute k from the third equation.

Summing up this discussion, we obtain the following proposition.

Proposition 2.1. *State variables X_t and Y_t follow equations (3) and (5), respectively. The volatility term of X_t, i.e., Σ_X is assumed to be a diagonal matrix with positive diagonal elements. We define r_t by equation (7). In addition, the dividend $D(t, Z_t)$ of the finite maturity stock is given by equation (17), and E and Φ are symmetric. Furthermore, we assume that $\Psi - \Phi$ is positive definite and the matrix pair $(K_X^{\mathbf{Q}}, \Sigma_X)$ is controllable. Then the finite maturity stock price $S^T(t, Z_t)$ has the following representation:*

$$S^T(t, Z_t) = \exp(kt + d'X_t + X_t'EX_t + c'Y_t),$$

where k, d and E are obtained from equation (20), whereas c is a free parameter.

Since $S^T(t, Z_t) = \exp(kt + d'X_t + X_t'EX_t + c'Y_t)$ under the condition indicated in Proposition 2.1, $S^T(t, X_t)$ does not depend on the time to maturity T. Hereafter, we denote by $S(t, Z_t)$ the finite maturity stock price.

Note that $\delta_0 + \delta_1'X_t + X_t'\Phi X_t$ is interpreted as the dividend yield because $D(t, Z_t)/S(t, Z_t) = \delta_0 + \delta_1'X_t + X_t'\Phi X_t$.

Next, imposing condition (15) on the finite maturity stock price leads to the following proposition.

Proposition 2.2. *Under the condition presented in Proposition 2.1, assuming the transversality condition (15), we obtain the following relationship:*

$$\lim_{T\to\infty} E_t^{\mathbf{Q}}\left[\int_t^T \exp\left(-\int_t^s r(u, X_u)du\right) D(s, Z_s)ds\right] = \exp(kt + d'X_t + X_t'EX_t + c'Y_t).$$

Proof. By equation (16),

$$\lim_{T\to\infty} E_t^{\mathbf{Q}}\left[\int_t^T \exp\left(-\int_t^s r(u,X_u)du\right)D(s,Z_s)ds\right] = \lim_{T\to\infty} S^T(t).$$

Since $\lim_{T\to\infty} S^T(t,Z_t) = S(t,Z_t)$, the above equation leads to the follwing equation:

$$\lim_{T\to\infty} E_t^{\mathbf{Q}}\left[\int_t^T \exp\left(-\int_t^s r(u,X_u)du\right)D(s,Z_s)ds\right] = \exp(kt + d'X_t + X_t'EX_t + c'Y_t). \square$$

2.3.2 Infinite Maturity Stock

In this subsection, let us consider an infinite maturity stock. As is the case with a finite maturity stock, we postulate that an infinite maturity stock continuously pays the dividend $D(t,Z_t)$ per time to holders. Let $S^\infty(t)$ denote the price of this stock at time t. The discounted gain process from time 0, $g^\infty(t)$ is given by

$$g^\infty(t) = \int_0^t \exp\left(-\int_0^s r_u du\right)D(s,Z_s)ds + \exp\left(-\int_0^t r_s ds\right)S^\infty(t). \tag{22}$$

Note that $g^\infty(t)$ must be a martingale under the **Q** measure. Hence, $g^\infty(t) = E_t^{\mathbf{Q}}[g^\infty(T)]$ holds. This relationship and equation (22) lead to the following equation:

$$S^\infty(t) = E_t^{\mathbf{Q}}\left[\int_t^T \exp\left(-\int_t^s r_u du\right)D(s,Z_s)ds + \exp\left(-\int_t^T r_s ds\right)S^\infty(T)\right]. \tag{23}$$

This relationship holds for $T \geq t$.

Now, imposing the transversality condition for the infinite maturity stock price given by

$$\lim_{T\to\infty} E_t^{\mathbf{Q}}\left[\exp\left(-\int_t^T r(s,X_s)ds\right)S^\infty(T)\right] = 0, \tag{24}$$

equation (23) reduces to

$$S^\infty(t) = \lim_{T\to\infty} E_t^{\mathbf{Q}}\left[\int_t^T \exp\left(-\int_t^s r(u,X_u)du\right)D(s,Z_s)ds\right]. \tag{25}$$

From this equation, we obtain the following proposition:

Proposition 2.3. *If two transversality conditions (15) and (24) hold under the condition presented in Proposition 2.1, then the infinite maturity stock price has the following representation:*

$$S^\infty(t) = \exp(kt + d'X_t + X_t'EX_t + c'Y_t).$$

Accordingly, the infinite maturity stock price representation is obtained when the two transversality conditions hold.

2.3.3 Theorem on Stock Price Representation

Given Proposition 2.3, when the two transversality conditions hold, the infinite maturity stock price has a closed-form representation. Here, we discuss the sufficient condition for the terminal transversality.

According to [12], as for the transversality condition for the dividend, we prove the following proposition:

Proposition 2.4. *If the dividend yield $\delta_0+\delta_1' X_t+X_t'\Phi X_t > 0$, then the transversality condition for the terminal dividend (15) holds.*

Proof. First, we denote by $\delta(t, X_t)$ the dividend yield $\delta_0 + \delta_1' X_t + X_t'\Phi X_t$. Let us define ζ_t as

$$\zeta_t = \exp\left(\int_0^t (\delta(u, X_u) - r(u, X_u))du\right)\overline{D}(t, Z_t).$$

By applying Ito's lemma into this equation, we obtain the stochastic process of ζ_t as follows:

$$\begin{aligned} d\zeta_t &= \exp\left(\int_0^t (\delta(u, X_u) - r(u, X_u))du\right)(\delta(u, X_u) - r(u, X_u))\overline{D}(t, Z_t)dt \\ &\quad + \exp\left(\int_0^t (\delta(u, X_u) - r(u, X_u))du\right)\left(\tfrac{\partial \overline{D}(t,Z_t)}{\partial t} + \mathcal{D}_Z\overline{D}(t, Z_t)\right)dt \\ &\quad + \exp\left(\int_0^t (\delta(u, X_u) - r(u, X_u))du\right)\tfrac{\partial \overline{D}(t,Z_t)}{\partial Z_t}\Sigma_Z dW_t^{\mathbf{Q}}, \end{aligned}$$

where we assume that the stochastic process of $Z_t = (X_t', Y_t')'$ is represented as $dZ_t = K_Z dt + \Sigma_Z dW_t^{\mathbf{Q}}$ and denote by $\mathcal{D}_Z$ the infinitesimal generator of Z_t; in other words, $\mathcal{D}_Z\overline{D}(t, Z_t) = K_Z' \frac{\partial \overline{D}(t,Z_t)}{\partial Z} + \frac{1}{2}\mathrm{Tr}\left(\Sigma_Z\Sigma_Z' \frac{\partial^2 \overline{D}(t,Z_t)}{\partial Z^2}\right)$.

Here, applying Ito's lemma into equation (13), we find that the drift term of the finite maturity stock's cumulative discounted gain $g^T(t)$ reduces to $\left(\frac{\partial}{\partial t} + \mathcal{D}_Z - (r(t, X_t) - \delta(t, X_t))\right)\overline{D}(t, Z_t)$. This means the drift term of $d\zeta_t$ vanishes because the cumulative discounted gain follows a martingale under $\mathbf{Q}$. Hence, we obtain the following equation:

$$d\zeta_t = \exp\left(\int_0^t (\delta(u, X_u) - r(u, X_u))du\right)\frac{\partial \overline{D}(t, Z_t)}{\partial Z_t}\Sigma_Z dW_t^{\mathbf{Q}}.$$

This equation leads to the following equation:

$$\zeta_T = \zeta_t + \int_t^T \exp\left(\int_0^s (\delta(u, X_u) - r(u, X_u))du\right)\frac{\partial \overline{D}(s, Z_s)}{\partial Z_s}\Sigma_Z dW_s^{\mathbf{Q}}.$$

Let us denote by $I_t(T)$ the second term of the right-hand side of the above equation. $I_t(T)$ is a local martingale because it is a stochastic integral with respect to a Brownian motion. In addition, $I_t(T)$ is more than $-\zeta_t$ because ζ_t is positive by definition. Hence, we find that $I_t(T)$ is a supermartingale because it is a local martingale with a lower bound.

We assume that $\delta(t, X_t) \geq \epsilon > 0$. From this, we have the following inequality:

$$\exp\left(\int_0^T (\delta(u, X_u) - r(u, X_u))du\right)\overline{D}(T, Z_T) > e^{\epsilon T} \exp\left(\int_0^T -r(u, X_u)du\right)\overline{D}(T, Z_T).$$

Hence,

$$e^{\epsilon T} \exp\left(\int_0^T -r(u, X_u)du\right)\overline{D}(T, Z_T) < \zeta_t + I_t(T).$$

Taking the expecataion for both sides of the above inequality,

$$e^{\epsilon T} E_t^{\mathbf{Q}}\left[\exp\left(\int_0^T -r(u, X_u)du\right)\overline{D}(T, Z_T)\right] < \zeta_t + E_t^{\mathbf{Q}}[I_t(T)] \leq \zeta_t + I_t(t) = \zeta_t,$$

where the last inequality is given by the fact that $I_t(T)$ is a supermartingale. This leads to the following inequality:

$$E_t^{\mathbf{Q}}\left[\exp\left(\int_0^T -r(u, X_u)du\right)\overline{D}(T, Z_T)\right] < e^{-\epsilon T}\zeta_t.$$

Therefore, the left-hand side of this inequality approaches zero as T approaches infinity. □

Note that the dividend yield $\delta(t, X_t) = \delta_0 + \delta_1' X_t + X_t' \Phi X_t > 0$ is equivalent to $\delta_0 > \frac{1}{4}\delta_1' \Phi^{-1}\delta_1$.

Let us consider the transversality condition for the infinite stock price.

Proposition 2.5. *If the infinite maturity stock price exists, the finite maturity stock price uniquely exists, and the transversality condition for the terminal dividend holds, then the transversality condition for the infinite maturity stock price holds.*

Proof. We assume that there exists a t such that $S^\infty(t) \neq S(t)$. Equation (23) means that $S^\infty(t)$ satisfies equation (14). However, this contradicts the assumption of the unique existence of the finite maturity stock price. Hence, $S^\infty(t) = S(t) = \overline{D}(t, Z_t)$ for any t. This leads to the transversality condition for the infinite maturity stock price as follows:

$$\lim_{T\to\infty} E_t^{\mathbf{Q}}\left[\exp\left(-\int_t^T r_s ds\right)\overline{D}(T, Z_T)\right] = \lim_{T\to\infty} E_t^{\mathbf{Q}}\left[\exp\left(-\int_t^T r_s ds\right)S^\infty(T, Z_T)\right] = 0.$$

Therefore, the transversality condition for the infinite maturity stock price holds. □

Summing up our discussion up to this point, we have the following theorem.

Theorem 2.1. *We assume the following:*

- *Φ and $\Psi - \Phi$ are positive definite,*
- *a matrix pair $(K_X^{\mathbf{Q}}, \Sigma_X)$ is controllable,*
- *$\delta_0 > \frac{1}{4}\delta_1'\Phi^{-1}\delta_1$.*

Then, the non-defaultable stock price is well defined and has the following representation:

$$S(t, Z_t) = exp(kt + d'X_t + X_t'EX_t + c'Y_t),$$

where the coefficients of X_t and Y_t satisfy the stock price matrix equation given by

$$-2EK_X^{\mathbf{Q}} + 2E\Sigma_X\Sigma_X'E + \Phi - \Psi = 0_{N\times N},$$
$$(-K_X^{\mathbf{Q}\prime} + 2E\Sigma_X\Sigma_X')d + 2EK_X^{\mathbf{Q}}\theta^{\mathbf{Q}} + \delta_1 = 0_{N\times 1},$$
$$k + d'K_X^{\mathbf{Q}}\theta^{\mathbf{Q}} + \frac{1}{2}\mathrm{Tr}(\Sigma_X\Sigma_X'(2E + dd')) + \delta_0 = 0.$$

2.4 Processes of Stock and Bond Prices, Correlation, and Risk Premiums

In this subsection, we first note the stochastic processes of stock and bond prices, which provide the instantaneous excess returns and volatilities of these prices. In addition, we derive the correlation representation between stock and bond prices. These representations evolve depending on the state variable X_t. We find that our model provides more flexible structures to the dynamics of stocks and bonds and their dependencies.

Furthermore, we provide the definitions of term premium and equity risk premium. In section 4, we estimate these premiums using Japanese data.

First, we write the stochastic process of the n-year zero-coupon bond price under the **P** measure. Using equation (10) and the fact that the drift term of the price return is equal to r_t under **Q**, the stochastic process under **Q** is given by

$$\frac{dP^n(t)}{P^n(t)} = r_t dt + \{(A_n + A_n')X_t + b_n\}'\,\Sigma_X dW_{t,1}^{\mathbf{Q}}. \tag{26}$$

Remembering the setting of the essentially affine risk premium, or $\Lambda_{t,1} = \lambda_1 + \Lambda_1 X_t$, we obtain the bond price process under **P** as

$$\begin{aligned}\frac{dP^n(t)}{P^n(t)} &= (r_t + \{(A_n + A_n')X_t + b_n\}'\,\Sigma_X(\lambda_1 + \Lambda_1 X_t))dt \\ &\quad + \{(A_n + A_n')X_t + b_n\}'\,\Sigma_X dW_{t,1}^{\mathbf{P}}.\end{aligned} \tag{27}$$

From equation (27), we find that the instantaneous excess return and the volatility of the bond price return depend on the state variable X_t.

Next, let us derive the stock price process under **P**. The cumulative gain associated with holding the stock from time 0, $\tilde{g}(t)$ is given by

$$\tilde{g}(t) = \int_0^t D(s)ds + S(t) = \int_0^t (\delta_0 + \delta_1'X_s + X_s'\Phi X_s)S(s)ds + S(t). \tag{28}$$

Hence, from equation (28) and Ito's lemma, the process of the return associated with holding the stock is given under **Q** by

$$
\begin{aligned}
\frac{d\tilde{g}(t)}{S(t)} &= (\delta_0 + \delta_1' X_t + X_t' \Phi X_t)dt + \frac{dS(t)}{S(t)} \\
&= r_t dt + ((d + 2EX_t)' \Sigma_X + c' \Sigma_{Y,1}) dW_{t,1}^{\mathbf{Q}} + c' \Sigma_{Y,2} dW_{t,2}^{\mathbf{Q}}.
\end{aligned} \tag{29}
$$

The second equality is attributable to the fact that the drift term must be a risk-free rate under **Q**.

Using equation (29) and by the assumption of the essentially affine risk premium, we obtain the return process associated with holding the stock under **P**:

$$
\begin{aligned}
\frac{d\tilde{g}(t)}{S(t)} &= (r_t + ((d + 2EX_t)' \Sigma_X + c' \Sigma_{Y,1})(\lambda_1 + \Lambda_1 X_t) + c' \Sigma_{Y,2}(\lambda_2 + \Lambda_2 X_t))\, dt \\
&\quad + ((d + 2EX_t)' \Sigma_X + c' \Sigma_{Y,1}) dW_{t,1}^{\mathbf{P}} + c' \Sigma_{Y,2} dW_{t,2}^{\mathbf{P}}.
\end{aligned} \tag{30}
$$

Similarly to the bond return, the excess return and the volatility of a stock holding's return depend on the state variable X_t.

Equations (27) and (30) lead to the following correlation between the bond return and the stock holding return:

$$
\frac{(D'\Sigma_X + c'\Sigma_{Y,1})\Sigma_X' \left\{\tilde{A}_n X_t + b_n\right\}}{\sqrt{(D'\Sigma_X + c'\Sigma_{Y,1})(D'\Sigma_X + c'\Sigma_{Y,1})' + c'\Sigma_{Y,2}\Sigma_{Y,2}' c}\ \sqrt{\left\{\tilde{A}_n X_t + b_n\right\}' \Sigma_X \Sigma_X' \left\{\tilde{A}_n X_t + b_n\right\}}}, \tag{31}
$$

where $\tilde{A}_n = A_n + A_n'$ and $D = d + 2EX_t$.

Next, let us define the bond and equity risk premiums. With respect to the bond risk premium, the term premium subsequently defined is estimated and analyzed in many empirical studies. The term premium TP_t^n of the n-year zero-coupon yield at time t is defined as

$$
\begin{aligned}
TP_t^n &= \frac{1}{n} \log E_t^{\mathbf{P}} \left[\exp\left(- \int_t^{t+n} r(u, X_u) du \right) \right] - \frac{1}{n} \log E_t^{\mathbf{Q}} \left[\exp\left(- \int_t^{t+n} r(u, X_u) du \right) \right] \\
&= y_t^n + \frac{1}{n} \log E_t^{\mathbf{P}} \left[\exp\left(- \int_t^{t+n} r(u, X_u) du \right) \right],
\end{aligned} \tag{32}
$$

where y_t^n represents the n-year zero-coupon yield.

The equity risk premium EP_t^n for n years is defined as the excess expected holding return of the stock for n years over the n-year zero-coupon yield. Expressing the mathematical form, we define it as follows:

$$
EP_t^n = \frac{1}{n} E_t^{\mathbf{P}} \left[\int_t^{t+n} D(s) ds + S(t+n) - S(t) \right] / S(t) - y_t^n. \tag{33}
$$

In section 4, we estimate and analyze the term premium and the equity risk premium based on Japanese stock and bond data.

3. Estimation Methodology

In this section, we explain the estimation methodology used to conduct our empirical studies. Our model can be regarded as the state-space model. Since the observation equation becomes a nonlinear function, we cannot use the Kalman filter in order to estimate the latent state variables. Thus, we apply the unscented Kalman filter, a nonlinear filter, to actual financial market data. Furthermore, we estimate the model parameter using the quasi-maximum likelihood method.

3.1 State Space Representation

Moving toward our empirical study, we approximate our continuous time model to a discrete time model. We set the unit of time as one month. We write X_t's and Y_t's processes in discrete time, respectively, as follows:

$$X_{t+1} = \exp(-K_X^{\mathbf{P}})X_t + (I - \exp(-K_X^{\mathbf{P}}))\theta^{\mathbf{P}} + \sqrt{V}\epsilon_{1,t+1}^{\mathbf{P}},$$
$$Y_{t+1} = Y_t + \mu^{\mathbf{P}} + K_Y^{\mathbf{P}}X_t + \Sigma_{Y,2}\epsilon_{2,t+1}^{\mathbf{P}},$$

where $V = \int_{-1}^{0} e^{K_X^{\mathbf{P}}u}\Sigma_X\Sigma_X' e^{K_X^{\mathbf{P}}u'}du$ and $\sqrt{V}$ represents the Cholesky decomposition of V. In addition, $\epsilon_{1,t+1}^{\mathbf{P}}$ and $\epsilon_{2,t+1}^{\mathbf{P}}$ are random variables, each with a standard normal distribution and each is independent from the other. For simplicity, we assume that $\Sigma_{Y,1}$ in equation (2) is a zero matrix. The above equations indicate the transition equation of the state variable Z_t. We group the X_t's and Y_t's processes together and write them as $Z_{t+1} = f(Z_t) + \Sigma\epsilon_{t+1}$.

We specify the setting of this transition equation. This specification imposes some restrictions on the model parameters. As a result, we are able to estimate the parameters without ending up with overfitting or underfitting.

In our empirical analysis, the dimension N of X_t is 3, and we ignore Y_t, i.e., we let M be 0. This setting aims to reduce the computational burden associated with an estimation. Note that X_t with three factors is flexible enough to represent a variety of shapes that the yield curve generally adopts.

Considering X_t's correlation structure, we find that it is determined by $K_X^{\mathbf{P}}$ and Σ_X. However, since the correlation matrix is symmetric, $K_X^{\mathbf{P}}$ and Σ_X are unidentifiable in an estimation. Hence, we assume that $K_X^{\mathbf{P}}$ is a lower triangular matrix and Σ_X is a diagonal matrix.

Furthermore, according to the invariant transformation by [6] and [1] in order to exclude an arbitrariness associated with a nonsingular linear transformation of X_t, we assume that Σ_X is the following matrix:

$$\Sigma_X = \begin{bmatrix} 0.1 & 0 & 0 \\ 0 & 0.1 & 0 \\ 0 & 0 & 0.1 \end{bmatrix}.$$

If we deal with the case in which $M > 1$, by the same argument as Σ_X, we assume that $\Sigma_{Y,2}$ is the following matrix:

$$\Sigma_{Y,2} = \begin{bmatrix} 0.1 & 0 \\ 0 & 0.1 \end{bmatrix},$$

where we let M be 2.

We impose a restriction on Λ_t, the market price of risk, in order that $K_X^{\mathbf{Q}}$ becomes a lower triangular matrix. That is, $K_X^{\mathbf{Q}}$ and $K_X^{\mathbf{P}}$ have the same matrix form. If Λ_1 is a lower triangular matrix, then $K_X^{\mathbf{Q}}$ becomes a lower triangular matrix from equation (4) because Σ_X is diagonal. Hence, we assume that Λ_1 is a lower triangular. As for λ_1, we have no restrictions.

The measurement equation is defined by

$$T_t = \begin{bmatrix} zyield_t^{(n_1)} \\ \vdots \\ zyield_t^{(n_l)} \\ logS(t) \\ dyield_t \end{bmatrix} = \begin{bmatrix} g_1(X_t) \\ \vdots \\ g_l(X_t) \\ g_{l+1}(X_t) \\ g_{l+2}(X_t) \end{bmatrix} + \begin{bmatrix} \eta_{t,1} \\ \vdots \\ \eta_{t,l} \\ \eta_{t,l+1} \\ \eta_{t,l+2} \end{bmatrix},$$

where $zyield_t^{(n_i)}$ is the zero-coupon yield with the n_i-month time to maturity at time t, $logS(t)$ is the log stock price, and $dyield_t$ is the dividend yield of the stock. The function $g_i(X_t)$ $(i = 1, \ldots, l)$ is given by

$$g_i(X_t) = -\frac{1}{n_i}\left(X_t' A_{n_i} X_t + b_{n_i}' X_t + c_{n_i}\right),$$

from equation (10). The coefficients A_{n_i}, b_{n_i}, and c_{n_i} are solutions to equation (12).

The function $g_{l+1}(X_t)$ is given by

$$g_{l+1}(X_t) = kt + d'X_t + X_t'EX_t,$$

where k, d, and E are solutions to equation (20).

The function $g_{l+2}(X_t)$ is given by

$$g_{l+2}(X_t) = \delta_0 + \delta_1' X_t + X_t'\Phi X_t.$$

Furthermore, $(\eta_{t,1}, \ldots, \eta_{t,l}, \eta_{t,l+1}, \eta_{t,l+2})$ is the measurement error following a multivariate normal distribution. This variance-covariance matrix is given by

$$Cov(\eta_{t,1}, \ldots, \eta_{t,l}, \eta_{t,l+1}, \eta_{t,l+2}) = diag(h_1, \cdots, h_1, h_2, h_3) \equiv H.$$

3.2 Unscented Kalman Filter

As we can see in the previous subsection, because the form of $g_i(X_t)$ is nonlinear, the measurement equation is a nonlinear function. Hence, we cannot rely on the Kalman filter to estimate the latent state variable X_t. We instead apply the unscented Kalman filter developed by [9] to actual financial market data.

The first step of the unscented Kalman filter is initialization:

$$\hat{X}_0 = E^{\mathbf{P}}[X_0], \quad P_0 = E^{\mathbf{P}}[(X_0 - \hat{X}_0)(X_0 - \hat{X}_0)'].$$

In this initialization step, we calculate the unconditional expectation and variance-covariance matrix of the state variable X_t.

Next, we explain the filter's prediction step. We calculate the prediction of X_t using "sigma points." For $k = 1, \ldots, S$, where S is the number of observation dates, we calculate $2N + 1$ points called sigma points given by

$$\chi_{k-1} = [\hat{X}_{k-1}, \hat{X}_{k-1} + \gamma\sqrt{P_{k-1}}, \hat{X}_{k-1} - \gamma\sqrt{P_{k-1}}],$$

where χ_{k-1} is an $N \times (2N + 1)$ matrix and $\sqrt{P_{k-1}}$ represents the square root matrix of P_{k-1}. $\hat{X}_{k-1} \pm \gamma\sqrt{P_{k-1}}$ in the above equation is defined as

$$\hat{X}_{k-1} \pm \gamma\sqrt{P_{k-1}} = (\hat{X}_{k-1}, \ldots, \hat{X}_{k-1}) \pm \gamma\sqrt{P_{k-1}},$$

where $\gamma = \sqrt{\alpha^2(N + \kappa)}$ is called the scaling parameter. In this study, we assume that $\alpha = 1$ and $\kappa = 0$.

The next step is the time update of the state variable X_t. We transform the aforementioned sigma points by function f as follows:

$$\chi^*_{k|k-1} = [f(\hat{X}_{k-1}), f(\hat{X}_{k-1} + \gamma\sqrt{P_{k-1}}), f(\hat{X}_{k-1} - \gamma\sqrt{P_{k-1}})],$$

where $\chi^*_{k|k-1}$ has an $N \times (2N + 1)$ matrix.

Using each column of $\chi^*_{k|k-1}, \chi^*_{i,k|k-1}$, we calculate the "mean" and "covariance" of the state variable given by

$$\hat{X}_k^- = \sum_{i=0}^{2N} W_i^m \chi^*_{i,k|k-1},$$
$$P_k^- = \sum_{i=0}^{2N} W_i^c (\chi^*_{i,k|k-1} - \hat{X}_k^-)(\chi^*_{i,k|k-1} - \hat{X}_k^-)' + \Sigma_X \Sigma_X',$$

where weights are assigned as follows:

$$W_0^m = \frac{\alpha^2(N + \kappa) - N}{\alpha^2(N + \kappa)}, \quad W_0^c = W_0^m + 1 - \alpha^2 + \beta,$$
$$W_i^m = W_i^c = \frac{1}{2\alpha^2(N + \kappa)}, \quad i = 1, \ldots, 2N,$$

where β is a non-negative weighting parameter used to incorporate knowledge of the higher order distribution moments. In this study, we set $\beta = 0$. We define the sigma points of the predicted state variables as follows:

$$\chi_{k|k-1} = [\hat{X}_k^-, \hat{X}_k^- + \gamma\sqrt{P_k^-}, \hat{X}_k^- - \gamma\sqrt{P_k^-}].$$

Next, we transform these sigma points $\chi_{k|k-1}$ with a g function that provides the yields and the log stock price:

$$T_{k|k-1} = [g(\hat{X}_k^-), g(\hat{X}_k^-) + \gamma\sqrt{P_k^-}, g(\hat{X}_k^-) - \gamma\sqrt{P_k^-}],$$

$$T_k^- = \sum_{i=0}^{2N} W_i^m T_{i,k|k-1}.$$

Finally, we obtain the following measurement update equations,

$$P_{T_k,T_k} = \sum_{i=0}^{2N} W_i^c (T_{i,k|k-1} - \hat{T}_k^-)(T_{i,k|k-1} - \hat{T}_k^-)',$$

$$P_{X_k,T_k} = \sum_{i=0}^{2N} W_i^c (\chi_{i,k|k-1} - \hat{X}_k^-)(T_{i,k|k-1} - \hat{T}_k^-)' + H.$$

Using these variance-covariance matrices, we compute the Kalman gain K_k given by

$$K_k = P_{X_k,T_k} P_{T_k,T_k}^{-1}.$$

Given the Kalman gain, the time update of the conditional expectation and variance of X_t is

$$X_k = \hat{X}_k^- + K_k(T_k - \hat{T}_k^-), \quad P_k = P_k^- - K_k P_{T_k,T_k} K_k'.$$

With the model parameters given, we estimate the latent state variables through the above steps of the unscented Kalman filter.

3.3 Quasi-Maximum Likelihood Method

We estimate the model parameters using the quasi-maximum likelihood method. The log likelihood function in this model is given by

$$\log L(\Theta) = -\frac{NS}{2}\log 2\pi - \frac{1}{2}\sum_{k=1}^{S}\left(\log|P_{T_k,T_k}| + (T_k - T_k^-)' P_{T_k,T_k}^{-1}(T_k - T_k^-)\right),$$

where Θ is a set of model parameters and S is the number of our observation dates.

The optimal model parameter Θ is estimated as the solution to the maximization of $\log L(\Theta)$.

3.4 Data

We use monthly data from January 1996 to September 2013 for the JGB zero-coupon yields, the Topix, and the dividend yield of the Topix. Zero-coupon yields are computed from the Broker's JGB prices using the method presented in [13]. The maturities included are six months and two, five, 10, and 20 years. Data for the JGB prices, the Topix, and its monthly dividend yield are downloaded from the Nikkei NEEDS Financial Quest.

4. Estimation Result

4.1 Model Fit

In this subsection, we illustrate the fit of our model to actual financial market data. The estimation of the state variable X_t and the model parameters is conducted based on the methodology described in section 3. The Appendix presents estimates of the model parameters. Estimates of the state variable X_t are shown in Figure 1.

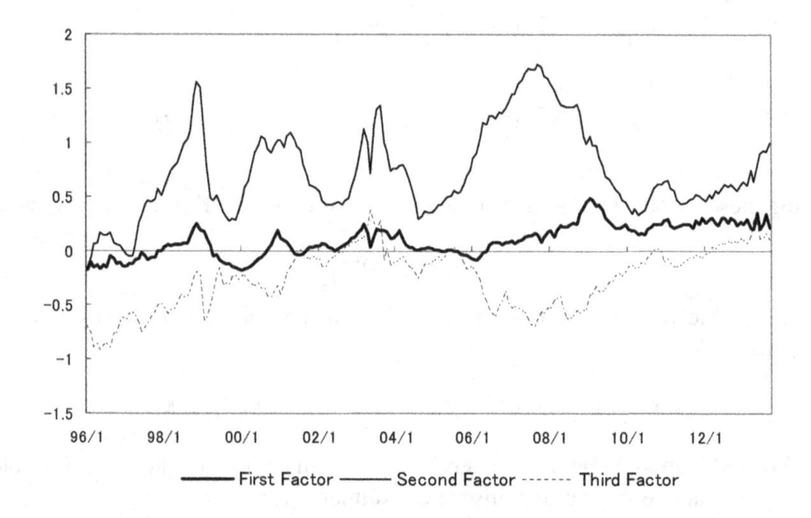

Figure 1. Estimates of the state variable X_t

Based on the estimated state variable and parameters, we can examine the performance of our model. Figure 2 displays the comparison between the observation data and the estimates of the zero-coupon yields, the stock index, and the dividend yield as calculated from the filtered variable $X_{t|t}$. For the zero-coupon yields and stock price, the estimation results show a good fit to the market data. Table 1 reports the summary statistics of the estimation errors. The table indicates that the

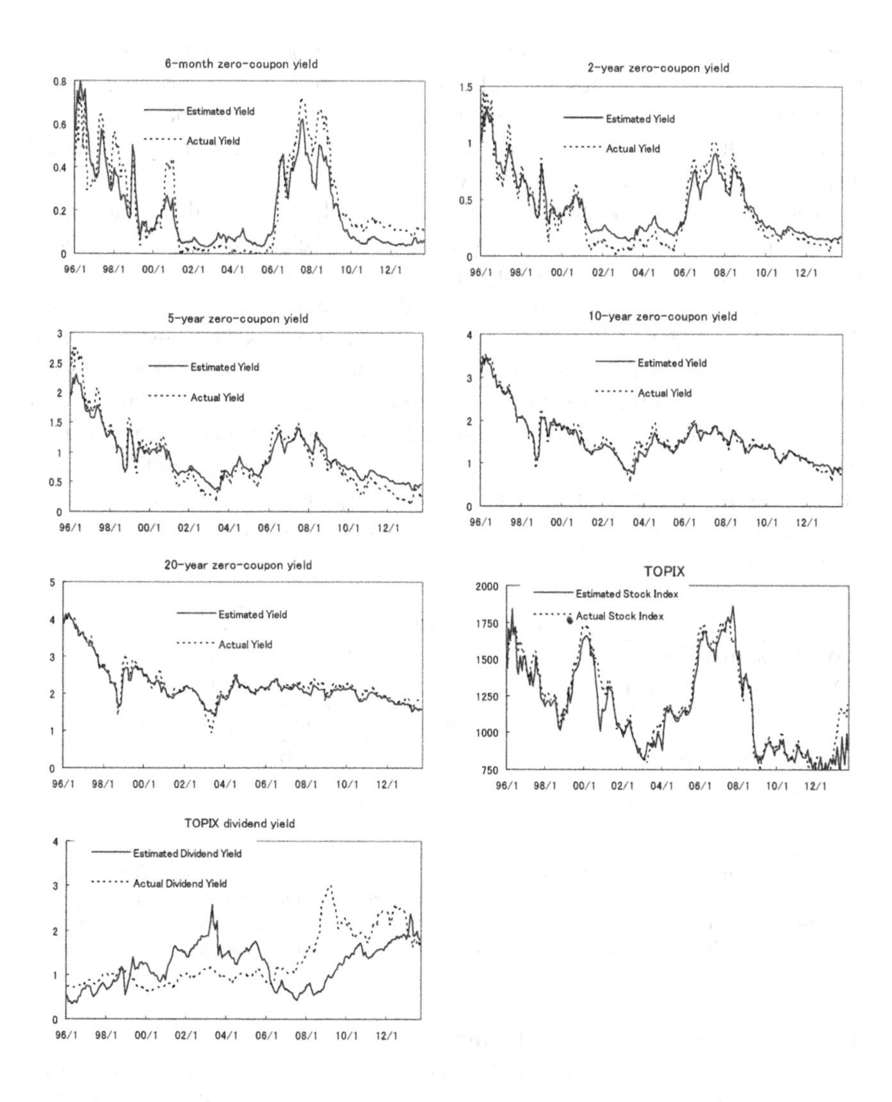

Figure 2. Model Fit: The graphs show a comparison between estimated values and actual data. Yields, including zero coupon and dividend yields, are shown in percentage terms. Two values for Topix are shown in points.

mean absolute errors for almost all zero-coupon yields are below ten basis points. For the Topix, the mean absolute relative error is about 5 %. Measuring the explained percentage variation for the Topix, defined as one minus the ratio of error variance to the observation variance, we obtain 91.8 %. Accordingly, our model displays a good fit to the zero-coupon yields and stock index.

The model's performance for the dividend yield is worse than that for the zero-coupon yields and stock index. This can most likely be attributed to the misspecification of our model. In reality, investors value a company's stock not only by the dividend amount but also the internal reserve amount. With this in mind, the estimated dividend yield of our model can be regarded as the adjustment of the realized dividend yield by internal reserves. Hence, the estimated dividend yield could take a value different from the observed dividend yield. Conducting an estimation with the exclusion of the dividend yield from the measurement data is possible; however, the approach can lead to an unrealistic estimate of the dividend yield. Thus, to obtain a realistic estimate, we incorporate the actual dividend yield into the observation data, although this might introduce measurement errors for the dividend yield to some extent.

	0.5-year	2-year	5-year	10-year
Mean Estimated Error	−1.713	4.186	5.044	−2.400
Mean Absolute Error	7.677	8.320	14.48	9.593

	20-year	Stock	dividend yield
Mean Estimated Error	−1.852	−1.949	−12.60
Mean Absolute Error	8.887	4.946	56.33

Table 1 Summary statistics of the estimation errors on the zero-coupon yields, the Topix, and its monthly dividend yield: The estimation errors on the zero-coupon yields and dividend yield are defned as the difference between the observed data and estimates. These are indicated in basis points. As for the Topix, the estimation error is defined as the relative error, indicated in percentage terms.

4.2 Correlation between Stocks and Bonds

When investors make decisions regarding their asset allocations, they need correlations among financial assets as the input data. However, the use of correlations computed directly from historical returns is at a risk of deterioration in the portfolio diversification in case where correlations of the future largely vary from those of the past. Hence, the correlations should contain forward-looking information on asset prices. As shown in equation (31), our model enables us to compute the implied correlation between stocks and bonds once we estimate the filtered state variable.

Figure 3 displays the implied correlations between the Topix and the Japanese government bond prices. One is the correlation between the Topix and bonds with a six-month time to maturity, and the other is the Topix's correlation with bonds with a 10-year time to maturity.

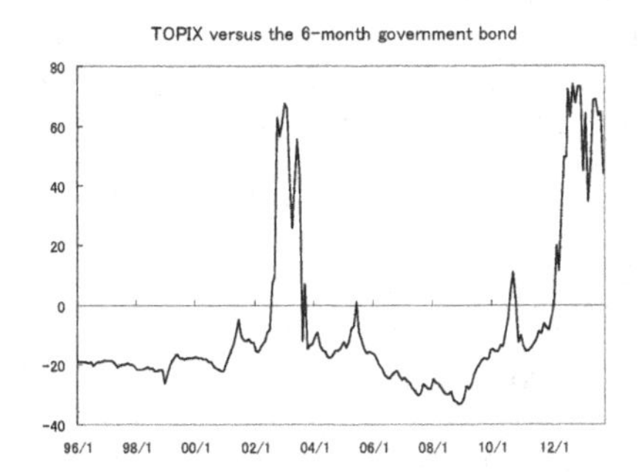

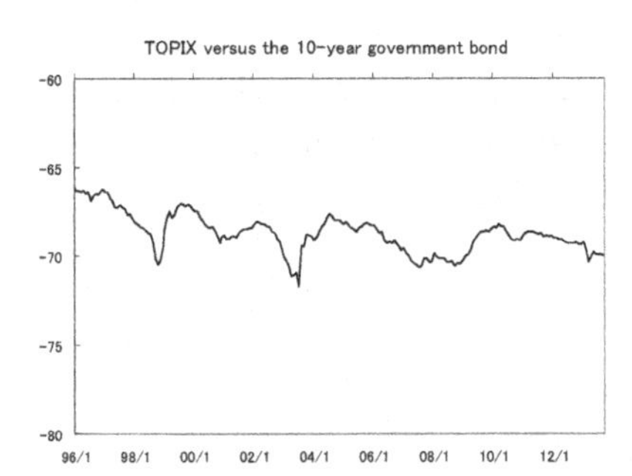

Figure 3. Correlations between the Topix and the Japanese government bonds: the graphs are shown in percentage terms.

As for the correlation between the Topix and the longer-term to maturity bonds, no large fluctuation continued to have negative values over the sample period, as the right-hand graph indicates. On the other hand, the correlation between the Topix and the shorter-maturity term bonds takes mild negative values for most of the sample period; however, it takes positive values during a certain interval of the sample period. Since the spring of 2012 when the BOJ decided to further enhance the CME introduced in October 2010, the correlation has taken significant positive values. This decision by the BOJ is likely to cause widespread expectations that prolonged quantitative easing may increase interest in the liquidity-driven market among investors. Accordingly, from the above discussion, it can be said that the sign of the implied correlation changed from negative to positive because of the BOJ's action.

4.3 Risk Premium

In this subsection, we analyze the estimates of bond and stock risk premiums. In particular, we are interested in how the unconventional monetary policies that the BOJ introduced after the financial crisis beginning in the fall of 2008 have affected risk premiums.

Figure 4 illustrates estimates of the term premiums for the shorter- and the longer-term maturity bonds, defined as equation (32). Focusing on developments after the beginning of 2009, we can observe the contrast between two developments. For the shorter term maturity bonds, premiums have remained approximately constant. On the other hand, the term premium for longer-term maturity bonds has largely continued to decline since the beginning of 2011. This might reflect the fact that the BOJ began the CME in October 2010. Before introducing this policy, the BOJ had primarily bought the Japanese government bonds with relatively shorter durations. However, the asset purchase program established via the CME attempted to buy longer-term maturity bonds on a large scale and lower both the longer-term interest rates and their term premiums. Notably, the BOJ's purchase of longer-term bonds could reduce the investors' concerns about a deteriorated balance of supply and demand and thus lead to a decline in the longer-term maturity bonds' risk premium.

Figure 4. Term Premiums (shown in percentage terms)

Figure 5 displays the equity risk premiums for two, five, and 10 years, defined as equation (33). From this figure, we can observe that the equity risk premiums rose sharply when Lehman Brothers collapsed. Looking back at past stress events in Japan, for example, in November 1997 when Hokkaido Takushoku Bank, one of the city banks, and Yamaichi Securities Company went bankrupt and in early 2003 when the Resona Holdings capital adequacy ratio fell drastically low, equity risk premiums showed large increases. In this manner, the equity risk premiums estimated based on our model appear to capture the developments following the stress events.

Turning focus to developments in the equity risk premiums after the Lehman crash, we can observe that the risk premiums have remained high. Until at least the end of our sample period, September 2013, derailing the course of the BOJ, the decline in the term premium of the longer-term maturity bonds has not impacted equity risk premiums.

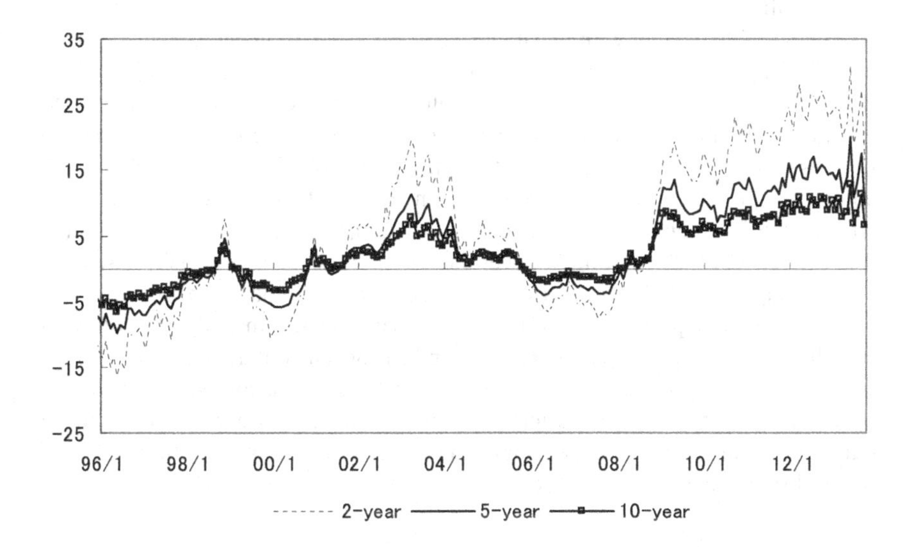

Figure 5. Equity Risk Premiums (shown in percentage terms)

5. Conclusion

This paper proposed a joint pricing model for stocks and bonds in the no-arbitrage framework. Specifically, our pricing model is based on the quadratic Gaussian term structure model studied in [1] and [10]. This setting ensures a positive nominal interest rate. On the other hand, our stock price is defined as the dividend discount cash flow model incorporating the quadratic Gaussian term structure model in the no-arbitrage condition. Specifying the dividend as a function using the quadratic form of the state variables leads to a stock price representation that is exponential-quadratic in the state variables. We proved that the coefficients determining the stock price have to satisfy some matrix equations, including an algebraic Riccati equation. Moreover, in general, these matrix equations do not have any solutions; however, we specified the sufficient condition in which the matrix equations do have a unique solution.

In an empirical analysis using Japanese data, we estimated the latent state variables and the model parameters based on the quasi-maximum likelihood method with an unscented Kalman filter. As a result, we obtained a good fit to the actual financial market data. This could be because our model, which ensures a positive nominal interest rate, works well with Japanese data, which contain a lengthy low interest rate environment. Using estimated filtered state variables, we computed the implied correlation between stocks and bonds. While the correlation between the Topix and longer-term maturity government bond price evolves in a relatively stable manner with negative values, the correlation between the shorter-term bonds and the stock index takes positive values since the beginning of the BOJ's CME, although it takes negative values for most of the sample period. As for the risk premiums, the term premium of bonds with the longer-term maturities has continued to decrease since the introduction of CME. On the other hand, equity risk premiums have not decreased since the collapse of Lehman Brothers. At least until September 2013, the end of our sample period, we did not observe spillover effects of prompting to lower the risk premiums of more risky assets.

Although our study focused on Japanese data, our model can be applied to other countries. In particular, our model should be effective in analyzing countries with low interest rates, such as the U.S. and European countries. In the U.S, the quantitative easing policy ended in October 2014 and investors have now developed a strong interest in the period of the next policy rate rise. An analysis using our model would clarify the development of the risk premiums for the stocks and bonds during the exit from an accommodative monetary policy.

Acknowledgement
The author is grateful to participants at TMU Finance Workshop 2014 for their useful comments. He also expresses his thanks to the editors and the anonymous referee for helpful suggestions.

References

1. Ahn, D.H., Dittmar, R.F, and Gallant, A.R. (2002) "Quadratic Term Structure Models: Theory and Evidence," *The Review of Financial Studies*, **15**(1), pp.243-288.
2. Bäuerle, N. and Pfeiffer, R. (2013), "A Joint Stock and Bond Market based on the Hyperbolic Gaussian Model," *European Actuarial Journal*, **3**(1), pp.229-248.
3. Bekaert, G. and Grenadier, S.R. (2002) "Stock and Bond Pricing in Affine Economy," NBER Working Paper.
4. Cox, J.C, Ingersoll, J.E. and Ross, S.A. (1985) "A Theory of the Term Structure of Interest Rates, " *Econometrica*, **53**(2), pp.385-408.
5. d'Addona, S. and Kind, A.H. (2006) "International Stock-Bond Correlations in a Simple Affine Asset Pricing Model," *Journal of Banking and Finance*, **30**(10), pp.2747-2765.
6. Dai, Q. and Singleton, K.J. (2000) "Specification Analysis of Affine Term Strucuture Models," *Journal of Finance*, **55**(5), pp.1943-1978.
7. Duffee, G.R. (2002) "Term Premia and Interest Forecast in Affine Models," *Journal of Finance*, **57**(1), pp.405-443.
8. Gorovoi, V. and Linetsky, V. (2004) "Black's Model of Interest Rates as Options, Eigenfunction Expansions and Japanese Interest Rates," *Mathematical Finance*, **14**(1), pp.49-78.
9. Julier, S.J. and Uhlmann, J.K. (1997), "A New Extension of the Kalman Filter to Nonlinear Systems," *Proceedings of Aero Sense: 11th Int. Symp. Aerospace/Defense Sensing, Simulation and Controls*, pp.182-193.

10. Leippold, M. and Wu, L. (2002) "Asset Pricing under the Quadratic Class," *Journal of Financial and Quantitative Analysis*, **37**(2), pp.271-295.
11. Lemke, W. and Werner, T. (2009) "The Term Structure of Equity Premia in an Affine Arbitrage-Free Model of Bond and Stock Market Dynamics," ECB Working Paper.
12. Mamaysky, H. (2002) "On the Joint Pricing of Stocks and Bonds: Theory and Evidence," Yale ICF Working Paper.
13. McCulloch, H.J. (1971) "Measuring the Term Structure of Interest Rates," *Journal of Business*, **44**(1), pp.19-31.
14. Rogers, L.C.G. (1997), "The Potential Approach to the Term Structure of Interest Rates and Foreign Exchange Rates," *Mathmatical Finance*, **7**(2), pp.157-176.

Appendix:. Estimated Model Parameters

$\Phi_{1,1}$	$\Phi_{2,1}$	$\Phi_{2,2}$	$\Phi_{3,1}$	$\Phi_{3,2}$	$\Phi_{3,3}$
1.130×10^{-5}	4.471×10^{-6}	8.449×10^{-6}	-1.571×10^{-5}	-1.818×10^{-5}	5.534×10^{-4}
$K^{\mathbf{P}}_{X,1,1}$	$K^{\mathbf{P}}_{X,2,1}$	$K^{\mathbf{P}}_{X,2,2}$	$K^{\mathbf{P}}_{X,3,1}$	$K^{\mathbf{P}}_{X,3,2}$	$K^{\mathbf{P}}_{X,3,3}$
0.05021	0.01031	0.00541	−0.01226	−0.01643	0.06484
$\Psi_{1,1}$	$\Psi_{2,1}$	$\Psi_{2,2}$	$\Psi_{3,1}$	$\Psi_{3,2}$	$\Psi_{3,3}$
1.464×10^{-5}	5.319×10^{-6}	1.250×10^{-5}	-1.489×10^{-5}	-1.796×10^{-5}	5.895×10^{-4}
$\Lambda_{1,1,1}$	$\Lambda_{1,2,1}$	$\Lambda_{1,2,2}$	$\Lambda_{1,3,1}$	$\Lambda_{1,3,2}$	$\Lambda_{1,3,3}$
−0.4928	0.1552	−0.02693	−0.02056	0.1191	−0.6327
$\theta^{\mathbf{P}}_1$	$\theta^{\mathbf{P}}_2$	$\theta^{\mathbf{P}}_3$	$\lambda_{1,1}$	$\lambda_{1,2}$	$\lambda_{1,3}$
0.2055	0.2522	−0.1511	−0.1199	−0.0915	0.1486
$\delta_{1,1}$	$\delta_{1,2}$	$\delta_{1,3}$			
-4.262×10^{-5}	-1.074×10^{-4}	0.001731			

Option Pricing with Ambiguous Correlation and Fast Mean-reverting Volatilities*

Man Hau Leung and Hoi Ying Wong

Department of Statistics, The Chinese University of Hong Kong, Shatin, N.T., Hong Kong.

In pricing an option on multiple assets, volatilities of individual assets and the correlations among different assets are necessary inputs. In practice, implied volatility surface for individual underlying asset can be used to calibrate the marginal distribution, but information on correlations is generally difficult to obtain. By regarding the correlation between two assets as an ambiguous parameter, we obtain the worst case price to option seller. The corresponding pricing problem is formulated as a stochastic optimal control problem in which the uncertain correlation takes the role as a control function. Consequently, the Black-Scholes equation is replaced by an HJB equation for deriving of the option price bounds. We solve this problem using the framework of stochastic volatility asymptotics and explain how volatility surfaces of individual assets can be robustly pulled together to estimate the price bounds of an option on multiple assets. Empirical study with foreign exchange option data illustrates the practical use of the proposed approach.

Key words: Ambiguous correlation; Asymptotic; HJB equation; Stochastic volatility

1. Introduction

The determination of option price has been studied extensively over the past decades. Black and Scholes (1973) laid down a cornerstone in option pricing theory, resulted in the celebrated Black-Scholes formula. Their result is based on the assumption of a constant volatility level of stock return. This assumption is inconsistent with the observed volatility smile implied by market data.

*This research is supported by the Direct Research Grant of The Chinese University of Hong Kong. Send all correspondence to Hoi Ying Wong, Department of Statistics, The Chinese University of Hong Kong, Shatin, N.T., Hong Kong. Email:hywong@cuhk.edu.hk.

Stochastic volatility (SV) models have been proposed to capture the volatility structure implied by the market, such as Hull and White (1987) and many others. SV models essentially postulate a parametric stochastic process for the volatility movement. A very comprehensive account for the asymptotic analysis of mean-reverting SV models can be found in the book by Fouque et al. (2011). A major finding is the use of a linear regression of volatility against log-moneyness to maturity ratio (LMMR). The scheme extracts effective parameters for pricing various derivative products, for example see Ilhan et al. (2004).

While SV models are successful in capturing the volatility smile implied by options on single asset, its extension to options on multiple assets is not obvious. Multi-asset options require the correlations among underlying assets. However, the market generally lacks of liquidly traded multi-asset options for calibrating the correlation. A straight forward method uses the historical correlations but it often produces inconsistent and unstable correlation estimates. Therefore, the estimated confidence level of the correlation is regarded as a more robust statistics.

Motivated by this, Fouque, Pun and Wong (2014) introduce the concept of ambiguous correlation into portfolio selection so that only the upper and lower bounds of the correlation is known. We combine their ambiguous correlation concept and SV model to option pricing with multiple underlying assets. More specifically, individual underlying assets follow a stochastic volatility model but a correlation ρ between two asset returns is ambiguous in the way that $\rho \in [\underline{\rho}, \overline{\rho}]$.

The ambiguous correlation approach belongs to the class of parameter ambiguity models. A popular ambiguity model in finance is the ambiguous volatility approach such as those in Avellaneda et al. (1995) and Peng (2007a). Similar to the ambiguous correlation approach, the ambiguous volatility model assumes a known bounds for the volatility such that $\sigma \in [\underline{\sigma}, \overline{\sigma}]$), but does not specify the stochastic process of the volatility. Chapter 4 in Pham (2009) contains a comprehensive formulation of the ambiguous volatility model using the stochastic optimal control framework. Investigation on the behaviour of option pricing under ambiguous volatility can be found in Frey and Sin (1999) , Nicolato and Venardos (2003) , and El Karoui et al. (1998). Empirical results however suggest that the option bounds derived from ambiguous volatility are often too wide for practical usage.

The role of ambiguity (uncertainty) in influencing market behaviour also attracts attention in financial literature. Chen and Epstein (2002) demonstrate that on top of risk aversion, ambiguity aversion constitutes another source of asset return premium. Epstein and Ji (2013a, 2013b) further formulate a utility model in continuous time to capture ambiguity about both the drift and volatility, illustrating its use in asset pricing.

Instead of regarding all volatilities and correlations to be ambiguous parameters, our ambiguous correlation approach incorporates market information contained in volatility surfaces through calibrating parameters to SV models while leaving the correlation as an ambiguous parameter. With more information taken into consideration, the option price bounds obtained are hopefully narrower than the trivial bounds, that are deduced from the purely ambiguous model.

Under ambiguous correlation, we solve the option pricing problem under both the Black-Scholes model and the fast-mean reverting SV model. It is shown that the valuation is related to the HJB framework. We show that the optimal choice of the correlation ρ admits a simple representation that takes either the upper bound or lower bound depending on the cross Gamma. An asymptotic approximations of the price bounds are then obtained for the SV model with an ambiguous correlation. In the two-asset case, the zeroth order term is found to be the ambiguous correlation version of the Black-Scholes price. To illustrate the potential use of the model, we derive the analytical price bounds for an exchange option.

The rest of the paper is organized as follows. Section 2 presents option pricing under uncertain correlation in the Black-Scholes world. Section 3 deals with option pricing under ambiguous correlation with a fast mean reverting stochastic volatility model. Section 4 contains the numerical and empirical study on foreign exchange options. Section 5 concludes the paper.

2. The Black-Scholes Model with Ambiguous Correlation

To simply mathematical notation, consider an option on two Black-Scoles assets S_1 and S_2, where the underlying asset prices satisfy the stochastic differential equations (SDE)

$$
\begin{aligned}
dS_i &= (\mu_i - q_i)S_i dt + \sigma_i S_i dW_i^{\mathbb{P}}, \quad i = 1,2, \\
\mathbb{E}[dW_1^{\mathbb{P}} dW_2^{\mathbb{P}}] &= \rho dt,
\end{aligned} \tag{1}
$$

where $W_i^{\mathbb{P}}$ are standard Brownian motions under the physical probability measure $\mathbb{P}$ and the constants are interpreted as

μ_i = expected return of asset i, q_i = dividend yield of asset i;
σ_i = volatility of asset i, ρ = correlation between $W_1^{\mathbb{P}}$ and $W_2^{\mathbb{P}}$.

We are interested in the case of ambiguous correlation. The correlation matrix Ω then varies arbitrarily in the set

$$
\Theta = \left\{ \Omega = \begin{pmatrix} 1 & \rho \\ \rho & 1 \end{pmatrix}, \ \rho \in [\underline{\rho}, \overline{\rho}], \quad -1 \leq \underline{\rho} \leq \overline{\rho} \leq 1, \right\}.
$$

In other words, the correlation coefficient can be any number within $[\underline{\rho}, \overline{\rho}]$. Note that we also allow ρ to be stochastic but its support is $[\underline{\rho}, \overline{\rho}]$.

Under such a setting, the market becomes *incomplete* and there exists a family of risk neutral measures $\mathbb{Q}_\rho$ characterized by ρ. Option price can only be obtained when a particular $\mathbb{Q}_\rho$, or ρ, is chosen. Therefore, the upper and lower bounds of the option price are the interesting quantities for investors as it is no way to determine the value of the option accurately.

Consider a European contract $\Phi(S_1(T), S_2(T))$ that depends on asset prices through their terminal values only. We require Φ to be non-path dependent, bounded and uniformly continuous in (S_1, S_2). The price upper bound (P_{sup}) and lower bound (P_{inf}) can be written as the extreme values of the expectation of discounted payoff

$$
\begin{aligned}
P_{sup} &= \sup_{\rho\in[\underline{\rho},\overline{\rho}]} \mathbb{E}^{\mathbb{Q}_\rho}\left[e^{-r\tau}\Phi(S_1(T), S_2(T)) \mid \mathcal{F}_t\right], \\
P_{inf} &= \inf_{\rho\in[\underline{\rho},\overline{\rho}]} \mathbb{E}^{\mathbb{Q}_\rho}\left[e^{-r\tau}\Phi(S_1(T), S_2(T)) \mid \mathcal{F}_t\right] \\
&= -\sup_{\rho\in[\underline{\rho},\overline{\rho}]} \mathbb{E}^{\mathbb{Q}_\rho}\left[-e^{-r\tau}\Phi(S_1(T), S_2(T)) \mid \mathcal{F}_t\right].
\end{aligned} \tag{2}
$$

The upper bound is sometimes known as the super-replication cost in the literature. The Black-Scholes market is a complete market if the correlation is a known parameter. The option can be fully replicated by the underlying assets S_1 and S_2 in the complete Black-Scholes market. However, when the correlation is uncertain and market practitioners only know the region that the correlation belongs to, the cost of the replication portfolio is constructed under the worst case scenario of the possible correlation. It turns out that this super-replication cost is greater than the replication cost with a known correlation.

The formulation of ambiguous correlation problem in the Black-Scholes market is very similar to the classical ambiguous volatility model for option pricing on a single asset. Therefore, the following Theorem 2.1 can be straightforwardly established using viscosity solution of an HJB equation.

Theorem 2.1. *The price upper bound is viscosity solution to the following nonlinear PDE, which is a type of HJB equation.*

$$
\begin{cases} \mathcal{L}_{BS,\sup} P_{\sup} = 0 \\ P_{\sup}(T) \quad = \Phi(S_1(T), S_2(T)) \end{cases}, \tag{3}
$$

where

$$
\begin{aligned}
\mathcal{L}_{BS,\sup} &= \frac{\partial}{\partial t} + \sum_{i=1}^{2}(r-q_i)S_i\frac{\partial}{\partial S_i} + \frac{1}{2}\sum_{i=1}^{2}\sigma_i^2 S_i^2\frac{\partial^2}{\partial S_i^2} \\
&\quad + \sup_{\rho\in[\underline{\rho},\overline{\rho}]}\left[\rho\sigma_1\sigma_2 S_1 S_2\frac{\partial^2}{\partial S_1\partial S_2}\right] - r\cdot.
\end{aligned} \tag{4}
$$

A similar result holds on the price lower bound with the "sup" appearing in (3) *and* (4) *replaced by "inf".*

Proof. The proof mimics that of Theorem 4.6.9 in Chapter 4 of Pham (2009). It is because the volatilities σ_1 and σ_2 are known finite quantities while $\rho \in [\underline{\rho}, \overline{\rho}] \subset [-1, 1]$ is a bounded region.

Alternatively, the computation of the non-linear expectation for the price bound can make use of the G-Brownian motion (BM) introduced by Peng (2007, 2008). Appendix A provides the definition of G-BM. Once we can treat the asset price processes to be driven by the G-BM defined above, result follows by the non-linear Feynman-Kac formula associated with G-BM due to Peng (2010). □

General audiences in finance and economics may not be familiar to the notion of G-expectation[1] although Epstein and Ji (2013a) do mention the connection between G-expectation and ambiguity. Therefore, we link the proof of Theorem 2.1 to the classical optimal control framework. However, the proof for the model combining ambiguous correlation and SV is much more involved while the G-expectation approach conveniently brings us the theoretical result. Appendix A then provides some brief information for the construction of a multi-dimensional G-Brownian motion.

Theorem 2.1 asserts that the determination of option price bounds under ambiguous correlation can be achieved through solving the HJB equation subjected to terminal condition of different payoff functions. The properties of HJB equation are well studied in many literatures such as the book by Young and Zhou (1999). We establish a property concerning (3).

Lemma 2.1. *The HJB equation* (3) *admits at most one viscosity solution.*

Proof. By Theorem 6.1 in Chapter 4 of the book by Young and Zhou, we need to verify the following conditions to establish the lemma.

1. $(r - q_i)S_i, \sigma_i S_i$ and Φ as functions of $(t, (S_1, S_2), \rho)$ are uniformly continuous.

2. Denote φ as any one of the functions $(r - q_i)S_i, \sigma_i S_i$ and Φ. We have a constant $L > 0$ such that for all $t < T, (S_1, S_2), (\widehat{S_1, S_2}), \rho \in [\underline{\rho}, \overline{\rho}]$,

$$\begin{cases} |\varphi(t, (S_1, S_2), \rho) - \varphi(t, (\widehat{S_1, S_2}), \rho)| \leq L|(S_1, S_2) - (\widehat{S_1, S_2})|, \\ |\varphi(t, (0, 0), \rho)| \leq L \end{cases}$$

It is easy to verify these conditions since r, q_i and σ_i are constants and by the assumptions on Φ. □

2.1 Optimal choice of ρ

The control process in (3) is the correlation coefficient ρ. Denote the optimal choice of ρ as $\rho^*_{\sup}$ ($\rho^*_{\inf}$) for the price upper bound (lower bound). The optimal

[1] We thank an anonymous referee for this comment and for directing us to the book by Pham (2009)

ρ is solved simultaneously with the corresponding price bound as a solution pair $(\rho^*_{\text{sup}}, P_{\text{sup}})$ and $(\rho^*_{\text{inf}}, P_{\text{inf}})$.

Theorem 2.2. *The optimal choice of ρ can be expressed as*

$$\rho^*_{\text{sup}} = \overline{\rho}\mathbb{I}\left\{\frac{\partial^2 P_{\text{sup}}}{\partial S_1 \partial S_2} \geq 0\right\} + \underline{\rho}\mathbb{I}\left\{\frac{\partial^2 P_{\text{sup}}}{\partial S_1 \partial S_2} < 0\right\}, \tag{5}$$

so that the equation (3) *becomes*

$$\begin{cases} \mathcal{L}_{BS,\text{sup}} P_{\text{sup}} = 0 \\ P_{\text{sup}}(T) \quad = \Phi(S_1(T), S_2(T)) \end{cases}, \tag{6}$$

where

$$\mathcal{L}_{BS,\text{sup}} = \frac{\partial}{\partial t} + \sum_{i=1}^{2}(r - q_i)S_i \frac{\partial}{\partial S_i} + \frac{1}{2}\sum_{i=1}^{2}\sigma_i^2 S_i^2 \frac{\partial^2}{\partial S_i^2} + \rho^*_{sup}\sigma_1\sigma_2 S_1 S_2 \frac{\partial^2}{\partial S_1 \partial S_2} - r \cdot .$$

The expression for ρ^*_{inf} is obtained by reverting the inequality signs in (5) *and the associated operator is obtained similarly.*

Proof. If $\frac{\partial^2 P_{\text{sup}}}{\partial S_1 \partial S_2} = 0$, the option price is independent of ρ, implying the choice of ρ^*_{sup} is arbitrary and is taken to be $\overline{\rho}$ here. Otherwise, by noticing σ_i and S_i in (4) are positive, the representation of ρ^*_{sup} is justified. Similar arguments for ρ^*_{inf}. □

Theorem 2.2 asserts that the optimal choice of the correlation depends on the sign of the cross Gamma. This result can be viewed as a generalization of Avellaneda et al. (1995) in which the optimal choice of the volatility parameter depends on the sign of the gamma.

Solving the above HJB equation analytically is not always an easy task in general. Numerical schemes are available to solve HJB equations related to financial problems, for example see Wang and Forsyth (2008).

Despite the general difficulty in solving (3), there are cases where the solution is available.

Theorem 2.3. *Let $P_{BS}(S_1, S_2; \rho)$ be the Black-Scholes pricing formula of an option with payoff $\Phi(S_1, S_2)$, where ρ is a known constant. Then, the following results hold.*

- *If $\frac{\partial^2 P_{BS}(S_1, S_2; \rho)}{\partial S_1 \partial S_2} \geq 0$ for any fixed ρ and for all r, q_i, S_i, σ_i, t, then $\rho^*_{\text{sup}} \equiv \overline{\rho}$ and the price upper bound is given by $P_{\text{sup}}(S_1, S_2) = P_{BS}(S_1, S_2; \overline{\rho})$.*

- *If $\frac{\partial^2 P_{BS}(S_1, S_2; \rho)}{\partial S_1 \partial S_2} < 0$ for any fixed ρ and for all r, q_i, S_i, σ_i, t, then $\rho^*_{\text{sup}} \equiv \underline{\rho}$ and the upper price bound is $P_{\text{sup}}(S_1, S_2) = P_{BS}(S_1, S_2; \underline{\rho})$.*

The result for $P_{\inf}$ and $\rho^*_{\inf}$ is obtained by reverting the inequality signs and the form of $P_{\inf}$ follows naturally.

Proof. By Lemma 2.1, the HJB equation (3) admits no more than one viscosity solution. If $\frac{\partial^2 P_{BS}(\rho)}{\partial S_1 \partial S_2} \geq 0$ for any fixed ρ and for all r, q_i, S_i, σ_i, t, we can easily verify that the pair $(\overline{\rho}, P_{BS}(\overline{\rho}))$ satisfies (3), which implies it is the only solution to (3). Other cases follow a similar argument. □

In general, the Black-Scholes price of an option is not the solution $P_{\sup}$ to equation (3). Theorem 2.3 asserts that under certain conditions depending only on the derivative of Black-Scholes price, we can obtain the solution pair $(\rho^*_{\sup}, P_{\sup})$ in terms of the original Black-Scholes pricing formula. Verifying these conditions is an easier task than to look for a solution to (3) itself. The following shows some examples that Theorem 2.3 is typically useful.

1. Exchange option with payoff $\Phi(S_1, S_2) = (S_1 - S_2)^+$: The cross Gamma takes the form,
$$\frac{\partial^2 P^E_{BS}}{\partial S_1 \partial S_2} = -\frac{e^{-q_1 \tau} n(a)}{S_2(t) \sigma \sqrt{\tau}} < 0,$$
where τ is the time to maturity, a is some constant, σ is the combined positive volatility and $n(\cdot)$ is the probability density function (pdf) of the standard normal distribution. The negativity of the cross Gamma implies that the price bounds of the exchange option under uncertain correlation is $P^E \in [P^E_{BS}(\overline{\rho}), P^E_{BS}(\underline{\rho})]$.

2. Quanto option with payoff $\Phi(S_1, S_2) = (S_1 S_2 - K)^+$: The cross Gamma takes the form,
$$\frac{\partial^2 P^F_{BS}}{\partial S_1 \partial S_2} = e^{-q_1 \tau} \left(N(a) + \frac{n(a)}{\sigma \sqrt{\tau}} \right) > 0,$$
where τ is the time to maturity, a is some constant, σ is the combined positive volatility, $n(\cdot)$ is the normal pdf and $N(\cdot)$ is the normal cumulative distribution function (cdf). Therefore, the price bounds of option on foreign asset under uncertain correlation between the asset price and foreign exchange rate is given by $P^F \in [P^F_{BS}(\underline{\rho}), P^F_{BS}(\overline{\rho})]$.

3. Minimum call option with payoff $\Phi(S_1, S_2) = (\min(S_1, S - 2) - K)^+$: The cross Gamma takes the form
$$\frac{\partial^2 P_m}{\partial S_1 \partial S_2} = \frac{e^{-q_1 \tau} n(a) N\left(\frac{b - \rho a}{\sqrt{1 - \rho^2}} \right)}{S_2 \sigma \sqrt{\tau}} > 0,$$

where τ is the time to maturity, a and b are some constants, σ is the combined positive volatility, $n(\cdot)$ is the normal pdf and $N(\cdot)$ is the normal cdf. Theorem 2.3 implies that $P^m \in [P^m_{BS}(\underline{\rho}), P^m_{BS}(\overline{\rho})]$. Again by parity relationships, the corresponding solutions for call on maximum and put on maximum/minimum options can be easily derived.

For more complex option positions, say a straddle strategy of max/min options, the cross Gamma fluctuates across positive and negative values. The conditions in Theorem 2.3 are not satisfied and hence one has to resort to numerical methods.

3. Fast Mean-reverting Stochastic Volatility Model

This section extends the option pricing framework under the Black-Scholes model with an ambiguous correlation to the fast mean-reverting stochastic volatility setting. We will show that the result associated with the Black-Scholes assets is the zeroth order approximation of the model considered in this section.

Suppose the assets S_1, S_2 follow a mean-reverting stochastic volatility model with two factors. Under the risk neutral measure $\mathbb{Q}$, the SDE describing the dynamics of the asset price S_i and the volatility factors Y and Z are as follows.

$$
\begin{aligned}
dS_i &= (r - q_i)S_i dt + f_i(Y,Z)S_i dW_i, \quad i = 1,2, \\
dY &= \left[\frac{1}{\epsilon}(m - Y) - \frac{\sqrt{2}v}{\sqrt{\epsilon}}\Lambda(Y,Z)\right]dt + \frac{\sqrt{2}v}{\sqrt{\epsilon}}dW_y, \\
dZ &= \left[\delta c(Z) - \sqrt{\delta}g(Z)\Gamma(Y,Z)\right]dt + \sqrt{\delta}g(Z)dW_z,
\end{aligned} \tag{7}
$$

where W_i, W_y and W_z are standard Brownian motions. The correlation coefficients are, for $i = 1, 2$,

$$
\begin{aligned}
\mathbb{E}[dW_1 dW_2] &= \rho dt, \quad \mathbb{E}[dW_i dW_y] = \rho_{iy} dt, \\
\mathbb{E}[dW_i dW_z] &= \rho_{iz} dt, \quad \mathbb{E}[dW_y dW_z] = \rho_{yz} dt.
\end{aligned}
$$

The correlation matrix Ω can be written as

$$
\Omega = \begin{pmatrix} 1 & \rho & \rho_{1y} & \rho_{1z} \\ \rho & 1 & \rho_{2y} & \rho_{2z} \\ \rho_{1y} & \rho_{2y} & 1 & \rho_{yz} \\ \rho_{1z} & \rho_{2z} & \rho_{yz} & 1 \end{pmatrix}.
$$

The constants are

r	= risk-free interest rate,	q_i	= dividend yield of asset i,
ϵ, δ	= small positive constants,	m, v^2	= long term mean, variance of Y.

The functions $f_i(Y,Z), \Lambda(Y,Z), c(Z), g(Z)$ and $\Gamma(Y,Z)$ are assumed to be deterministic, bounded and Lipschitz continuous in all arguments. $f_i(Y,Z)$ is also non-negative and bounded away from zero representing the instantaneous volatility of asset i.

The small positive constants ϵ and δ control the speed of mean-reversion of volatility factors Y and Z respectively. We can observe that Y is fast mean-reverting and Z is slow mean-reverting. Combination of both creates a moderate mean-reverting effect in volatility which fits in market observations.

Under the assumption of uncertain correlation, each off diagonal entry of Ω can take value arbitrarily in a predefined region while keeping Ω semi-positive definite. However, with the aid of information from individual volatility surfaces, the degree of uncertainty can be reduced.

Fouque et al. (2003) derived an asymptotic approximation to the option price for option products on single asset. The influence of correlation coefficients ρ_{iy} and ρ_{iz} on the option price approximation is reflected in the effective parameters, which can be obtained through calibration from market data. ρ_{yz} does not play a role in the approximation. From these observations, the uncertainty in correlation structure left is on ρ only. We can abuse notation to say Ω is restricted in the set Θ,

$$\Theta = \left\{ \Omega = \begin{pmatrix} 1 & \rho & \rho_{1y} & \rho_{1z} \\ \rho & 1 & \rho_{2y} & \rho_{2z} \\ \rho_{1y} & \rho_{2y} & 1 & \rho_{yz} \\ \rho_{1z} & \rho_{2z} & \rho_{yz} & 1 \end{pmatrix}, \quad \rho \in [\underline{\rho}, \overline{\rho}], \quad -1 \leq \underline{\rho} \leq \overline{\rho} \leq 1, \right\}.$$

The price bounds therefore admit the same form as in (2). We can again define the appropriate G-normal distribution, G-BM as well as the G-expectation for the following theorem. The construction of G-BM associated with our problem can refer to Appendix A.

Theorem 3.1. *The price upper bound is a viscosity solution to the following HJB equation.*

$$\begin{cases} \mathcal{L}_{SV,\mathrm{sup}} P_{\mathrm{sup}} = 0 \\ P_{sup}(T) = \Phi(S_1(T), S_2(T)) \end{cases}, \tag{8}$$

where

$$\mathcal{L}_{SV,\mathrm{sup}} = \frac{1}{\epsilon}\mathcal{L}_0 + \frac{1}{\sqrt{\epsilon}}\mathcal{L}_1 + \mathcal{L}_{2,sup} + \sqrt{\delta}\mathcal{M}_1 + \delta\mathcal{M}_2 + \sqrt{\frac{\delta}{\epsilon}}\mathcal{M}_3,$$

with the operators defined as

$$
\begin{aligned}
\mathcal{L}_0 &= (m-y)\frac{\partial}{\partial y} + v^2\frac{\partial^2}{\partial y^2},\\
\mathcal{L}_1 &= -\sqrt{2}v\Lambda\frac{\partial}{\partial y} + \sum_{i=1}^{2}\sqrt{2}\rho_{iy}vf_iS_i\frac{\partial^2}{\partial S_i\partial y},\\
\mathcal{L}_{2,sup} &= \frac{\partial}{\partial t} + \sum_{i=1}^{2}(r-q_i)S_i\frac{\partial}{\partial S_i} + \frac{1}{2}\sum_{i=1}^{2}f_i^2S_i^2\frac{\partial^2}{\partial S_i^2}\\
&\quad + \sup_{\rho\in[\underline{\rho},\overline{\rho}]}\left[\rho f_1f_2S_1S_2\frac{\partial^2}{\partial S_1\partial S_2}\right] - r\cdot,\\
\mathcal{M}_1 &= -g\Gamma\frac{\partial}{\partial z} + \sum_{i=1}^{2}\rho_{iz}gf_iS_i\frac{\partial^2}{\partial S_i\partial z},\\
\mathcal{M}_2 &= c\frac{\partial}{\partial z} + \frac{1}{2}g^2\frac{\partial^2}{\partial z^2},\\
\mathcal{M}_3 &= \sqrt{2}\rho_{yz}vg\frac{\partial^2}{\partial y\partial z}.
\end{aligned}
\tag{9}
$$

A similar result holds on the price lower bound with the "sup" appearing in (8) *and* (9) *replaced by "inf".*

Proof. We can treat the asset price processes to be driven by the G-BM defined in (22). Result follows by the non-linear Feynman-Kac formula associated with G-BM. □

Again, we see that the optimal correlation is determined by the sign of the cross Gamma. However, the cross Gamma is much more difficult to compute under the SV model. While numerical methods can be applied to solve (8), we use an asymptotic expansion.

3.1 Asymptotic expansions

We focus on the price upper bound here although similar result holds for the price lower bound.

A combination of regular and singular perturbations is used to analyze the HJB equation (8) to obtain an asymptotic approximation to the solution.

$$P_{\text{sup}} = \sum_{i,j=0}^{\infty}\epsilon^{\frac{i}{2}}\delta^{\frac{j}{2}}P_{i,j}. \tag{10}$$

The objective is to determine the approximation up to order $\sqrt{\epsilon}$ and $\sqrt{\delta}$.

$$P_{\text{sup}} \approx P_{sup}^{\epsilon,\delta} = P_{0,0} + \sqrt{\epsilon}P_{1,0} + \sqrt{\delta}P_{0,1}. \tag{11}$$

Bear in mind that the solution to the HJB equation (8) is also a pair $(\rho^*_{\text{sup}}, P_{\text{sup}})$. In this section, it is supposed that the optimal ρ^* is available and plugged into (9) so that $\mathcal{L}_{2,\text{sup}}$ reads

$$\mathcal{L}_{2,\text{sup}} = \frac{\partial}{\partial t} + \sum_{i=1}^{2}(r - q_i)S_i\frac{\partial}{\partial S_i} + \frac{1}{2}\sum_{i=1}^{2} f_i^2 S_i^2 \frac{\partial^2}{\partial S_i^2} + \rho^*_{sup} f_1 f_2 S_1 S_2 \frac{\partial^2}{\partial S_1 \partial S_2} - r\cdot.$$

The actual choice of ρ^*_{sup} will be handled shortly.

Theorem 3.2. *The components in the desired expansion* (11) *are obtained by solving the following equations*

$$(12) \qquad \begin{cases} \langle \mathcal{L}_{2,\text{sup}} \rangle P_{0,0} & = 0 \\ P_{0,0}(T) & = \Phi(S_1(T), S_2(T)) \end{cases},$$

$$(13) \qquad \begin{cases} \langle \mathcal{L}_{2,sup} \rangle P_{1,0} & = \mathcal{A}_{\text{sup}} P_{0,0} \\ P_{1,0}(T) & = 0 \end{cases},$$

$$(14) \qquad \begin{cases} \langle \mathcal{L}_{2,\text{sup}} \rangle P_{0,1} & = -\langle \mathcal{M}_1 \rangle P_{0,0} \\ P_{0,1}(T) & = 0 \end{cases},$$

where $\langle\cdot\rangle$ *denotes the expectation with respect to the invariant distribution of the operator* $\mathcal{L}_0$, *i.e.* $\mathcal{N}(m, v^2)$, *under the risk neutral measure* $\mathbb{Q}_\rho$ *determined by* ρ^*_{sup} *and* $\mathcal{A}_{\text{sup}} = \langle \mathcal{L}_1 \mathcal{L}_0^{-1}(\mathcal{L}_{2,\text{sup}} - \langle \mathcal{L}_{2,\text{sup}} \rangle)\rangle$. *Detailed notations can be found in Appendix B.*

Proof. Put the expansion (11) into (8) and equate successive order of $\sqrt{\epsilon}$ and $\sqrt{\delta}$ of the result with the true price.

1. Considering $O(\frac{1}{\epsilon})$,

$$\mathcal{L}_0 P_{0,0} = 0.$$

   Since $\mathcal{L}_0$ only takes partial derivatives with respect to y, $P_{0,0}$ should be independent of y.
   The next order is $O(\frac{1}{\sqrt{\epsilon}})$,

$$\begin{aligned} \mathcal{L}_1 P_{0,0} + \mathcal{L}_0 P_{1,0} &= 0 \\ \Rightarrow \qquad \mathcal{L}_0 P_{1,0} &= 0. \end{aligned}$$

   Similar argument shows that $P_{1,0}$ is independent of y.
   Continue with $O(1)$,

$$(15) \qquad \begin{aligned} \mathcal{L}_{2,sup} P_{0,0} + \mathcal{L}_1 P_{1,0} + \mathcal{L}_0 P_{2,0} &= 0 \\ \Rightarrow \qquad \mathcal{L}_{2,sup} P_{0,0} + \mathcal{L}_0 P_{2,0} &= 0. \end{aligned}$$

This is a Poisson equation on $P_{2,0}$ given $P_{0,0}$. By the Fredholm solvability condition, to ensure the existence of a solution in $P_{2,0}$ with at most polynomial growth to infinity and noticing $P_{0,0}$ is independent of y_i, equation (12) must hold.

Remark: This is the same as (3) after the replacement of σ_i^2 by $\langle f_i^2 \rangle$, which we rename as $\overline{\sigma_i}^2$.

2. Considering $O(\sqrt{\epsilon})$,

$$\mathcal{L}_{2,sup}P_{1,0} + \mathcal{L}_1 P_{2,0} + \mathcal{L}_0 P_{3,0} = 0.$$

The Fredholm solvability condition on $P_{3,0}$ gives

$$\langle \mathcal{L}_{2,sup}P_{1,0} + \mathcal{L}_1 P_{2,0} \rangle = 0. \tag{16}$$

From (15) we can write explicitly $P_{2,0} = -\mathcal{L}_0^{-1}(\mathcal{L}_{2,sup} - \langle \mathcal{L}_{2,sup} \rangle)P_{0,0}$. Together with the fact that $P_{1,0}$ is independent of y, (16) can be rewritten as equation (13).

3. Taking $O(\sqrt{\frac{\delta}{\epsilon}})$ and $O(\frac{\sqrt{\delta}}{\epsilon})$ respectively,

$$\begin{aligned} \mathcal{L}_0 P_{0,1} &= 0, \\ \mathcal{L}_1 P_{0,1} + \mathcal{L}_0 P_{1,1} + \mathcal{M}_3 P_{0,0} &= 0 \\ \Rightarrow \qquad \mathcal{L}_0 P_{1,1} &= 0. \end{aligned}$$

One can conclude that both $P_{0,1}$ and $P_{1,1}$ are independent of y.
Considering $O(\sqrt{\delta})$,

$$\begin{aligned} & \mathcal{L}_{2,sup}P_{0,1} + \mathcal{L}_1 P_{1,1} + \mathcal{L}_0 P_{2,1} + \mathcal{M}_1 P_{0,0} + \mathcal{M}_3 P_{1,0} = 0 \\ \Rightarrow \qquad & \mathcal{L}_{2,sup}P_{0,1} + \mathcal{L}_0 P_{2,1} + \mathcal{M}_1 P_{0,0} = 0. \end{aligned} \tag{17}$$

The Fredholm solvability condition on $P_{2,1}$ gives

$$\langle \mathcal{L}_{2,sup}P_{0,1} + \mathcal{M}_1 P_{0,0} \rangle = 0.$$

Equation (14) follows by the fact that both $P_{0,0}$ and $P_{0,1}$ are independent of y. □

Lemma 3.1. *The HJB equation* (12) *admits at most one viscosity solution.*

Proof. In the proof of Lemma 2.1, the conditions that the coefficients of the HJB equation have to satisfy in order to ensure the number of viscosity solution does not exceed one are presented.

Regarding (12), the functions under investigation are $(r-q_i)S_i$, f_iS_i and Φ. By the constant nature of r and q_i and the assumptions on f_i and Φ, the conditions are easily verified. □

With constant correlation, the accuracy of such an approximation was studied by Fouque et al. (2011). The accuracy of the approximation (11) can be determined following a similar argument.

Theorem 3.3.

Assume the payoff function is smooth in both S_1 and S_2. For fixed (t, S_1, S_2, y, z) and any $\epsilon \leq 1, \delta \leq 1$, there exists a constant $C > 0$ such that

$$|P - P^{\epsilon,\delta}| \leq C(\epsilon + \delta + \sqrt{\epsilon\delta}).$$

For payoff function which is continuous but only piecewise smooth, the accuracy of the approximation is

$$|P - P^{\epsilon,\delta}| \leq C(\epsilon|\ln(\epsilon)| + \delta + \sqrt{\epsilon\delta}).$$

Proof. We prove the theorem for when the payoff function is smooth. The case of a piecewise smooth payoff can be proven by a regularization argument similar to that introduced in Papanicolaou et al. (2003).

From (17), we deduce that

$$P_{2,1} = -\mathcal{L}_0^{-1}((\mathcal{L}_{2,sup} - \langle\mathcal{L}_{2,sup}\rangle)P_{0,1} + (\mathcal{M}_1 - \langle\mathcal{M}_1\rangle)P_{0,0}). \tag{18}$$

Consider order $O(\sqrt{\epsilon\delta})$,

$$\mathcal{L}_0P_{3,1} + \mathcal{L}_1P_{2,1} + \mathcal{L}_{2,sup}P_{1,1} + \mathcal{M}_1P_{1,0} + \mathcal{M}_3P_{2,0} = 0.$$

The solvability condition on $P_{3,1}$ gives

$$\langle\mathcal{L}_{2,sup}\rangle P_{1,1} = \mathcal{A}_{sup}P_{0,1} + \mathcal{B}P_{0,0} - \langle\mathcal{M}_1\rangle P_{1,0} - \langle\mathcal{M}_3P_{2,0}\rangle,$$

where

$$\mathcal{B} = \langle\mathcal{L}_1\mathcal{L}_0^{-1}(\mathcal{M}_1 - \langle\mathcal{M}_1\rangle)\rangle.$$

If we consider the residual

$$R = P^{\epsilon,\delta}_{sup} + \epsilon P_{2,0} + \epsilon^{\frac{3}{2}}P_{1,0} + \sqrt{\epsilon\delta}P_{1,1} + \sqrt{\delta}\epsilon P_{2,1} - P,$$

which satisfies

$$\begin{aligned}
\mathcal{L}_{SV,sup}R &= \frac{1}{\epsilon}(\mathcal{L}_0 P_{0,0}) + \frac{1}{\sqrt{\epsilon}}(\mathcal{L}_0 P_{1,0} + \mathcal{L}_1 P_{0,0}) \\
&+ (\mathcal{L}_0 P_{2,0} + \mathcal{L}_1 P_{1,0} + \mathcal{L}_{2,sup} P_{0,0}) \\
&+ \sqrt{\epsilon}(\mathcal{L}_0 P_{3,0} + \mathcal{L}_1 P_{2,0} + \mathcal{L}_{2,sup} P_{1,0}) \\
&+ \sqrt{\delta}\left(\frac{1}{\epsilon}(\mathcal{L}_0 P_{0,1}) + \frac{1}{\sqrt{\epsilon}}(\mathcal{L}_0 P_{1,1} + \mathcal{L}_1 P_{0,1} + \mathcal{M}_3 P_{0,0})\right) \\
&+ \sqrt{\delta}(\mathcal{L}_0 P_{2,1} + \mathcal{L}_1 P_{1,1} + \mathcal{L}_{2,sup} P_{0,1} + \mathcal{M}_1 P_{0,0} + \mathcal{M}_3 P_{1,0}) \\
&+ \epsilon R_1 + \sqrt{\epsilon\delta} R_2 + \delta R_3, \qquad (19)
\end{aligned}$$

where

$$\begin{aligned}
R_1 &= \mathcal{L}_{2,sup} P_{2,0} + \mathcal{L}_1 P_{3,0} + \sqrt{\epsilon}\mathcal{L}_{2,sup} P_{3,0}, \\
R_2 &= \mathcal{L}_2 P_{1,1} + \mathcal{L}_1 P_{2,1} + \mathcal{M}_1 P_{1,0} + \mathcal{M}_3 P_{2,0} \\
&\quad + \sqrt{\epsilon}(\mathcal{L}_{2,sup} P_{2,1} + \mathcal{M}_1 P_{2,0} + \mathcal{M}_3 P_{3,0}) + \epsilon \mathcal{M}_1 P_{3,0}, \\
R_3 &= \mathcal{M}_1 P_{0,1} + \mathcal{M}_2 P_{0,0} + \mathcal{M}_3 P_{1,1} \\
&\quad + \sqrt{\epsilon}(\mathcal{M}_1 P_{1,1} + \mathcal{M}_2 P_{1,0} + \mathcal{M}_3 P_{2,1}) + \epsilon(\mathcal{M}_1 P_{2,1} + \mathcal{M}_2 P_{2,0}).
\end{aligned}$$

We can see that all the terms in (19) vanish except R_1, R_2 and R_3, i.e.

$$\mathcal{L}_{SV,sup}R = \epsilon R_1 + \sqrt{\epsilon\delta} R_2 + \delta R_3.$$

At the same time, R_1, R_2 and R_3 are smooth functions of t, S_1, S_2, y, z. For $\epsilon \leq 1, \delta \leq 1$, they are bounded by smooth functions of t, S_1, S_2, y, z independent of ϵ and δ, uniformly bounded in t, S_1, S_2, z and at most linearly growing in y through the solution of Poisson equations in B.

We have at maturity time T,

$$\begin{aligned}
R(T, S_1, S_2, y, z) &= \epsilon(P_{2,0} + \sqrt{\epsilon}P_{1,0})(T, S_1, S_2, y, z) \\
&\quad + \sqrt{\epsilon\delta}(P_{1,1} + \sqrt{\epsilon}P_{2,1})(T, S_1, S_2, y, z) \\
&\triangleq \epsilon G_1(S_1, S_2, y, z) + \sqrt{\epsilon\delta} G_2(S_1, S_2, y, z), \qquad (20)
\end{aligned}$$

where G_1 and G_2 are independent of t share similar properties as the R_i defined above.

It follows that by invoking the Feynman-Kac formula associated with G-BM due to Peng (2010), we have an expectation representation of the residual as

$$R = \epsilon \mathbb{E}^*\{e^{-r\tau} G_1(S_{1,T}, S_{2,T}, Y_T, Z_T) - \int_t^T e^{-r\tau} R_1(u, S_{1,u}, S_{2,u}, Y_u, Z_u) du | \mathcal{F}_t\}$$

$$+ \sqrt{\epsilon\delta} \mathbb{E}^*\{e^{-r\tau} G_2(S_{1,T}, S_{2,T}, Y_T, Z_T) - \int_t^T e^{-r\tau} R_2(u, S_{1,u}, S_{2,u}, Y_u, Z_u) du | \mathcal{F}_t\}$$

$$+ \delta \mathbb{E}^*\{- \int_t^T e^{-r\tau} R_3(u, S_{1,u}, S_{2,u}, Y_u, Z_u) du | \mathcal{F}_t\}.$$

This proves $P = P^{\epsilon,\delta} + O(\epsilon, \delta, \sqrt{\epsilon\delta})$. □

3.2 Optimal choice of ρ

Recall the representation of ρ^*_{sup} in (5),

$$\rho^*_{\text{sup}} = \overline{\rho}\mathbb{I}\left\{\frac{\partial^2 P_{\text{sup}}}{\partial S_1 \partial S_2} \geq 0\right\} + \underline{\rho}\mathbb{I}\left\{\frac{\partial^2 P_{sup}}{\partial S_1 \partial S_2} < 0\right\}.$$

Obviously this representation remains valid regarding the HJB equation (9).

Due to the complexity of (8), we have introduced an asymptotic expansion (10) to P_{sup} in the previous section to approximate its true value. If we manage to obtain ρ^*_{sup} without actually knowing P_{sup}, we can take ρ^*_{sup} and solve the corresponding asymptotic approximation according to (12) - (14). However, this is not always possible. For example substituting the approximation (10) into (5) directly does not guarantee a correct value of ρ^*_{sup}.

From a more practical of view, the determination of an exact ρ^*_{sup} might not be necessary. Instead, we need to obtain an expression of ρ^*_{sup} which depends on the order of accuracy we require for the approximation of P_{sup} in ϵ and δ . It is similar to the idea of introducing a "ρ^*_{sup} expansion" which matches the order of accuracy for both ρ^*_{sup} and P_{sup}.

We then investigate the relationship between ρ^*_{sup} and P_{sup} at each order of accuracy.

1. Zeroth order: $P_{0,0}$

   If we would like to obtain the approximation to P_{sup} up to the zeroth order only, i.e. the only concern is $P_{0,0}$, the approximation problem turns into solving (12). It is a particular case of the Black-Scholes setup where we take $\overline{\sigma_i}$ as the volatility of S_i. In view of this we may take $\rho^*_{\text{sup}} = \rho^0_{\text{sup}}$ where

$$\rho^0_{\text{sup}} = \overline{\rho}\mathbb{I}\left\{\frac{\partial^2 P_{0,0}}{\partial S_1 \partial S_2} \geq 0\right\} + \underline{\rho}\mathbb{I}\left\{\frac{\partial^2 P_{0,0}}{\partial S_1 \partial S_2} < 0\right\}.$$

2. First order: $P_{0,0} + \sqrt{\epsilon} P_{1,0} + \sqrt{\delta} P_{0,1}$

Approximation up to first order is the main objective of this chapter. This refers to a scenario where we retain terms of order $\sqrt{\epsilon}$ and $\sqrt{\delta}$ while terms with order $\epsilon^{\frac{i}{2}}\delta^{\frac{j}{2}}, i, j \geq 0, i+j \geq 2$ are truncated out.

We first introduce an expansion on the partial derivative $\frac{\partial^2 P_{\text{sup}}}{\partial S_1 \partial S_2}$ as

$$\frac{\partial^2 P_{\text{sup}}}{\partial S_1 \partial S_2} = \sum_{i,j=0}^{\infty} \epsilon^{\frac{i}{2}} \delta^{\frac{j}{2}} \xi_{i,j}.$$

Lemma 3.2. *Under the scenario of approximation up to first order, the following events are equivalent.*

$$\left\{ \frac{\partial^2 P_{\text{sup}}}{\partial S_1 \partial S_2} \geq 0 \right\} \iff \{\xi_{0,0} \geq 0\}.$$

Proof. Consider the indicator function $\hat{\mathbb{I}} \triangleq \mathbb{I}\left\{ \frac{\partial^2 P_{\text{sup}}}{\partial S_1 \partial S_2} \geq 0, \xi_{0,0} < 0 \right\}$.

$$\begin{aligned}
\hat{\mathbb{I}} &= \frac{\sum_{i,j=0}^{\infty} \epsilon^{\frac{i}{2}} \delta^{\frac{j}{2}} \xi_{i,j}}{\sum_{i,j=0}^{\infty} \epsilon^{\frac{i}{2}} \delta^{\frac{j}{2}} \xi_{i,j}} \hat{\mathbb{I}} \leq \frac{\sum_{i,j=0}^{\infty} \epsilon^{\frac{i}{2}} \delta^{\frac{j}{2}} \xi_{i,j} - \xi_{0,0}}{\sum_{i,j=0}^{\infty} \epsilon^{\frac{i}{2}} \delta^{\frac{j}{2}} \xi_{i,j}} \hat{\mathbb{I}} \\
&= \frac{\sum_{i=1}^{\infty} \epsilon^{\frac{i}{2}} \xi_{i,0} + \sum_{j=1}^{\infty} \delta^{\frac{j}{2}} \xi_{0,j} + \sum_{i,j=1}^{\infty} \epsilon^{\frac{i}{2}} \delta^{\frac{j}{2}} \xi_{i,j}}{\sum_{i,j=0}^{\infty} \epsilon^{\frac{i}{2}} \delta^{\frac{j}{2}} \xi_{i,j}} \hat{\mathbb{I}}
\end{aligned}$$

For the first term in the numerator, we have

$$\frac{\sum_{i=1}^{\infty} \epsilon^{\frac{i}{2}} \xi_{i,0}}{\sum_{i,j=0}^{\infty} \epsilon^{\frac{i}{2}} \delta^{\frac{j}{2}} \xi_{i,j}} \hat{\mathbb{I}} = \sqrt{\epsilon} \frac{\xi_{1,0}}{\sum_{j=0}^{\infty} \delta^{\frac{j}{2}} \xi_{0,j}} \hat{\mathbb{I}} + \text{terms of order } \epsilon \text{ or higher.}$$

Similarly for the second term,

$$\frac{\sum_{j=1}^{\infty} \delta^{\frac{j}{2}} \xi_{0,j}}{\sum_{i,j=0}^{\infty} \epsilon^{\frac{i}{2}} \delta^{\frac{j}{2}} \xi_{i,j}} \hat{\mathbb{I}} = \sqrt{\delta} \frac{\xi_{0,1}}{\sum_{i=0}^{\infty} \epsilon^{\frac{i}{2}} \xi_{i,0}} \hat{\mathbb{I}} + \text{terms of order } \delta \text{ or higher.}$$

The third term contains terms of order $\sqrt{\epsilon\delta}$ or higher. Combining these three observations,

$$\hat{\mathbb{I}} \leq \sqrt{\epsilon} \frac{\xi_{1,0}}{\sum_{j=0}^{\infty} \delta^{\frac{j}{2}} \xi_{0,j}} \hat{\mathbb{I}} + \sqrt{\delta} \frac{\xi_{0,1}}{\sum_{i=0}^{\infty} \epsilon^{\frac{i}{2}} \xi_{i,0}} \hat{\mathbb{I}} + \text{terms of order } \epsilon, \delta, \sqrt{\epsilon\delta} \text{ or higher.}$$

Invoking this argument repeatedly, we can conclude that $\hat{\mathbb{I}}$ is less than or equal to terms of order $\epsilon, \delta, \sqrt{\epsilon\delta}$ or higher. Under the scenario of approximation up to first order,we can conclude that $\hat{\mathbb{I}} = 0$. □

Lemma 3.3. *The zeroth order term in the derivative expansion coincides with the derivative of the zeroth order term of the price expansion, i.e.*

$$\frac{\partial^2 P_{0,0}}{\partial S_1 \partial S_2} = \xi_{0,0}.$$

Proof. Differentiating equation (8) with respect to S_1 and S_2, we have

$$\begin{cases} \mathcal{D}_{\text{sup}}\left(\dfrac{\partial^2 P_{\text{sup}}}{\partial S_1 \partial S_2}\right) & = 0 \\ \dfrac{\partial^2 P_{\text{sup}}}{\partial S_1 \partial S_2}(T) & = \dfrac{\partial^2 \Phi}{\partial S_1 \partial S_2}(T) \end{cases},$$

where

$$\mathcal{D}_{\text{sup}} = \frac{1}{\epsilon}\mathcal{L}_0 + \frac{1}{\sqrt{\epsilon}}\mathcal{D}_1 + \mathcal{D}_{2,sup} + \sqrt{\delta}\mathcal{E}_1 + \delta\mathcal{M}_2 + \sqrt{\frac{\delta}{\epsilon}}\mathcal{M}_3,$$

with $\mathcal{L}_0$, $\mathcal{M}_1$ and $\mathcal{M}_2$ defined in (9) and

$$\mathcal{D}_1 = \mathcal{L}_1 + \sum_{i=1}^{2} \sqrt{2}\rho_{iy} v f_i \frac{\partial}{\partial y_i},$$

$$\mathcal{D}_{2,\text{sup}} = \frac{\partial}{\partial t} + \sum_{i=1}^{2}(r - q_i)(1 + S_i\frac{\partial}{\partial S_i}) + \frac{1}{2}\sum_{i=1}^{2} f_i^2(2S_i\frac{\partial}{\partial S_i} + S_i^2\frac{\partial^2}{\partial S_i^2}),$$

$$+ \frac{\partial^2}{\partial S_1 \partial S_2}\left(\sup_{\rho\in[\underline{\rho},\overline{\rho}]} [\rho f_1 f_2 S_1 S_2 \cdot\ ]\right) - r\,\cdot,$$

$$\mathcal{E}_1 = \mathcal{M}_1 + \sum_{i=1}^{2} \rho_{iz} g f_i \frac{\partial}{\partial z}.$$

If the payoff function is piecewise smooth, $\frac{\partial^2 \Phi}{\partial S_1 \partial S_2}$ can be defined piece-wisely. Following the proof of (12), it can be shown that $\xi_{0,0}$ satisfies

$$\begin{cases} \langle \mathcal{D}_{2,\text{sup}}\rangle \xi_{0,0} & = 0 \\ \xi_{0,0}(T) & = \dfrac{\partial^2 \Phi}{\partial S_1 \partial S_2}(T) \end{cases}.$$

Differentiating equation (12) with respect to S_1 and S_2 shows that $\frac{\partial^2 P_{0,0}}{\partial S_1 \partial S_2}$ is the solution to the same equation. Result follows. □

Remark: In general the relationship $\frac{\partial^2 P_{i,j}}{\partial S_1 \partial S_2} = \xi_{i,j}$ does not hold for $i, j \geq 0$.

Combining these two results, we have the following theorem.

Period	March 2013, with a total of 21 trading days
Maturities	1mth, 2mth, 3mth, 4mth, 6mth, 9mth, 1yr
Strikes	ATM, Both Call and Put Delta of 0.35, 0.25, 0.15, 0.1, 0.05
Number of data point	1617

Table 1 Description of data collection

Theorem 3.4. *In determining the price bound approximation up to first order, i.e.* $P_{sup} = P_{0,0} + \sqrt{\epsilon}P_{1,0} + \sqrt{\delta}P_{0,1}$, *we can take* $\rho^*_{sup} = \rho^0_{sup}$ *where*

$$\rho^0_{\text{sup}} = \overline{\rho}\mathbb{I}\left\{\frac{\partial^2 P_{0,0}}{\partial S_1 \partial S_2} \geq 0\right\} + \underline{\rho}\mathbb{I}\left\{\frac{\partial^2 P_{0,0}}{\partial S_1 \partial S_2} < 0\right\}.$$

4. Empirical Study

Our empirical study uses an FX triangle as an illustrative example. In the currency option market, one can often infer the implied correlation from an FX triangle provided that option data are available in the three FX rates involved. In this case, correlation is not very difficult to estimate as in the technical note by Wystup (2000). However, there are situations that option data only available in two out of the three FX rates. When a financial institution wants to make the option market for the remaining FX rate, it is useful to obtain the price bounds by taking into account of volatility smiles of the two liquid FX option markets. A typical example appears in Hong Kong, where options on US dollar (USD) against offshore Chinese Yuan in Hong Kong (CNH) and options on USD against Hong Kong dollar (HKD) are liquidly traded. Although there is a demand for options on HKD against CNH, not much options on this are available for the time being.

To examine the quality of the ambiguous correlation with SV model, we apply the model to a FX triangle with options data available to all of the three FX rates involved.

4.1 The Data

Three chosen currencies are the Great British Pound (GBP), Japanese Yen (JPY) and the US dollar (USD). Trading of products on the exchange rates among any pair of them are more frequent in the market so that data are more available and reliable.

Define $S_{FOR-DOM}$ as the exchange rate representing the price of one unit of foreign currency in terms of the domestic currency. To simplify notations, denote $S_1 = S_{GBP-JPY}$, $S_2 = S_{USD-JPY}$, $S_3 = S_{GBP-USD}$, r_1, r_2 and r_3 to be the risk free interest rate in the UK, Japan and the USA respectively. Market data of the exchange rates and the bid-ask quotation of options on these currency pairs are collected for numerical study. Risk-free interest rate is chosen to be the LIBOR quotes of appropriate tenor.

The conversion of strike price in terms of delta to strike price in terms of exchange rate can be found in Reiswich and Wystup (2010).

Figures 1 and 2 display the volatility surfaces implied by options on GBP-JPY and USD-JPY, respectively. It can be seen from Figure 2 that the volatility surface of USD-JPY exchange rate is highly skewed and shows a "U-shape".

4.2 Equivalence of products

The product under investigation is call option on S_3, $\max(S_3(T) - K, 0)$. By the relationship among the exchange rates, i.e. $S_1/S_2 = S_3$, option on S_3 can be treated as product based on S_1 and S_2. Notice that S_1 and S_2 are expressed in terms of JPY while S_3 is expressed in terms of USD. Option data thus reveals market information from different market measures. The equivalence between the two products can be derived through a simple change of measure. The result is given below.

(21)

$$\mathbb{E}^{\mathbb{Q}_U}[e^{-r_3\tau}\max(S_3(T) - K, 0) \mid \mathcal{F}_t] = \frac{\mathbb{E}^{\mathbb{Q}_J}[e^{-r_2\tau}\max(S_1(T) - KS_2(T), 0) \mid \mathcal{F}_t]}{S_2(t)}$$

The price of product $\max(S_3(T) - K, 0)$ can be obtained directly from the market implied volatility. On the other hand, the price bound of product stated on right hand side of (21) can be obtained using the theory developed. Comparison between both results is carried out to evaluate the performance of the theory.

The payoff on the right hand side of (21) is similar to that of an exchange option. It is easy to verify that the optimal choice of ρ is $\underline{\rho}$ ($\overline{\rho}$) for price upper (lower) bound.

4.3 Performance of the asymptotic solution

The theoretical price bounds predicted by stochastic volatility asymptotic approximation are computed. The restricted region on the correlation coefficient is taken to be the 95 % (90 %) confidence interval (CI) of the sample correlation coefficient among the exchange rate. Specifically, the sample correlation is estimated at 0.7521 while the 90% and 95% CIs are [0.53, 0.88] and [0.47, 0.89], respectively.

For a fixed maturity, the price bounds predicted by theory are shown to contain market bid-ask spread across different strike as well as over calendar time as shown in Figure 3. The bounds predicted by theory are also much narrower than the trivial bound of the implied volatility of the option. Poor performance is observed when the maturity of option is very short. In addition, the convex curves against the moneyness for the short maturity is also due to the error associated with the asymptotic approximation.

The problem of very short maturity options was reported in existing literature. Fouque et al. (2003) mentioned the behaviour of option price with maturity of one month or less is much different from those with longer maturities. The asymptotic approximation alone is not sufficient to capture the volatility structure over a

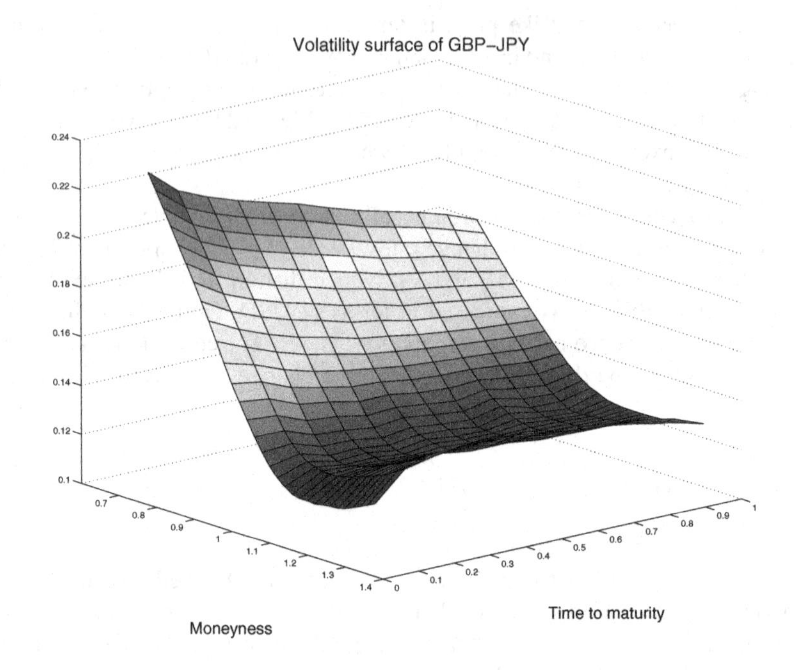

Figure 1. GBP-JPY exchange rate

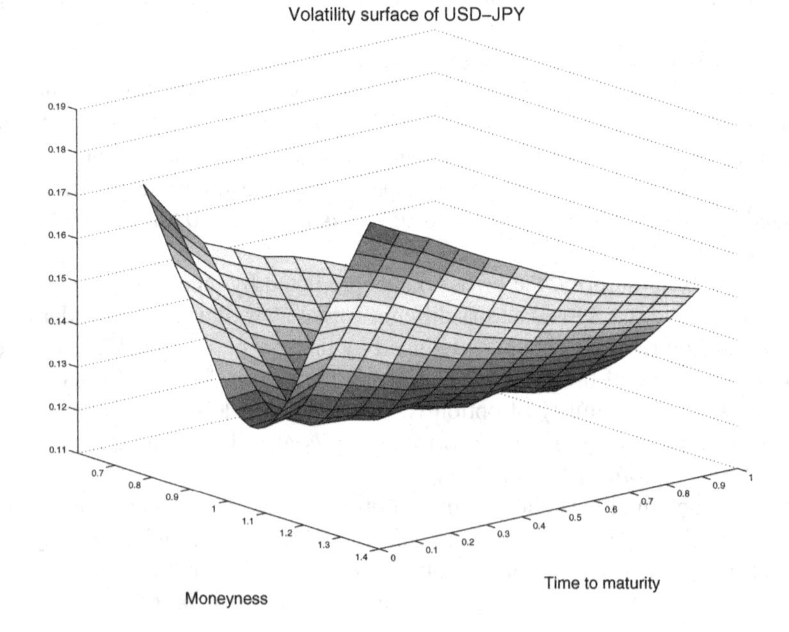

Figure 2. USD-JPY exchange rate

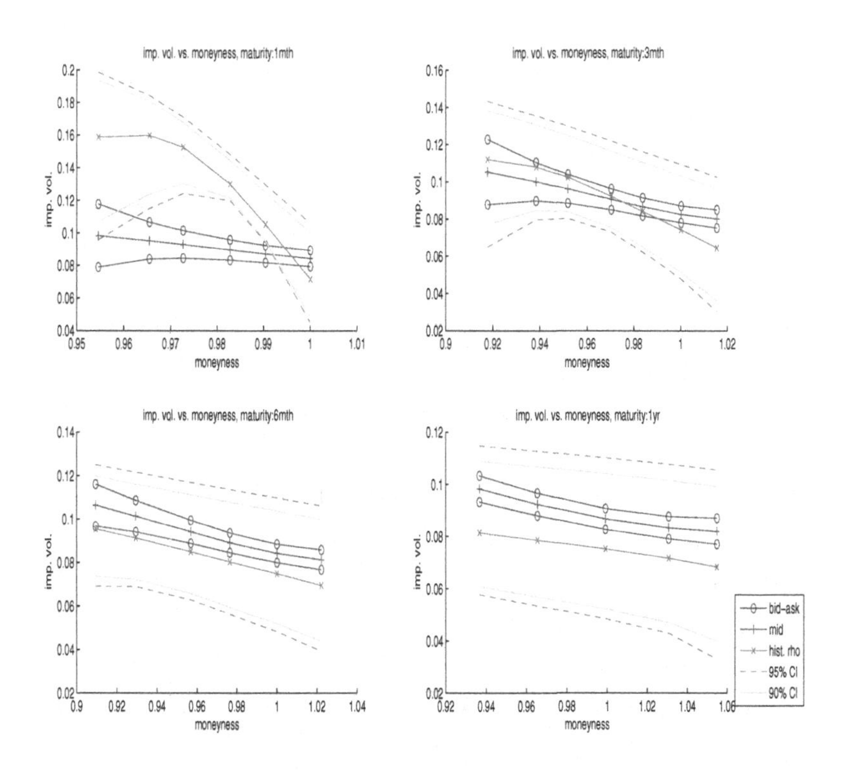

Figure 3. SV: Fixed maturity across different strike.

whole spectrum of maturities. Fouque et al. (2004) suggest the notion of maturity cycle to capture this feature.

Admittedly, the asymptotic solution may suffer from errors associated with the asymptotic expansion. Therefore, a future research should consider an alternative stochastic volatility model for which closed-form solution for the bounds can be obtained. The Heston is therefore on our research agenda.

5. Conclusion

This paper investigates the evaluation of option price on multiple assets by assuming the correlation among the assets to be uncertain. The HJB equations that governs the option price bounds are derived. In two asset case, the optimal choice of ρ can be determined for particular payoff functions.

The bounds on option price can be obtained under the stochastic volatility framework with mean-reverting feature, where the zeroth order term is found to

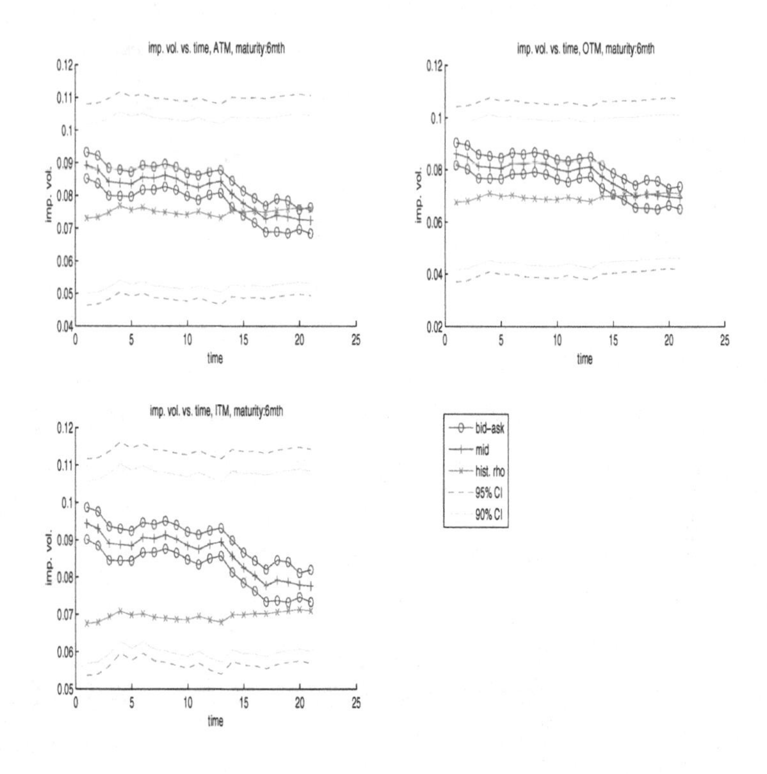

Figure 4. SV: Fixed maturity across calendar date.

be the Black-Scholes price with averaged volatility. The application involves a linear regression which is easy and fast to implement.

Empirical study with foreign exchange option data shows that the predicted price bounds is consistent with market bid-ask spread in general. The choice of a proper correlation bound is an interesting future empirical study.

References

1. Aït-Sahalia, Y., Lo, A. W.: Nonparametric estimation of state-price densities implicit in financial asset prices, *The Journal of Finance*, 53(2), 499-547, 1998.
2. Avellaneda, M., Levy, A., Parás, A.: Pricing and hedging derivative securities in markets with uncertain volatilities, *Applied Mathematical Finance*, 2(2), 73-88, 1995.
3. Björk, T.: *Arbitrage Theory in Continuous Time*, Oxford university press, 2004.
4. Black, F., Scholes, M.: The pricing of options and corporate liabilities, *The Journal of Political Economy*, 637-654, 1973.
5. Carr, P., Madan, D.: Option valuation using the fast Fourier transform,*Journal of Computational Finance*, 2(4), 61-73, 1999.

6. Chen, Z., Epstein, L.: Ambiguity, risk, and asset returns in continuous time, *Econometrica*, 70(4), 1403-1443, 2002.
7. El Karoui, N., Jeanblanc-Picqué, M., Shreve, S. E.: Robustness of the Black and Scholes formula, *Mathematical Finance*, 8(2), 93-126, 1998.
8. Elices, A., Fouque, J. P.: Perturbed Copula: Introducing the skew effect in the codependence, *arXiv preprint arXiv:1003.0041*, 2010.
9. Epstein, L. G., Ji, S.: Ambiguous volatility and asset pricing in continuous time, *Review of Financial Studies*, 26(7), 1740-1786, 2013a.
10. Epstein, L. G., Ji, S.: Ambiguous volatility, possibility and utility in continuous time, *Journal of Mathematical Economics*, 2013b.
11. Fouque, J. P., Papanicolaou, G., Sircar, R., Sølna, K.: Multiscale stochastic volatility asymptotics, *Multiscale Modeling & Simulation*, 2(1), 22-42, 2003.
12. Fouque, J. P., Papanicolaou, G., Sircar, R., Sølna, K.: Maturity cycles in implied volatility, *Finance and Stochastics*, 8(4), 451-477, 2004.
13. Fouque, J.P., Papanicolaou, G., Sircar, R., Sølna, K.: *Multiscale Stochastic Volatility for Equity, Interest Rate, and Credit Derivatives*, Cambridge: Cambridge University Press, 2011.
14. Fouque, J.P., Pun, C.S., Wong, H.Y. Portfolio optimization with ambiguous correlation and stochastic volatilities. Working paper of UC Santa Barbara, 2014.
15. Frey, R., Sin, C. A.: Bounds on European option prices under stochastic volatility, *Mathematical Finance*, 9(2), 97-116, 1999.
16. Heston, S. L.: A closed-form solution for options with stochastic volatility with applications to bond and currency options, *Review of Financial Studies*, 6(2), 327-343, 1993.
17. Hull, J., White, A.: The pricing of options on assets with stochastic volatilities, *The Journal of Finance*, 42(2), 281-300, 1987.
18. Ilhan, A., Jonsson, M., Sircar, R.: Singular perturbations for boundary value problems arising from exotic options, *SIAM Journal on Applied Mathematics*, 64(4), 1268-1293, 2004.
19. Johnson, H.: Options on the maximum or the minimum of several assets, *Journal of Financial and Quantitative Analysis*, 22(3), 277-283, 1987.
20. Kwok, Y. K.: *Mathematical Models of Financial Derivatives*. Springer, 2008.
21. Margrabe, W.: The value of an option to exchange one asset for another, *The Journal of Finance*, 33(1), 177-186, 1978.
22. Nicolato, E., Venardos, E.: Option pricing in stochastic volatility models of the Ornstein-Uhlenbeck type, *Mathematical Finance*, 13(4), 445-466, 2003.
23. Papanicolaou, G., Fouque, J. P., Sølna, K., Sircar, R.: Singular perturbations in option pricing, *SIAM Journal on Applied Mathematics*, 63(5), 1648-1665, 2003.
24. Peng, S.: G-Brownian motion and dynamic risk measure under volatility uncertainty, *arXiv preprint arXiv:0711.2834*, 2007a.
25. Peng, S.: G-expectation, G-Brownian motion and related stochastic calculus of Itô type, In *Stochastic Analysis and Applications*, (pp. 541-567). Springer Berlin Heidelberg, 2007b.
26. Peng, S.: Multi-dimensional G-Brownian motion and related stochastic calculus under G-expectation, *Stochastic Processes and their Applications*, 118(12), 2223-2253, 2008.

27. Peng, S.: Nonlinear expectations and stochastic calculus under uncertainty, *arXiv preprint arXiv:1002.4546*, 2010.
28. Pham, H. *Continuous-time Stochastic Control and Optimization with Financial Applications*. Springer, 2009.
29. Reiswich, D., Wystup, U.: A guide to FX options quoting conventions, *The Journal of Derivatives*, 18(2), 58-68, 2010.
30. Stulz, R.: Options on the minimum or the maximum of two risky assets: analysis and applications, *Journal of Financial Economics*, 10(2), 161-185, 1982.
31. Wang, J., Forsyth, P. A.: Maximal use of central differencing for Hamilton-Jacobi-Bellman PDEs in finance, *SIAM Journal on Numerical Analysis*, 46(3), 1580-1601, 2008.
32. Wystup, U. How the Greeks would have hedged correlation risk of foreign exchange options, (www.mathfinance.com/wystup/papers/rainbow.pdf), 2000.
33. Young, J., Zhou, X. Y.: *Stochastic Controls: Hamiltonian Systems and HJB Equations* (Vol. 43). Springer, 1999.

A. *G*-expectation

We digress here to introduce the tools in handling the above expectations. Define $lip(\mathbb{R}^2)$ to be the space of all bounded and Lipschitz real functions on $\mathbb{R}^2$. We define a G-normal distribution, which is a nonlinear expectation, on $lip(\mathbb{R}^2)$ as:

$$P_i^G(\phi) = u(1,0) : \phi \in lip(\mathbb{R}^2) \mapsto \mathbb{R}$$

where $u(t,x)$ is a bounded continuous function and is a viscosity solution of the nonlinear PDE:

$$\frac{\partial u}{\partial t} - \frac{1}{2} \sup_{\Omega \in \Theta} tr[\Omega D^2 u] = 0 \tag{22}$$

$$u(0,x) = \phi(x)$$

where $D^2 u = (\partial^2_{x_i,x_j} u)^2_{i,j=1}$ is the Hessian matrix of u.

The above definition is a particular case of the G-normal distribution suggested by Peng (2007b, 2008) where we have chosen Θ as the constraint set in the PDE (22). Following his work, we can define a canonical process based on the above G-normal distribution which mirrors the properties of ordinary Brownian motion and is named G-BM. The expectation associated to the G-BM is named G-expectation. Notice that $\sup_{\theta\in\Theta}$ is equivalent to $\sup_{\rho\in[\underline{\rho},\bar{\rho}]}$ by the definition of Θ. We can see that see that $\sup_{\rho\in[\underline{\rho},\bar{\rho}]} \mathbb{E}^{\mathbb{Q}_\rho}(\cdot)$ is indeed the G-expectation under interest.

B. Effective Parameters

Define the solution to the following Poisson equations as ϕ_{ij} and rename ϕ_{ii} as ϕ_i.

For $i, j = 1, 2$,

$$\mathcal{L}_0 \phi_{ij}(y,z) = f_i f_j - \langle f_i f_j \rangle.$$

Define the effective parameters.
For $i, j = 1, 2$,

$$\sigma_i^{*2} = \overline{\sigma_i}^2 + \sqrt{2\epsilon}\nu\left\langle\Lambda\frac{\partial\phi_i}{\partial y}\right\rangle$$
$$Q_{ij}^{\epsilon} = \frac{\sqrt{\epsilon}\rho_{iy}\nu}{\sqrt{2}}\left\langle f_i\frac{\partial\phi_j}{\partial y}\right\rangle \quad ; \quad R_i^{\epsilon} = \sqrt{2\epsilon}\nu\rho_{iy}\left\langle f_i\frac{\partial\phi_{12}}{\partial y}\right\rangle$$
$$Q_i^{2\delta} = \sqrt{\delta}g\langle\Gamma\rangle\frac{\partial\sigma_i^*}{\partial z} \quad ; \quad Q_{ij}^{3\delta} = \sqrt{\delta}\rho_{iz}g\langle f_i\rangle\frac{\partial\sigma_j^*}{\partial z}$$

B.1 $\mathcal{A}_{\text{sup}}$ and $\langle\mathcal{M}_1\rangle$

The explicit expression of $\mathcal{A}_{\text{sup}}$ is

$$\begin{aligned}\sqrt{\epsilon}\mathcal{A}_{sup} &= -\sum_{i=1}^{2}\frac{\sqrt{\epsilon}\nu}{\sqrt{2}}\left\langle\Lambda\frac{\partial\phi_i}{\partial y}\right\rangle S_i^2\frac{\partial^2}{\partial S_i^2} - \rho^*\sqrt{2\epsilon}\nu\left\langle\Lambda\frac{\partial\phi_{12}}{\partial y}\right\rangle S_1S_2\frac{\partial^2}{\partial S_1\partial S_2}\\ &+\sum_{i,j=1}^{2}Q_{ij}^{\epsilon}S_i\frac{\partial}{\partial S_i}S_j^2\frac{\partial^2}{\partial S_j^2} + \sum_{i=1}^{2}\rho^*R_i^{\epsilon}S_i\frac{\partial}{\partial S_i}S_1S_2\frac{\partial^2}{\partial S_1\partial S_2}\end{aligned} \tag{23}$$

The explicit expression of $\langle\mathcal{M}_1\rangle$ can be written as, after a chain rule,

$$\sqrt{\delta}\mathcal{M}_1 = -\sum_{i=1}^{2}Q_i^{2\delta}\frac{\partial}{\partial\sigma_i^*} + \sum_{i,j=1}^{2}Q_{ij}^{3\delta}S_i\frac{\partial^2}{\partial S_i\sigma_j^*} \tag{24}$$

C. Calibration

C.1 Regression coefficients

The term LMMR is defined as $\ln(\frac{K}{S_i})/\tau$. The regression coefficients are

$$a^* = \sigma_i^* + \frac{Q_{ii}^{\epsilon}}{\sigma_i^*}(\frac{1}{2} - \frac{r}{\sigma_i^{*2}}), \qquad b^{\epsilon} = \frac{Q_{ii}^{\epsilon}}{\sigma_i^{*3}},$$
$$a^{\delta} = -\frac{Q_i^{2\delta}}{2} + \frac{Q_{ii}^{3\delta}}{2}(\frac{1}{2} - \frac{r}{\sigma_i^{*2}}), \; b^{\delta} = \frac{Q_{ii}^{3\delta}}{2\sigma_i^{*2}}.$$

C.2 Conversion of regression coefficients to effective parameters

From the definition of regression coefficients in C.1, we have some useful facts.

1. $\sigma_i^* = a^* + O(\sqrt{\epsilon})$.
2. b^{ϵ} is of order $O(\sqrt{\epsilon})$.
3. b^{δ} is of order $O(\sqrt{\delta})$.

Combining the first two gives

$$Q_{ii}^{\epsilon} = b^{\epsilon}(a^* + O(\sqrt{\epsilon}))^3 = b^{\epsilon}a^{*3} + O(\epsilon) + \text{higher order terms.}$$

Put this back to the definition of a^*,

$$\begin{aligned}\sigma_i^* &= a^* - \frac{Q_{ii}^{\epsilon}}{\sigma_i^*}(\frac{1}{2} - \frac{r}{\sigma_i^{*2}}) = a^* - b^{\epsilon}\sigma_i^{*2}(\frac{1}{2} - \frac{r}{\sigma_i^{*2}}) = a^* - \frac{b^{\epsilon}(a^* + O(\sqrt{\epsilon}))^2}{2} + rb^{\epsilon} \\ &= a^* - \frac{b^{\epsilon}a^{*2}}{2} + rb^{\epsilon} + O(\epsilon) + \text{higher order terms.}\end{aligned}$$

From the property of b^{δ},

$$Q_{ii}^{3\delta} = 2b^{\delta}(a^* + O(\sqrt{\epsilon}))^2 = 2b^{\delta}a^{*2} + O(\sqrt{\delta\epsilon}) + \text{higher order terms.}$$

Finally for a^{δ},

$$\begin{aligned}Q_i^{2\delta} &= -2a^{\delta} + Q_{ii}^{3\delta}(\frac{1}{2} - \frac{r}{\sigma_i^{*2}}) = -2a^{\delta} + 2\sigma_i^{*2}b^{\delta}(\frac{1}{2} - \frac{r}{\sigma_i^{*2}}) \\ &= -2a^{\delta} + (a^* + O(\sqrt{\epsilon}))^2 b^{\delta} - 2rb^{\delta} \\ &= -2a^{\delta} + b^{\delta}a^{*2} - 2rb^{\delta} + O(\sqrt{\delta\epsilon}) + \text{higher order terms.}\end{aligned}$$

C.3 Multi-asset parameters

Under the additional assumptions, the solution to Poisson equations in B becomes, for $i, j = 1, 2$,

$$\mathcal{L}_0\phi(y,z) = f^2 - \langle f^2 \rangle, \quad \phi_{ij}(y,z) = \sigma_i\sigma_j\phi.$$

The effective parameters become

$$\begin{aligned}Q_{ij}^{\epsilon} &= \frac{\sqrt{\epsilon}\rho_{iy}\nu}{\sqrt{2}}\sigma_i\sigma_j^2\left\langle f\frac{\partial\phi}{\partial y}\right\rangle, \quad & R_i^{\epsilon} &= \sqrt{2\epsilon}\nu\rho_{iy}\sigma_i\sigma_1\sigma_2\left\langle f\frac{\partial\phi}{\partial y}\right\rangle, \\ Q_i^{2\delta} &= \sqrt{\delta}g\langle\Gamma\rangle\frac{\partial\sigma_i^*}{\partial z}, & Q_{ij}^{3\delta} &= \sqrt{\delta}\rho_{iz}g\sigma_i\langle f\rangle\frac{\partial\sigma_j^*}{\partial z}.\end{aligned}$$

From the definition of σ_i^{*2},

$$\sigma_i^{*2} = \left\langle f_i^2\right\rangle + \sqrt{2\epsilon}\nu\left\langle\Lambda\frac{\partial\phi_i}{\partial y}\right\rangle = \sigma_i^2\left(\left\langle f^2\right\rangle + \sqrt{2\epsilon}\nu\left\langle\Lambda\frac{\partial\phi}{\partial y}\right\rangle\right).$$

Simple calculation shows that

$$\frac{\partial\sigma_i^*}{\partial z} = \frac{\sigma_j^*}{\sigma_i^*}\frac{\sigma_i^2}{\sigma_j^2}\frac{\partial\sigma_j^*}{\partial z}.$$

The relationship between the parameters can be summarized as

$$
\begin{aligned}
Q_{12}^{\epsilon} &= \frac{\sigma_2^2}{\sigma_1^2} Q_{11}^{\epsilon} \quad ; \; Q_{21}^{\epsilon} = \frac{\sigma_1^2}{\sigma_2^2} Q_{22}^{\epsilon} \\
R_1^{\epsilon} &= 2\frac{\sigma_2}{\sigma_1} Q_{11}^{\epsilon} \quad ; \; R_2^{\epsilon} = 2\frac{\sigma_1}{\sigma_2} Q_{22}^{\epsilon} \\
Q_{12}^{3\delta} &= \frac{\sigma_1^*}{\sigma_2^*}\frac{\sigma_2^2}{\sigma_1^2} Q_{11}^{3\delta} \; ; \; Q_{21}^{3\delta} = \frac{\sigma_2^*}{\sigma_1^*}\frac{\sigma_1^2}{\sigma_2^2} Q_{22}^{3\delta}
\end{aligned}
$$

From the definition of σ_i^{*2} in B, we deduce that $\sigma_i^* - \sigma_i$ is of order $\epsilon^{\frac{1}{4}}$. Replacing σ_i by σ_i^* above leads to error of order $\epsilon^{\frac{3}{4}}$. Recall the objective of this paper is to obtain a price approximation that ignores terms of order higher than $\sqrt{\epsilon}$ and $\sqrt{\delta}$, the replacement of σ_i^* does not affect the desired accuracy.

D. Greeks of the Exchange Option

The price of the product $\max(S_1(T) - KS_2(T), 0)$ is

$$P = e^{-r_1\tau} S_1(t) N(d_+) - e^{-r_3\tau} K S_2(t) N(d_-)$$

with the notations

$$\sigma^2 = \sigma_1^2 - 2\rho\sigma_1\sigma_2 + \sigma_2^2, \quad d_\pm = \frac{\log(\frac{S_1(t)}{KS_2(t)}) + (r_3 - r_1 \pm \frac{\sigma^2}{2})\tau}{\sigma\sqrt{\tau}}.$$

Partial derivatives:

$$
\begin{aligned}
\frac{\partial^2 P}{\partial S_1^2} &= \frac{e^{-r_1\tau} n(d_+)}{S_1(t)\sigma\sqrt{\tau}}, & \frac{\partial^2 P}{\partial S_2^2} &= \frac{e^{-r_3\tau} n(d_-)}{S_2(t)\sigma\sqrt{\tau}}, \\
\frac{\partial^2 P}{\partial S_1 \partial S_2} &= -\frac{e^{-r_1\tau} n(d_+)}{S_2(t)\sigma\sqrt{\tau}}, & \frac{\partial^3 P}{\partial S_1^3} &= -\frac{e^{-r_1\tau} n(d_+)}{S_1(t)^2\sigma\sqrt{\tau}}\left(\frac{d_+}{\sigma\sqrt{\tau}} + 1\right), \\
\frac{\partial^3 P}{\partial S_2^3} &= \frac{e^{-r_3\tau} n(d_-)}{S_2(t)^2\sigma\sqrt{\tau}}\left(\frac{d_-}{\sigma\sqrt{\tau}} - 1\right), & \frac{\partial^3 P}{\partial S_1^2 \partial S_2} &= \frac{e^{-r_1\tau} n(d_+) d_+}{S_1(t) S_2(t)\sigma^2\tau}, \\
\frac{\partial^3 P}{\partial S_1 \partial S_2^2} &= \frac{e^{-r_1\tau} n(d_+)}{S_2(t)^2\sigma\sqrt{\tau}}\left(1 - \frac{d_+}{\sigma\sqrt{\tau}}\right), & & \\
\frac{\partial P}{\partial \sigma_1} &= S_1(t) e^{-r_1\tau} n(d_+)\sqrt{\tau}\frac{\sigma_1 - \rho\sigma_2}{\sigma}, & \frac{\partial P}{\partial \sigma_2} &= S_1(t) e^{-r_1\tau} n(d_+)\sqrt{\tau}\frac{\sigma_2 - \rho\sigma_1}{\sigma}, \\
\frac{\partial^2 P}{\partial S_1 \partial \sigma_1} &= e^{-r_1\tau} n(d_+)\sqrt{\tau}\frac{\sigma_1 - \rho\sigma_2}{\sigma}\left(1 - \frac{d_+}{\sigma\sqrt{\tau}}\right), & \frac{\partial^2 P}{\partial S_1 \partial \sigma_2} &= e^{-r_1\tau} n(d_+)\sqrt{\tau}\frac{\sigma_2 - \rho\sigma_1}{\sigma}\left(1 - \frac{d_+}{\sigma\sqrt{\tau}}\right), \\
\frac{\partial^2 P}{\partial S_2 \partial \sigma_1} &= \frac{S_1(t) e^{-r_1\tau} n(d_+) d_+ (\sigma_1 - \rho\sigma_2)}{S_2(t)\sigma^2}, & \frac{\partial^2 P}{\partial S_2 \partial \sigma_2} &= \frac{S_1(t) e^{-r_1\tau} n(d_+) d_+ (\sigma_2 - \rho\sigma_1)}{S_2(t)\sigma^2}.
\end{aligned}
$$

Callable Stock Loans

Chi Chung Siu[1]*, Sheung Chi Phillip Yam[2], and Wei Zhou

[1]Finance Discipline Group, UTS Business School, University of Technology, Sydney, Sydney, Australia. Email: chichung.siu@uts.edu.au
[2]Department of Statistics, The Chinese University of Hong Kong, Shatin, N.T., Hong Kong. Email: scpyam@sta.cuhk.edu.hk
[3]Quantitative Research, Equity Derivatives Group, J.P. Morgan, Central, Hong Kong. Email: wei.x.zhou@jpmorgan.com

In a non-recourse collateralized loan agreement, the lender's recovery is only limited to the market value of the asset as a collateral. In practice, a loan lender can protect himself against any loss by introducing either a margin requirement or a callable feature in the loan contract. In this respect, it is natural to ask: (1) which one of the two mentioned features should be more preferable to the loan manager; (2) how the borrower reacts towards either one of these two features. In the present work, we address these issues by solving explicitly the respective valuation problems of collateralized lending with margin requirement and callable feature under the Black-Scholes model. For the callable feature, we also provide systematic discussions on (1) how to identify whether the smooth-fit condition holds or fails at the optimal stopping boundaries of the associated Dynkin game, and (2) how to solve it when the smooth-fit condition fails at one or both boundaries. We finally utilize these explicit solutions to conduct detailed comparative analysis.

Key words: Collateralized lending; Call protection; Margin requirement; Dynkin games; Smooth-fit condition; Corner solution.
2000 Mathematics Subject Classification: 91A15, 60G40

1. Introduction

In this paper we extend the valuation problem of a non-recourse collatoralized loan under the Black-Scholes economy to the case with callable feature. In a

*Corresponding author

non-recourse collatoralized loan, the borrower approaches the lender for a loan by surrendering the ownership of the dividend-paying asset as a collateral and paying for the service fee to the lender. The borrower has an option, but not the obligation, to terminate the loan contract by redeeming the asset with a repayment of the principal and interest to the lender. The lender collect all dividend payments from the asset until it is redeemed by the borrower.

The redemption feature in the collateralized loan provides a limited downside protection, which is the initial market value of the asset as the collateral less the loan principal and the service fee, with unlimited upside benefit. Should the underlying asset deteriorate in value, the borrower can choose not to redeem collateral. On the other hand, the borrower can terminate the loan contract early by repaying the loan with interest when the market value of the underlying asset rises, as he could then engage in another contract with larger loan amount. Moreover, the collateralized loan can be also attractive to the lender with the rising value of the collateral for he/she is rewarded with larger dividend payments from the underlying asset for the duration of the loan contract, as opposed to the standard lending agreement, in which the lender seizes the ownership of the collateral from the borrower only when the borrower fails to repay the principal and interest. The natural question arises on what would be the fair value of the service fee charged by the lender for the counterparty to engage in the loan contract. In this respect, Jones and Nickerson [16] provided a game-theoretic model to analyze the fair valuation of the loan contract with the mortgage as a collateral in the economy with complete information and complete contingent claim market.

The collateralized loans discussed above are collectively known as the *non-recourse loans*. In a typical non-recourse loan, the lender's recovery is only limited to the market value of the underlying asset. In other words, lender can suffer a substantial loss when the market value of the underlying asset erodes, since he cannot request for further compensation from the borrower to cover the loss, which is the difference between the value of the loan and the value of the underlying asset. Initiated by Xia and Zhou [31], the valuation of the non-recourse stock loans, in which stock shares are used as the collateral, has been extended to various incomplete market settings (see, among others, [33], [6], [29], [30], [14], and [5]). In reality, lender can protect himself/herself by imposing either a margin requirement or a callable feature onto the collateralized loan. In a collateralized loan with margin requirement, when the market value of the underlying asset falls below a pre-specified fraction of the outstanding loan balance, the borrower has an *obligation* to redeem the collateral by repaying the fractional value of the outstanding loan balance, or else the lender has a right of selling the collateral to cover the margin requirement. On the other hand, in a collateralized loan with callable feature, the lender can choose to call back the loan. When the loan is called back by the lender, the borrower has an obligation to redeem the underlying asset by repaying certain portion of the outstanding loan balance, or else forfeit the asset to the lender.

With the consideration of the margin requirement and the callable feature in a collateralized loan, it becomes essential and natural to ask which feature, margin requirement or callable feature, should the lender impose onto the collateralized loan contract. Moreover, in entering the loan contract with either margin requirement or callable feature, it is also equally important to investigate the corresponding reaction from the borrower. Therefore, the problem of designing loan contract with margin requirement or callable feature fits naturally into the framework of the zero-sum Dynkin game. In this article, we provide a complete Dynkin-game analysis on the collateralized loan with margin requirement and callable feature in a Black-Scholes economy, i.e. when the underlying asset follows a geometric Brownian motion.

For the valuation of a collateralized loan with margin requirement, we follow the existing literature (see, for example, [11], [22], and [23]) to identify the contract as a perpetual American barrier option with possibly negative interest rate, and then apply a free boundary approach to establish its closed-form solution.

For the valuation of a callable collateralized loan, we identify the contract as a game option (or game contingent claim), a financial contract first introduced by Kifer [19] and later further investigated in Kyprianou[21], in which both the buyer and the seller of the option have the right to early exercise. In [19], Kifer introduced the notion of game options (or game contingent claims) in which both the buyer and the seller of the option have the right to early exercise. A super-replicating strategy for a game option is defined to be a pair (σ, Δ) of a stopping time σ and a self-financing strategy Δ, such that the option seller can use this portfolio Δ stopped at σ to super-replicate the payoff of the game option no matter what exercising time τ the option buyer chooses. The price of a game option is then defined to be the smallest possible capital required for the option seller to construct such a super-replicating strategy. The main result in Kifer's work is that (see Theorem 3.1 and Proposition 3.3 in [19]), the price of a game option is equal to the value of its associated Dynkin game. In view of this result, to solve for the valuation of a callable collateralized loan, it suffices to solve for the value of its associated Dynkin game; for classical theory of Dynkin game, see [8], [26], [3] and [4]. For the existing classes of the solvable Dynkin game models with applications, see, among others, [12], [1], [2], [9], and [32]. In particular, see Kifer[20] for a comprehensive overview on the recent development of the Dynkin games and their applications in finance.

As in the case of the collateralized loan with margin requirement, we use a free boundary approach to solve for the associated perpetual Dynkin game, namely, by solving a priori ordinary differential equation with a pair of free boundaries. To derive an explicit solution of the corresponding differential equation, one has to impose additional boundary conditions on the value function. Often, these conditions hinge on the principle of smooth-fit, which requires the associated payoff functions to be continuously differentiable across optimal stopping boundaries as long as payoff functions are continuously differentiable there. For an optimal stopping problem, the smooth-fit condition has been extensively discussed in Chapter IV in [28]. In particular, in [28] it has been proven that, if the payoff functions are continuously differentiable at the exercising boundary, and if the scale function of the underlying diffusion is also continuously differentiable at the exercising boundary, then the value function is also continuously differentiable at the exercising boundary. In the context of Dynkin game, the smooth-fit condition has also been rigorously proven in the literature when the payoff functions are continuously differentiable (see Lemma 4.1 in [1]). Therefore, if we know that the payoff function is continuously differentiable across the exercising boundary, then we can apply the smooth-fit condition to determine the exercising boundary.

However, the problem with the valuation of the callable collateralized loan is that the payoff functions are *not continuously differentiable*, and the smooth-fit condition *may fail* at one of the optimal stopping boundaries. Contrary to the literature on the optimal stopping problems with continuously differentiable payoff functions, there has not been much literature on developing the methods of solving the Dynkin games when (i) the payoff functions are not continuously differentiable and (ii) the smooth-fit conditions may fail at one or both optimal stopping boundaries. One noticeable exception is the work of Gapeev (2005) [13], where a Dynkin game version of the spread options was studied with the corresponding payoff functions that are clearly not continuously differentiable. In that work, a set of sufficient conditions (see equations (3.26) and (3.27) in [13]) were provided under which the smooth-fit conditions hold on both exercising boundaries; these smooth-fit conditions were then used to determine the closed form solutions to the game spread options. However, there is no complete characterization of the optimal stopping boundaries or the value function when that set of sufficient conditions fail to hold.

In this paper, we provide a complete solution to the Dynkin game associated to the callable collateralized loan contract. The diffusion we studied in this work is a geometric Brownian motion, whose scale function is smooth. Therefore, if the payoff function is differentiable at the optimal stopping boundary, then by the principle of smooth-fit, the value function should also be continuously differentiable there, i.e. the smooth-fit condition holds (this has been rigorously proven for perpetual optimal stopping problem, see Theorem 2.3 in [27]); to put this in another way, if the smooth-fit condition does not hold at the optimal stopping boundary, then the boundary should be located at the very point at which the corresponding payoff function is not differentiable. Our first contribution of this paper to the existing Dynkin game literature is to rigorously ascertain this claim. More specifically, for the Dynkin game associated to the callable collateralized loan, we find that, while the smooth-fit condition always holds at the upper stopping boundary, it could fail at the lower stopping boundary. If the smooth-fit condition fails at the lower boundary, then the lower boundary is indeed located at the point where the payoff function is not differentiable; if the smooth-fit condition holds at the lower boundary, then the locations of both optimal stopping boundaries can be found as interior solutions to a pair of transcendental equations. The second contribution of our present work is to provide explicit characterization, in terms of model parameters, of the cases where the smooth-fit condition holds or fails at the lower boundary.

The rest of the paper is organized as follows. In Section 2, we shall introduce our model for the price dynamics of the asset as the collateral, description of margin requirement and of callable feature for a collateralized loan. In Section 3, we state the main results of this paper, while Section 4 indicates the financial implication of our key results in Section 3. Section 5 is devoted to the proof of our main theorem. Conclusion and possible future research direction will be mentioned in Section 6, and technical proofs of two key lemmas will be put in Section 7.

2. Model Framework and Features of the Loan Contracts

We consider the continuous-time economy with complete contingent-claim market with two assets: a bond and a dividend-paying asset. The continuously-compounded interest rate r of the bond is assumed to be a constant.

Let $(\Omega, \mathcal{F}, \mathbb{F} = \{\mathcal{F}_t\}_{t\geq 0}, \mathbb{P})$ be the filtered risk-neutral probability space on which the one-dimensional Brownian motion $(W_t)_{t\geq 0}$ is defined with $W_0 = 0$ almost surely. Denote $\mathbb{P}_x$ to be the risk-neutral measure for the asset price process S_t under which $S_0 = x$, and $\mathbb{E}_x$ to be the expectation under $\mathbb{P}_x$.[1]

Under the risk-neutral measure $\mathbb{P}_x$, the asset price process S_t follows a geometric Brownian motion, i.e.

$$dS_t = (r - d) S_t dt + \kappa S_t dW_t,$$

where the dividend rate of the asset $d \geq 0$ and the volatility $\kappa > 0$ are both constants.

Let $0 < \pi < 1$, we begin our discussion with the contract specifications of the *non-recourse collateralized loans*, *collateralized loans with margin requirement*, and *callable collateralized loans*.

- Non-recourse collateralized loans:
  - At time $t = 0$, the borrower borrows an amount q ($q > 0$) from the lender with an asset S_0 as a collateral. At the same time, the lender

[1] The existence of the risk-neutral measure stems from our standing assumption that the contingent-claim market is complete, i.e. every asset can be replicated by the portfolio of the financial assets in the market. See also [16] for similar assumptions in the game-theoretic analysis on the mortgage contracts.

will charge a service fee c_n ($0 \leq c_n \leq q$). Therefore, the lender obtains an amount $S_0 + c_n - q$ from the borrower.

- The loan is continuously compounded at a rate γ. At any time $t \geq 0$, the borrower has an option, but not an obligation, to redeem his asset back by repaying the amount $qe^{\gamma t}$ to the lender. The dividends of the underlying asset are collected by the lender until the borrower redeems his asset back. In other words, the borrower pays an amount of $S_0 + c_n - q$ to enter a non-recourse collateralized loan with an option of redeeming his asset by paying back the loan amount with accrued interests.
- Under this contractual agreement, it is clear that the fair valuation of an non-recourse collateralized loan is equivalent to the fair valuation of an American call option with an initial value given by $S_0 + c_n - q$ and a time-dependent strike price $qe^{\gamma t}$, i.e. the payoff process Y_t of this option is given by[2]

$$Y_t \triangleq \left(S_t - qe^{\gamma t}\right)^+ . \tag{2.1}$$

From the specification of the non-recourse collateralized loan, it is clear that the optionality of redemption provides a downside risk protection to the borrower. Should the value of the asset erode, it would not be optimal for the borrower to redeem the asset back for such redemption would result in a loss. On the other hand, by shorting the American call option to the borrower, the lender have a possibility of facing unlimited downside risk when holding the deteriorated asset. To protect the lender from the erosion of the asset value, we now consider two additional clauses to the non-recourse collateralized loan: 1) margin requirement and 2) call protection.

- Collateralized loans with margin requirements:
  - In addition to the features of the non-recourse collateralized loan described above, a minimum margin requirement πq is posted. When the market price of the asset as the collateral falls below the margin amount accrued at the loan rate γ, i.e. $\pi qe^{\gamma t}$, the borrower has an *obligation* to redeem back the asset by repaying an amount $\pi qe^{\gamma t}$; if the borrower fails to do so, the lender can sell the asset to meet the margin requirement.

[2]As remarked in [31], the standard pricing techniques of the American call options can be not directly applied to the fair valuation of the non-recourse collateralized loan due to the possibility of "negative interest rate", i.e. when $\gamma \geq r$.

- Correspondingly, we denote c_m $(0 \leq c_m \leq q)$ to be the service fee charged by the lender in the collateralized loan contract with margin requirement.
- In view of the non-recourse collateralized loan, the fair valuation of a collateralized loan with margin requirement is equivalent to the fair valuation of an American call option with an initial value given by $S_0 + c_m - q$, a time-dependent strike price $qe^{\gamma}t$, and a time-dependent barrier $\pi qe^{\gamma t}$, i.e. a payoff process Y_t^m of this option is given by

$$Y_t^m \triangleq \left(S_t - qe^{\gamma t}\right)^+ 1_{\{t \leq \tau_{\pi q}\}} = Y_t 1_{\{t \leq \tau_{\pi q}\}}. \tag{2.2}$$

where

$$\tau_{\pi q} = \inf\left\{t > 0 : S_t \leq \pi qe^{\gamma t}\right\}. \tag{2.3}$$

In the collateralized loan with margin requirement, the mandatory redemption when the asset falls below a pre-specified value offers a downside risk protection to the lender by limiting his loss to be $\pi qe^{\gamma}t$ at the time of redemption.

Remark 2.1. In our collateralized loan with margin requirement, after the asset price falls below the accrued margin amount, the position of the borrower has to be closed out and the contract is then terminated. Different type of margin requirements has also been introduced in the literature of the stock loans. For example, [11] studied the value of the margin call stock loan in which the borrower is forced to pay back a fraction π of the loan once the stock falls below the accrued principal amount (not the accrued margin amount), and after which the contract is converted to a standard stock loan but with the loan amount reduced by a factor of $1 - \pi$. See also Liang et al. [22] and Liang and Wu [23] for the valuation of the margin call stock loan with an additional cap feature (cf. [24]) in which the stock loan retains the same margin call feature as in [11], and that the stock price also capped by a predetermined barrier.

- Callable collateralized loan:
  - In addition to features of the non-recourse collateralized loan, at any time $t \geq 0$, the lender can choose to call back the loan. Upon calling the loan at time t, the lender requires the borrower to immediately redeem the underlying asset by paying back a portion of the loan with an amount of $\pi qe^{\gamma t}$, or else surrender his collateral. If the borrower, upon receiving the call at t, chooses to repay a portion of the loan and redeems his collateral, the lender will then terminate the loan contract

by giving the borrower the amount $S_t - \pi q e^{\gamma t}$, otherwise the lender gives nothing to the borrower. Since the borrower will choose the payoff that gives him a higher value, the lender would pay the borrower the amount, Z_t,

$$Z_t \triangleq \left(S_t - (\pi q)\, e^{\gamma t}\right)^+ \; (\geq Y_t) \tag{2.4}$$

at the calling time t. Here, the main concern of the lender is to minimize his payment to the borrower by choosing the optimal call time.

- Correspondingly, we denote $c\ (0 \leq c \leq q)$ to be the service fee charged by the lender in the collateralized loan contract with callable feature.
- Let s be the time when the lender calls back the loan. Since the borrower would have the same payoff Y_t as that of the non-recourse collateralized loan for $t \leq s$ and the payoff Z_s for $s < t$, the fair valuation of a callable collateralized loan is equivalent to the fair valuation of a game option with an initial amount $S_0 + c - q$ and a payoff process $R_{s,t}$ given by

$$R_{s,t} \triangleq Y_t \mathbf{1}_{\{t \leq s\}} + Z_s \mathbf{1}_{\{s<t\}}. \tag{2.5}$$

Similar to the collateralized loan with margin requirement, the callable collateralized loan provides yet another downside risk protection to the lender by imposing a mandatory redemption at the call time. Different from the case with margin requirement, however, there is no specific characterization on what constitutes a call event at the inception of the loan contract. It only specifies the nature of the mandatory redemption at the call time. In this respect, the main objective of this paper is to study the scopes of the business opportunities offered by the collateralized loans with margin requirement and call protection. To understand the differences between the collateralized loans with margin requirement and call protection, we now turn to the study of their fair valuations.

As discussed in the introduction of the paper, our objectives are (i) to evaluate the value of the collateralized loan with margin requirement and (ii) to evaluate the value of a callable collateralized loan.

To this end, let us first consider the following optimal stopping problem

$$v_m(x) \triangleq \sup_{\tau \in \mathcal{T}} \mathbb{E}_x \left[e^{-r\tau} Y_\tau \mathbf{1}_{\{\tau \leq \tau_{\pi q}\}} \right] \tag{2.6}$$

where $\mathcal{T}$ denotes all $(\mathcal{F}_t)_{t\geq 0}$-stopping time and $\tau_{\pi q}$ is defined in (2.3), and the following Dynkin game

$$v(x) \triangleq \sup_{\tau \geq 0} \inf_{\sigma \geq 0} \mathbb{E}_x \left[e^{-r(\sigma \wedge \tau)} R_{\sigma,\tau} \right]. \tag{2.7}$$

In addition, consider the following two statements:

(a) The supremum and infimum in (2.7) are interchangeable, i.e. (σ, τ) is a saddle point,

$$\sup_{\tau \geq 0} \inf_{\sigma \geq 0} \mathbb{E}_x \left[e^{-r(\sigma \wedge \tau)} R_{\sigma,\tau} \right] = \inf_{\sigma \geq 0} \sup_{\tau \geq 0} \mathbb{E}_x \left[e^{-r(\sigma \wedge \tau)} R_{\sigma,\tau} \right]; \tag{2.8}$$

(b) The value of a collateralized loan with margin requirement, and respectively the value of a callable collateralized loan, are given by the optimal stopping problem (2.6) and the Dynkin game (2.7) respectively.

By [17], [19] and [10], Statements (a) and (b) can be validated, if we can verify the following conditions:

$$\mathbb{P}_x \left(\lim_{t \to \infty} e^{-rt} Z_t = 0 \right) = 1, \tag{2.9}$$

and

$$\mathbb{E}_x \left[\sup_{t > 0} e^{-rt} Z_t \right] < \infty. \tag{2.10}$$

Now, by the law of the iterated logarithm for Brownian motion, we see that $\lim_{t \to \infty} e^{-rt} Z_t = 0$ a.s., so (2.9) always holds.

The major hurdle comes in validating the integrability condition (2.10). If $d > 0$, or $d = 0$ and $\gamma - r > \frac{\kappa^2}{2}$, (2.10) holds true by the same arguments as in Lemma 3.1 in [31]. On the other hand, if $d = 0$ and $\gamma - r \leq \frac{\kappa^2}{2}$, one can no longer use Lemma 3.1 in [31] to verify (2.10). In other words, for $d > 0$, or $d = 0$ and $\gamma - r > \frac{\kappa^2}{2}$, statements (a) and (b) are immediately true, we can then find the fair values of the collateralized loan with margin requirement and the callable collateralized loan by solving (2.6) and (2.7), respectively; while if $d = 0$ and $\gamma - r \leq \frac{\kappa^2}{2}$, it is *not* clear that the interchangeability condition in (2.8) can be met, and hence we cannot take for granted that the values of these two contracts under consideration are indeed given by solving (2.6) and (2.7).

To circumvent this hurdle, in this paper, we shall first find out closed-form solutions to both the optimal stopping problem (2.6) and the Dynkin game (2.7), and then we utilize these explicit expressions of (2.6) and (2.7) to show *directly* that both Statements (a) and (b) remain valid even under the case when $d = 0$ and $\gamma - r \leq \frac{\kappa^2}{2}$.

3. Main Result

Let $\widetilde{r} \triangleq r - \gamma$ and $X_t \triangleq e^{-\gamma t} S_t$. X_t then satisfies the equation

$$dX_t = (\widetilde{r} - d) X_t dt + \kappa X_t dW_t$$

and the infinitesimal generator of the process $\left(e^{-\widetilde{r}t}X_t\right)_{0\leq t<\infty}$ is given by

$$\mathcal{A} \triangleq \frac{\kappa^2}{2}x^2\frac{d^2}{dx^2} + (\widetilde{r} - d)\, x\frac{d}{dx} - \widetilde{r}.$$

Define $g_1(x) \triangleq (x-q)^+$ and $g_2(x) \triangleq (x - \pi q)^+$. In terms of g_1, g_2, and X_t, the payoff processes (2.1), (2.4) and (2.5) can be rewritten as follows

$$Y_t = e^{\gamma t} g_1(X_t), \tag{3.1}$$

$$Z_t = e^{\gamma t} g_2(X_t), \tag{3.2}$$

$$R_{s,t} = e^{\gamma(s\wedge t)}\left(g_1(X_t)\mathbf{1}_{\{t\leq s\}} + g_2(X_s)\mathbf{1}_{\{s<t\}}\right). \tag{3.3}$$

Correspondingly, the value functions in (2.6) and (2.7) can be rewritten as

$$v_m(x) = \sup_{\tau\in\mathcal{T}} \mathbb{E}_x\left[e^{-\widetilde{r}\tau} g_1(X_\tau)\mathbf{1}_{\{\tau\leq\tau_{\pi q}\}}\right] \tag{3.4}$$

and

$$v(x) = \sup_{\tau\geq 0}\inf_{\sigma\geq 0} \mathbb{E}_x\left[e^{-\widetilde{r}(\sigma\wedge\tau)}\left[g_1(X_\tau)\mathbf{1}_{\{\tau\leq\sigma\}} + g_2(X_\sigma)\mathbf{1}_{\{\sigma<\tau\}}\right]\right]. \tag{3.5}$$

Let $\psi(x)$ and $\varphi(x)$ be two solutions to the differential equation $\mathcal{A}f = 0$, such that the function $x \longmapsto F(x) \triangleq \frac{\psi}{\varphi}(x)$ is increasing in x. One can then check that

$$(\psi(x), \varphi(x)) = \begin{cases} \left(x^{\lambda_1}, x^{\lambda_2}\right) & \text{if } d > 0 \text{ or } \widetilde{r} \neq -\frac{\kappa^2}{2} \\ (x\ln x, x) & \text{if } d = 0 \text{ and } \widetilde{r} = -\frac{\kappa^2}{2} \end{cases}. \tag{3.6}$$

where $\lambda_1 > \lambda_2$ are the roots of the quadratic equation

$$\frac{\kappa^2}{2}\lambda^2 + \left(\widetilde{r} - d - \frac{\kappa^2}{2}\right)\lambda - \widetilde{r} = 0, \tag{3.7}$$

and

$$F(x) = \begin{cases} x^{\lambda_1-\lambda_2} & \text{if } d > 0 \text{ or } \widetilde{r} \neq -\frac{\kappa^2}{2} \\ \ln x & \text{if } d = 0 \text{ and } \widetilde{r} = -\frac{\kappa^2}{2} \end{cases}. \tag{3.8}$$

Before listing out the key results of the present work, let us first provide a motivation leading to the solution of the Dynkin game (3.5). Recall the following observation as stated in Introduction: since a geometric Brownian motion has a smooth scale function, if the smooth-fit condition fails at a boundary, intuitively this boundary should be located at the point where the corresponding payoff function is not differentiable. For the lower optimal stopping boundary, the corresponding payoff function is g_2, which is not differentiable at $x = \pi q$. Therefore, we expect that, depending on model parameters, either the lower boundary is located at $x = \pi q$, or it lies above πq in which case the optimal boundary can be found by imposing a smooth-fit condition.

Moreover, note that if the lower stopping boundary is located at $x = \pi q$, the value of the callable collateralized loan (3.5) coincides with that of the collateralized loan with margin requirement (3.4), i.e. $v_m(x) = v(x)$. Indeed, a key observation is that, as long as $v_m(x) \le g_2(x) = (x-\pi q)^+$ for $x > \pi q$, then one can verify, by the same argument as in Part (ii) of Theorem 2 in Kyprianou [21], that $v_m(x) = v(x)$ for all x. In addition, another key observation is that, if $\frac{\partial}{\partial x} v_m(\pi q+) < 1$, then $v_m(x) \le g_2(x) = (x-\pi q)^+$ for all $x > \pi q$ (a detailed verification will be provided in the **Proof of Theorem 3.1**). By combining these two observations, we see that if $\frac{\partial}{\partial x} v_m(\pi q+) < 1$, then $v_m(x) = v(x)$ for all x. Therefore, we expect that the smooth-fit condition fails at the lower boundary whenever $\frac{\partial}{\partial x} v_m(\pi q+) < 1$. With this motivation in mind, we define

$$d^* = \inf\left\{d > 0 : \frac{d}{dx} v_m((\pi q)+) \triangleq \lim_{x \downarrow (\pi q)} \frac{v_m(x)}{x - (\pi q)} < 1\right\}, \tag{3.9}$$

which is the smallest dividend rate of the underlying asset such that the delta of the collateralized loan with margin requirement is smaller than unity. In the sequel, we shall show that d^* is the threshold dividend rate that determines whether the smooth-fit condition holds or fails at the lower exercising boundary.

In what follows, we state the solutions to our pricing problems under three mutually exclusive cases according to the values of $\widetilde{r}$ and d :

(C1) $\widetilde{r} \ge 0$ and $d = 0$,

(C2) (i) $\widetilde{r} = -\frac{\kappa^2}{2}$ and $d = 0$,

(ii) $0 > \widetilde{r} > -\frac{\kappa^2}{2}$ and $d = 0$,

(iii) $\{d \ge \widetilde{r}\} / \left\{0 \ge \widetilde{r} \ge -\frac{\kappa^2}{2}, d = 0\right\}$,

(iv) $\widetilde{r} > 0$ and $\widetilde{r} > d \ge d^*$.

(C3) $\widetilde{r} > 0$ and $d^* > d > 0$.

Remark 3.1. Since

$$\{d \ge \tilde{r}, \tilde{r} > 0\} \subset \left(\{d \ge \widetilde{r}\} / \left\{0 \ge \widetilde{r} \ge -\frac{\kappa^2}{2}, d = 0\right\}\right),$$

and

$$\{\widetilde{r} > 0, d^* > d > 0\} = \{\tilde{r} > 0, d^* > d \ge \tilde{r}\} \cup \{\tilde{r} > 0, d^* > r \ge d\} \cup \{\tilde{r} > 0, \tilde{r} \ge d^* > d\},$$

it is not immediate clear that Cases (C2)(iii) and (C3) are mutually exclusive. The fact that they are mutually exclusive is confirmed by Lemma 5.2 that d^* is bounded above by $\tilde{r}$ when $\tilde{r} > 0$, which implies that

$$\{\tilde{r} > 0, d^* > d \geq \tilde{r}\} = \{\tilde{r} > 0, d^* > r \geq d\} = \emptyset.$$

The reason why we split our problems into a number of cases is because Case (C2) precisely corresponds to the case $\frac{\partial}{\partial x} v_m (\pi q+) \leq 1$, whereas Cases (C1) and (C3) together correspond to the case $\frac{\partial}{\partial x} v_m (\pi q+) > 1$. Because solutions of (3.4) and (3.5) take trivial forms under (C1), we also distinguish between (C1) and (C3); the details will be given in Section 5. Case (C2) is further partitioned into four subcases for the sake of presentation.

With Cases (C1)-(C3) defined, we are now in position to state the results of the paper. First, the following lemma gives closed form solution for the optimal stopping problem (3.4) and relates it to the fair value of the collateralized loan with margin requirement.

Lemma 3.1. *Let v_m be the value of a collateralized loan with margin requirement in (3.4), we have*

1. *Under (C1),*

$$v_m(x) = \begin{cases} 0 & x \leq \pi q \\ x - \pi q \left(\frac{x}{\pi q}\right)^{-\frac{2\tilde{r}}{\kappa^2}} & x \geq \pi q \end{cases}.$$

2. *Under (C2) and (C3),*

$$v_m(x) = \begin{cases} 0 & x \leq \pi q \\ \frac{a_0 - q}{\varphi(a_0)} \frac{F(x) - F(\pi q)}{F(a_0) - F(\pi q)} \varphi(x) & \pi q \leq x \leq a_0 \\ x - q & x > a_0 \end{cases}. \tag{3.10}$$

where $a_0 \triangleq \alpha_0 q$ and α_0 is the unique solution to either one of the following equations

(a) *under (C2)(i),*

$$\alpha - \ln \alpha + \ln \pi - 1 = 0. \tag{3.11}$$

(b) *under (C2)(ii)-(iv) and (C3),*

$$(1 - \lambda_1)\alpha^{1-\lambda_2} + \lambda_1 \alpha^{-\lambda_2} = \left[(1 - \lambda_2)\alpha^{1-\lambda_1} + \lambda_2 \alpha^{-\lambda_1}\right]\pi^{\lambda_1 - \lambda_2}, \tag{3.12}$$

When $\widetilde{r} \geq 0$, the result of Lemma 3.1 is well-known in literature. In particular, the case $\widetilde{r} \geq 0$ and $d = 0$ can be found in [25]. In Section 5, we shall provide a proof of Lemma 3.1 for completeness, where the existence of d^* will also be proven.

The following definition and theorem give the closed-form solution to the Dynkin game (3.5).

Definition 3.1. Define the function v as follows:

1. Under (C1), $v(x) \triangleq g_2(x) = (x - \pi q)^+$ for all x.

   (a) Under (C2),

$$v(x) \triangleq v_m(x) \text{ for all } x.$$

   (b) Under (C3),

(3.13)
$$v(x) \triangleq \begin{cases} (x-\pi q)^+ & \text{if } x \le b_1 \\ \left(\frac{F(x)-F(b_1)}{F(a_1)-F(b_1)}\frac{a_1-q}{\varphi(a_1)} + \frac{F(a_1)-F(x)}{F(a_1)-F(b_1)}\frac{b_1-\pi q}{\varphi(b_1)}\right)\varphi(x) & \text{if } b_1 \le x < a_1 \\ x-q & \text{if } x \ge a_1 \end{cases}$$

with $(b_1, a_1) \triangleq (\beta_1 q, \alpha_1 q)$ and (β_1, α_1) is the unique pair of solutions to the system of equations on the interval $\left(\pi q, \frac{\overline{r}}{d}\pi q\right) \times \left(\frac{\overline{r}}{d}q, \infty\right)$:

(3.14)
$$\begin{cases} (1-\lambda_1)\alpha^{1-\lambda_2} + \lambda_1\alpha^{-\lambda_2} + (\lambda_1-\lambda_2)\left(\beta^{1-\lambda_2} - \pi\beta^{-\lambda_2}\right) = \beta^{\lambda_1-\lambda_2}\left((1-\lambda_2)\alpha^{1-\lambda_1} + \lambda_2\alpha^{-\lambda_1}\right), \\ (1-\lambda_1)\beta^{1-\lambda_2} + \lambda_1\pi\beta^{-\lambda_2} + (\lambda_1-\lambda_2)\left(\alpha^{1-\lambda_2} - \alpha^{-\lambda_2}\right) = \alpha^{\lambda_1-\lambda_2}\left((1-\lambda_2)\beta^{1-\lambda_1} + \lambda_2\pi\beta^{-\lambda_1}\right). \end{cases}$$

Theorem 3.1. *The function v defined in Definition 3.1 is C^1 on $(\pi q, \infty)$, piecewisely twice differentiable and is the value function for the Dynkin game with payoff functions g_1 and g_2, i.e.,*

$$\begin{aligned} v(x) &= \sup_{\tau\ge 0}\inf_{\sigma\ge 0}\mathbb{E}_x\left[e^{-\overline{r}(\sigma\wedge\tau)}\left[g_1(X_\tau)\mathbf{1}_{\{\tau\le\sigma\}} + g_2(X_\sigma)\mathbf{1}_{\{\sigma<\tau\}}\right]\right] \\ &= \inf_{\sigma\ge 0}\sup_{\tau\ge 0}\mathbb{E}_x\left[e^{-\overline{r}(\sigma\wedge\tau)}\left[g_1(X_\tau)\mathbf{1}_{\{\tau\le\sigma\}} + g_2(X_\sigma)\mathbf{1}_{\{\sigma<\tau\}}\right]\right]. \end{aligned} \tag{3.15}$$

The proof of the Theorem 3.1 will be given in Section 5. From Theorem 3.1, we see that, under (C1), the stopping region for the loan lender is the whole real line and so the lower exercising boundary effectively equals to infinity. Under (C2), the value of the Dynkin game in (3.5) coincides with the value of the optimal stopping problem in (3.4), and therefore the smooth-fit condition does not hold at the lower boundary located at πq, which can be interpreted as a 'corner solution'. Finally, under (C3), the smooth-fit condition of the Dynkin game in (3.5) holds at the lower boundary, which can be found as the interior solution of the transcendental equation (3.14). An important insight drawn from the Theorem 3.1 is that (2.8) is true even for the case $d = 0$ and $\gamma - r \le \frac{\kappa^2}{2}$.

Using the explicit solution of (3.5) given in Theorem 3.1, the following theorem completes the remaining puzzle in that the value function v in Definition 3.1 and Theorem 3.1 *is* indeed the fair value of the callable collateralized loan with initial market value of the collateral x, thereby confirming Kifer's result remains valid in the case when $d = 0$ and $\gamma - r \leq \frac{\kappa^2}{2}$.

Theorem 3.2. *Let $d = 0$ and $\gamma - r \leq \frac{\kappa^2}{2}$. Then $v(x)$ is equal to the initial value of the callable collateralized loan when the initial market price of the underlying asset is x, i.e. $v(x)$ is equal to the minimal capital that enables the lender of the loan to invest it into a self-financing portfolio, which will cover the payment the lender gives to the borrower up to a calling time σ, independent of the time τ the borrower may choose.*

Proof. Let $p(x)$ be the initial value of the callable collateralized loan when the initial stock price is x. We shall show that $p(x) = v(x)$.

1. First consider the case $d = 0$ and $-\frac{\kappa^2}{2} \leq r - \gamma < 0$. This case is contained in (C2). In this case, the optimal stopping time for the lender and the borrower are respectively given by $\sigma^* = \inf\{t \geq 0 : X_t \leq \pi q\}$ and $\tau^* = \inf\{t \geq 0 : X_t \geq a_0\}$. Since $v(X_t) \geq g_1(X_t)$ and $v(X_{\sigma^*}) = g_2(X_{\sigma^*}) = 0$ on $\{\sigma^* < \infty\}$, thus $v(X_{\sigma^* \wedge t}) \geq g_1(X_t)\mathbf{1}_{\{t \leq \sigma^*\}} + g_2(X_{\sigma^*})\mathbf{1}_{\{\sigma^* < t\}} = e^{-\gamma\sigma^* \wedge t} R_{\sigma^* \wedge t}$. Multiplying $e^{-\bar{r}(\sigma^* \wedge t)}$ on both sides, we obtain $e^{-\bar{r}(\sigma^* \wedge t)} v(X_{\sigma^* \wedge t}) \geq e^{-r(\sigma^* \wedge t)} R_{\sigma^* \wedge t}$. In accordance to Theorem 3.1, v is C^1 on $(\pi q, \infty)$ and is piecewisely twice differentiable, thus by applying Ito's formula, upon using that $e^{-\bar{r}u} X_u = e^{-ru} S_u$ for all u, we obtain

$$e^{-\bar{r}(\sigma^* \wedge t)} v(X_{\sigma^* \wedge t}) = v(x) + \int_0^{\sigma^* \wedge t} e^{-\bar{r}u} \mathcal{A} v(X_u)\, du + \int_0^{\sigma^* \wedge t} v'(X_u) e^{-ru} \kappa S_u dW_u.$$

For $x \in (\pi q, a_0)$, we clearly have $\mathcal{A} v(x) = 0$. For $x \in (a_0, \infty)$, by the optimality of a_0 we must have $\mathcal{A} v(x) \leq 0$. Thus $\mathcal{A} v(X_u) \leq 0$ for $u \leq \sigma^* \wedge t$, which implies

$$e^{-r\sigma^* \wedge t} R_{\sigma^* \wedge t} \leq e^{-\bar{r}\sigma^* \wedge t} v(X_{\sigma^* \wedge t}) \leq v(x) + \int_0^{\sigma^* \wedge t} v'(X_u) e^{-ru} \kappa S_u dW_u.$$

Since $v'(x)$ is bounded above, as can be directly verified upon using (3.13) and (3.10), the stochastic integral in the right hand side is actually a martingale. This suggests that if we define $\Delta_u \triangleq v'(X_u) S_u$, then (σ^*, Δ) is a hedge against the callable collateralized loan. We obtain $p(x) \leq v(x)$.

On the other hand, suppose (σ, Δ) is a hedge against the callable collateralized loan with initial value y, i.e.

$$e^{-r\sigma \wedge t} R_{\sigma \wedge t} \leq y + \int_0^{\sigma \wedge t} e^{-ru} \Delta_u \kappa dW_u.$$

Then by the continuity of $t \mapsto e^{-r\sigma\wedge t}R_{\sigma\wedge t}$ and the stochastic integral, for any stopping time τ we have

$$e^{-r(\sigma\wedge\tau)}R_{\sigma\wedge\tau} \leq y + \int_0^{\sigma\wedge\tau} e^{-ru}\Delta_u \kappa dW_u.$$

Since by the definition of a hedge, the process $t \to \int_0^{\sigma\wedge t} e^{-ru}\Delta_u \sigma dW_u$ is a local martingale bounded below, therefore it is a $\mathbb{P}_x$-supermartingale, therefore upon using optional sampling theorem,

$$y \geq \mathbb{E}_x\left[y + \int_0^{\sigma\wedge\tau} e^{-ru}\Delta_u \kappa dW_u\right] \geq \mathbb{E}_x\left[e^{-r(\sigma\wedge\tau)}R_{\sigma\wedge\tau}\right].$$

Therefore

$$\begin{aligned} y &\geq \sup_\tau \mathbb{E}_x\left[e^{-r(\sigma\wedge\tau)}R_{\sigma\wedge\tau}\right] \\ &\geq \inf_\sigma \sup_\tau \mathbb{E}_x\left[e^{-r(\sigma\wedge\tau)}R_{\sigma\wedge\tau}\right] = v(x). \end{aligned}$$

This shows $p(x) \geq v(x)$. Thus we have $p(x) = v(x)$.

2. For the case $d = 0$ and $\widetilde{r} \geq 0$, the theorem can be proven by similar arguments as above. □

4. Financial Implications

In this section, we present the financial interpretations of Lemma 3.1 and Theorem 3.1. Let us first restate the two key questions raised in the abstract of the paper:

(Q1) With the options of imposing callable feature and margin requirement to the non-recourse collateralized loans, which one should the lender prefer?

(Q2) Does the adoption of the callable feature *always* increase the business opportunity over that of the non-recourse collateralized loan?

Recall that c_m (resp. c) is the service fee charged by the loan lender for a collateralized loan with margin requirement (resp. the callable collateralized loan). The valuation of collateralized loans with margin requirement (resp. callable collateralized loans) can be formulated as the borrower initially buying at price $(S_0 - q + c_m)$ (resp. $S_0 - q + c$) an American contingent claim with payoff Y and barrier πq (resp. a game contingent claim (GCC) with payoffs Y and Z). Together with Lemma 3.1 and Theorem 3.1, the fair values of service fees should be such that $S_0 - q + c_m = v_m(S_0)$ and $S_0 - q + c = v(S_0)$.

4.1 Margin Requirement versus Call Protection

To answer (Q1), we shall identify the explicit cases (depending on the model parameters) under which the adoption of the callable feature is favored over the imposition of the margin requirement, and vice versa.

(C1) Assume that $\widetilde{r} \geq 0$ and $d = 0$.[3]

- Collateralized loan with margin requirement:
  If $S_0 \leq \pi q$, i.e. the initial market price of the collateral is less than the margin requirement, Lemma 3.1 states that there is no business between the two parties. On the other hand, if $S_0 > \pi q$, i.e. $\frac{q}{S_0} < \frac{1}{\pi}$, Lemma 3.1 gives $v_m(S_0) = S_0 - \pi q\left(\frac{S_0}{\pi q}\right)^{-\frac{2\tilde{r}}{\kappa^2}}$, which translates into $c_m = q - \pi q\left(\frac{S_0}{\pi q}\right)^{-\frac{2\tilde{r}}{\kappa^2}}$. Note that since $S_0 > \pi q$, $\tilde{r} \geq 0$, we have $\pi\left(\frac{S_0}{\pi}\right)^{-\frac{2\tilde{r}}{\kappa^2}} < 1$, which means that lender can earn non-zero service fee $c_m > 0$ by engaging in a collateralized loan contract with the margin requirement with the borrower. Therefore, as long as the loan-to-value (LTV) ratio is less than $\frac{1}{\pi}$, there is business opportunity for a collateralized loan with margin requirement. Furthermore, the result also reveals that imposing margin requirement to the non-recourse collateralized loans *increases* the lender's incentive to carry out a business than it would otherwise.
- Collateralized loan with callable feature:
  Now suppose that the lender has the right to call back the loan, and suppose that the LTV is less that $\frac{1}{\pi}$, i.e. $S_0 > \pi q$, then Theorem 3.1 states that the value of the callable collateralized loan becomes $v(S_0) = S_0 - \pi q$. Hence, we have $c = q - \pi q < c_m$. Thus, including the callable feature helps to *reduce* the service fee. However, in this case $v(S_0) = S_0 - \pi q$ implies that the optimal calling time of the lender is $\sigma^* = \inf\{t > 0 : v(X_t) = g_2(X_t)\} = 0$. This means that, at inception, the lender receives a total amount of $S_0 + c - q = S_0 - \pi q$ from the borrower and then *immediately calls the loan back*, delivering $S_0 - \pi q$ back to the borrower. That is, there is actually *no physical exchange between the two parties* in this case. Therefore, a business opportunity exists for a collateralized loan with margin requirement is *ruled out* when the callable feature is included in the contract, suggesting

[3] When the dividend rate $d = 0$, the loan rate γ is usually set at a higher rate than the risk-free interest r; while if the cash rate $d > 0$, the lender has already taken the advantage of getting all cash payments from the collateral, in order to boost the business opportunity, the lender may be willing to reduce the loan rate γ to be even lower than the prevailing interest rate r. This argument may suggest that, Case (C1), i.e. the case $r - \gamma \geq 0$ and $d = 0$, might not be likely to occur in practice; nonetheless, we still include (C1) in the discussion below for the purpose of completeness.

that a loan manager should *not* include the callable feature into the collateralized loan contract.

(C2) Assume that (i) $\widetilde{r} = -\frac{\kappa^2}{2}$ and $d = 0$, (ii) $0 > \widetilde{r} > -\frac{\kappa^2}{2}$ and $d = 0$, (iii) $\{d \geq \widetilde{r}\} / \left\{0 \geq \widetilde{r} \geq -\frac{\kappa^2}{2}, d = 0\right\}$ or (iv) $\widetilde{r} > 0$ and $\widetilde{r} > d \geq d^*$.

- Collateralized loan with margin requirement and with callable feature: Theorem 3.1 and Definition 3.1 imply that $v = v_m$, i.e. the fair values of the collateralized loans with callable feature and the margin requirement *coincide*. If $S_0 \leq \pi q$, the initial stock price is too low and there is no business opportunity. If $S_0 \geq a_0$ the optimal redeem time is $\tau^* = \inf\{t > 0 : v(X_t) = g_1(X_t)\} = 0$, which means at initial time the borrower pays a total amount $S_0 - q + c = v(S_0) = S_0 - q$ to the lender and then immediately redeem the asset back by paying $q - S_0$, therefore there is no actual physical exchange between the lender and the borrower. In the case $S_0 \in (\pi q, a_0)$, i.e. $\frac{q}{S_0} \in \left(\frac{1}{\alpha_0}, \frac{1}{\pi}\right)$, $v = v_m$ and hence $c = c_m$, which means that in this case it is *indifferent* between including the margin requirement and including the callable feature into the contract, and a business opportunity exists as long as the LTV lies in $\left(\frac{1}{\alpha_0}, \frac{1}{\pi}\right)$.

(C3) Assume that $\widetilde{r} > 0$ and $d^* > d > 0$.

- Collateralized loan with margin requirement.
  For a collateralized loan with margin requirement, as in (C2), a business opportunity exists as long as the LTV lies in $\left(\frac{1}{\alpha_0}, \frac{1}{\pi}\right)$.
- Collateralized loan with callable feature:
  Now suppose that the lender has the right to call back the loan, and suppose that the LTV lies in $\left(\frac{1}{\alpha_0}, \frac{1}{\pi}\right)$, i.e. $\pi q < S_0 < a_0$. Theorem 3.1 and Definition 3.1 state that $b_1 < a_1$ are the optimal stopping boundaries in this case. In Lemma 5.3 below it is proved that $\pi q < b_1$ and $a_1 < a_0$. It remains to discuss the business opportunity for the following three cases (i) $\pi q < S_0 \leq b_1 < a_1 < a_0$, (ii) $\pi q < b_1 < S_0 < a_1 < a_0$, and (iii) $\pi q < b_1 < a_1 \leq S_0 < a_0$.

  Under (C3)(i), $\pi q < S_0 \leq b_1$ implies that the optimal calling time is

  $$\sigma^* = \inf\{t > 0 : v(X_t) = g_2(X_t)\} = 0,$$

  i.e. no actual exchange occurs at the initial time.
  Under (C3)(ii), one can check that $v(S_0) < v_m(S_0)$ and consequently $c = q + v(S_0) - S_0 < q + v_m(S_0) - S_0 = c_m$. Hence, there is a business opportunity as long as the LTV lies in $\left(\frac{1}{\alpha_1}, \frac{1}{\beta_1}\right) \subset \left(\frac{1}{\alpha_0}, \frac{1}{\pi}\right)$.

Under (C3)(iii), $a_1 \le S_0 < a_0$, so the optimal redeem time is

$$\tau^* = \inf\{t > 0 : v(X_t) = g_1(X_t)\} = 0,$$

neither is there any actual exchange between the two parties at initial time.

In sum, the business opportunity exists when $\pi q < b_1 < S_0 < a_1 < a_0$. In particular, we see that including the callable feature helps to *reduce* the service fee of the loan, but at the cost of a *narrower* range of marketable LTV ratio, i.e. the callable feature rules out some business opportunity that would have existed otherwise. In this case, it is suggested that a loan manager should provide two types of products, one with the margin requirement and another one with the call option. If a client's LTV ratio lies in $\left(\frac{1}{\alpha_0}, \frac{1}{\pi}\right) \setminus \left(\frac{1}{\alpha_1}, \frac{1}{\beta_1}\right)$, then collateralized loan with margin requirement is the only product he can choose. On the other hand, if a client's LTV lies in $\left(\frac{1}{\alpha_1}, \frac{1}{\beta_1}\right)$, whether choosing a collateralized loan with margin requirement (which charges higher service fee and it also takes longer time for the contract to be terminated) or a callable collateralized loan (which charges lower service fee but the contract terminates earlier than a collateralized loan with margin requirement) is decided based on client's preference.

4.2 Call Protection versus Non-recourse Feature

We next provide a comparison between the callable collateralized loan and the non-recourse collateralized loan.

In particular, we shall answer (Q2) in the following two cases:

- $d = 0$, and $0 > \tilde{r} \ge -\frac{\kappa^2}{2}$:

  We first consider the non-recourse collateralized loan. In this case, the lender receives no dividend payment ($d = 0$) from the underlying asset as collateral and the excessive loan rate over the risk-free interest rate is sufficiently low with respect to the volatility of the asset κ ($0 > r - \gamma \ge -\frac{\kappa^2}{2}$). By applying the same arguments in Case(a) of Theorem 3.1 in [31], there would not be any incentive for the lender to do business.

  In contrast, with the additional call feature, Section 4.1 tells us that the lender is willing to start up the business as long as the loan-to-value (LTV) ratio is between $\left(\frac{1}{\alpha_0}, \frac{1}{\pi}\right)$. Thus, the additional call feature provides the lender *more incentive* to do business.

- $d > 0$ or $\tilde{r} < -\frac{\kappa^2}{2}$:

  In this case, it is easy to see that λ_1 in (3.7) is greater than 1. Following the same arguments in Case(b) of Theorem 3.1 in [31], the borrower is willing to borrow the loan as long as the LTV ratio is greater than $\frac{\lambda_1-1}{\lambda_1}$, i.e. $\frac{q}{S_0} > \frac{\lambda_1-1}{\lambda_1}$.

When the lender has an additional right to call back the loan, our Lemma 5.3 states that $y_1 \triangleq F(a_1) < y_0 \triangleq F(a_0) < F\left(\frac{\lambda_1}{\lambda_1-1}q\right)$, which implies that $\frac{q}{a_1} > \frac{q}{a_0} > \frac{\lambda_1-1}{\lambda_1}$. Hence, a *higher* loan-to-value ratio is needed for the borrower to ask for a loan. In other words, adding the callable feature to the non-recourse collateralized loan *diminishes* borrower's incentive to borrow money.

Remark 4.1. From the analysis of Cases (C1)-(C3) in Section 4.1 reveals that the scope of business generated by the imposition of the margin requirement is *at least as great* as the one generated by the adoption of the call protection. Moreover, the analysis in Section 4.2 indicates that the adoption of callable feature to the non-recourse collateralized loan does not *always* imply greater business opportunity over the adoption of the margin requirement. On one hand, the inclusion of callable feature protects the lender from the eroded market value of the asset as a collateral and hence such inclusion induces the lender higher incentive to make the (callable) loan. On the other hand, knowing that inclusion of the call protection serves as a more stringent requirement to the borrower that is absent in the non-recourse loan, the borrower would not agree to ask for the callable collateralized loan unless they can borrow *more* than they would under the non-recourse collateralized loan. The main result of the paper (Theorem 3.1) delineates explicitly this conflicting resolution between the lender and borrower in terms of the model parameters.

4.3 Numerical examples

In this section, we provide examples to numerically illustrate the financial interpretations in the last section. To facilitate comparison with the existing literature, we revisit the example of non-recourse collateralized loan when the collateral is a share of a stock, as discussed extensively in [31].

Example 4.1.

Consider a market model with $r = 0.05$, $\kappa = 0.15$, $\gamma = 0.07$, $d = 0$, $S_0 = \$100$ and $\pi = 0.8$.

Let us first recall the results of the non-recourse stock loan in [31] under the same set model parameters. In this case, $\lambda_1 = 1.7778 > 1$ and the borrower is willing to engage in an non-recourse stock loan as long as the loan-to-value (LTV) ratio is greater than $\frac{\lambda_1-1}{\lambda_1} = 0.4376$, i.e. $\frac{q}{S_0} > 43.76\%$. In other words, borrower is not willing to borrow unless the loan amount is more than 43.76% of the current stock price, S_0. The following is a table of fee charged (c_n) as a function of the principal amount of the non-recourse stock loan, by invoking Theorem 3.1 in [31]:

q	\$50	\$60	\$70	\$80	\$90	\$100	\$110
c_n	\$0.7010	\$3.9976	\$9.0264	\$15.1764	\$22.0971	\$29.5716	\$37.4587

Consider now that the lender can impose either margin requirement or callable feature. Since $\widetilde{r} = -0.02 < -\frac{\kappa^2}{2} = -0.001125$, we are in Case (C2)(iii), which means that the lender is indifferent between imposing margin requirement or callable feature. Hence, it suffices to discuss the fee charged (c) for the stock loan with callable feature only.

By solving (3.12) numerically, we obtain $\alpha_0 = 1.49525$, and hence $1/\alpha_0 = 0.6688$. Therefore, a callable stock loan is marketable if its loan-to-value lies in the interval $\left(\frac{1}{\alpha_0}, \frac{1}{\pi}\right) = (66.88\%, 125\%)$. With the initial value of the stock $S_0 = 100$, the interval $(66.88\%, 125\%)$ translates into the interval $(\$66.88, \$125)$. Note that the borrower would only borrow for the loan amount that is more than 66.88% of the current stock price S_0 in the case of the stock loan with callable feature, as opposed to 43.76% of the current stock price S_0 in the case of non-recourse stock loan. In other words, the borrower demands for larger loan amount in face of the callable feature. In addition, the presence of the call protection ($\pi = 0.8$) also puts an upper bound to the maximum loan amount to be no more than 125% of the current stock price S_0 to protect the lender. In this respect, it is worthy to mention that the non-recourse stock loan offers no such protection.

The following is a table of fee charged (c) as a function of the principal amount of the loan for the callable stock loan:

q	\$70	\$80	\$90	\$100	\$110	\$120
c	\$0.1242	\$1.9376	\$5.3895	\$10.0193	\$15.5263	\$21.7054

By comparing the two tables above, it is clear that the fee charged by the lender in the callable stock loan to be *significantly less* than the fee in the non-recourse stock loan. This is logical as the margin requirement already provides protection to the lender in face of eroded stock price, and hence would correspondingly charge less fee than the non-recourse stock loan of equal loan amount.

Figure 1 provides a graphical illustration of $v(x)$, the initial value of the callable stock loan, computed using (3.10), under Case (C2)(iii). Note that $v(x)$ is always bounded above by $f_1(x) \triangleq (x-(q-\delta))^+$ and below by $f_2(x) \triangleq (x-q)^+$, where $\delta = \pi q = \$64$. Moreover, observe that $\frac{\partial}{\partial x}v(x) = \frac{\partial}{\partial x}v_m(x)$ is not continuous at $x = \pi q = \$64$, i.e. smooth-fit condition fails at $x = \$64$.

Example 4.2.

Consider a market model with $r = 0.05$, $\kappa = 0.15$, $\gamma = 0.02$, $d = 0.015$, $S_0 = \$100$ and $\pi = 0.8$.

Let us discuss the non-recourse stock loans. In this case, $\lambda_1 = 1.4748 > 1$ and the borrower is willing to engage in an non-recourse stock loan as long as the loan-to-value (LTV) ratio is greater than $\frac{\lambda_1-1}{\lambda_1} = 0.3219$, i.e. $\frac{q}{S_0} > 32.19\%$. In other words, borrower is not willing to borrow unless the loan amount is more than 32.19% of the current stock price, S_0. The following is a table of fee charged (c_n) as a function of the principal amount of the non-recourse stock loan, by invoking

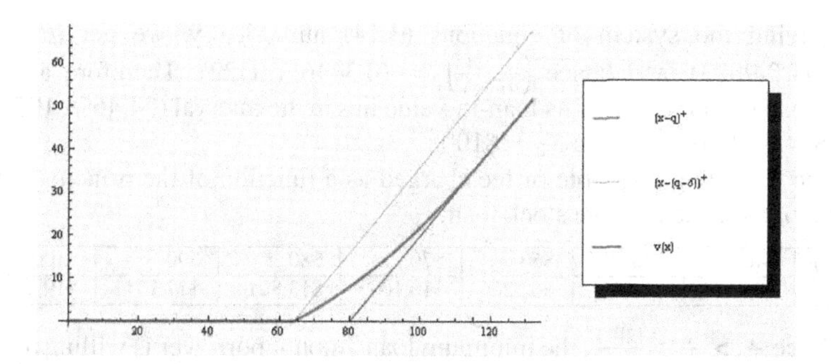

Figure 1. : Graphical illustration of the price of the initial value of the callable stock loan (v) with respect to initial stock price (x) under the set of the model parameters: $r = 0.05$, $\kappa = 0.15$, $\gamma = 0.07$, $d = 0$, $\pi = 0.8$, $q = 80$, and $\delta = 64$.

Theorem 3.1 in [31]

q	\$40	\$50	\$60	\$70	\$80	\$90	\$100
c_n	\$1.1647	\$5.0157	\$10.4534	\$16.8926	\$24.0118	\$31.6180	\$39.5872

Note that this example is different from Example 4.1 in that the excessive loan rate over the risk-free interest rate is negative $\tilde{r} = r - \gamma < 0$ and the lender receives (small) dividend payment $d > 0$. Although the lender receives the dividend payments from the stock, the dividend rate is relatively small. Together with the loan rate he charges to be significantly lower than the risk-free interest rate, the lender would then charge higher fee than the non-recourse stock loan in Example 4.1.

Consider first the fee charged for the stock loan with margin requirement. Since $\widetilde{r} > 0$ and $d = 0.015 \neq 0$, we are under Case (C2)(ii), (C2)(iv), or (C3). Using (3.10), we have

$$\frac{\partial}{\partial x} v_m((\pi q)+, d) = \widetilde{g}_1'(F(a_0))(\lambda_1 - \lambda_2)(\pi q)^{\lambda_1 - 1}$$
$$= \left((1-\lambda_2) a_0^{1-\lambda_1} + \lambda_2 q a_0^{-\lambda_1}\right)(\pi q)^{\lambda_1 - 1} = \pi^{\lambda_1 - 1}\left((1-\lambda_2)\alpha_0^{1-\lambda_1} + \lambda_2 \alpha_0^{-\lambda_1}\right)$$

with α_0 the unique solution to the equation (3.12). Starting with the initial point $d_0 = \frac{\widetilde{r}}{2}$, using bisection method we obtain that $d^* = 0.0183$. Since $d = 0.015 < 0.0183 = d^*$, we are in Case (C3).

By solving (3.12) numerically, we obtain $\alpha_0 = 2.9133$, and hence $1/\alpha_0 = 0.3433$. Therefore a stock loan with margin requirement is marketable if its loan-to-value lies in the interval $(\frac{1}{\alpha_0}, \frac{1}{\pi}) = (34.33\%, 125\%)$, i.e. $(\$34.33, \$125)$ with $S_0 = \$100$.

The following is a table of fee charged (c_m) as a function of the principal amount of the loan for the stock loan with margin requirement:

q	\$40	\$50	\$60	\$70	\$80	\$90	\$100	\$110	\$120
c_m	\$0.4922	\$2.9815	\$6.5195	\$10.4244	\$14.2848	\$17.8214	\$20.8271	\$23.1369	\$24.6129

Solving the system of equations (3.14) numerically, we get $(\beta_1, \alpha_1) = (0.9504, 2.9023)$, and hence $\left(\frac{1}{\alpha_1}, \frac{1}{\beta_1}\right) = (0.3446, 1.0522)$. Therefore a callable stock loan is marketable if its loan-to-value lies in the interval (34.46%, 105.22%), i.e. (\$34.46, \$105.22) with $S_0 = \$100$.

The following is a table of fee charged as a function of the principal amount of the loan for the callable stock loan:

q	\$40	\$50	\$60	\$70	\$80	\$90	\$100
c	\$0.4634	\$2.8841	\$6.3266	\$10.1043	\$13.8014	\$17.1341	\$19.8911

Since $\frac{1}{\alpha_1} > \frac{1}{\alpha_0} > \frac{\lambda_1 - 1}{\lambda_1}$, the minimum loan amount borrower is willing to undertake is highest in the stock loan with callable feature, followed by the stock loan with margin requirement, with the lowest loan amount in the non-recourse stock loan, *ceteris paribus*. It is worthy to mention that the scope of the business opportunity in the stock loan with margin requirement is larger than that in the stock loan with callable feature, i.e. $(\frac{1}{\alpha_1}, \frac{1}{\beta_1}) = (34.46\%, 105\%) \subset (34.33\%, 125\%) = (\frac{1}{\alpha_0}, \frac{1}{\pi})$.

Figure 2 provides a graphical illustration of $v(x)$, the initial value of the callable stock loan, computed using (3.13), under Case (C3). As in Example 4.1, $v(x)$ is always bounded above by $f_1(x) \triangleq (x - (q - \delta))^+$ and below by $f_2(x) \triangleq (x - q)^+$, where $\delta = \pi q = \$64$. While $\frac{\partial}{\partial x} v(x)$ is not continuous at $x = \pi q = \$64$, i.e. smooth-fit condition fails at $x = \$64$, the smooth-fit conditions remain to hold at $x = b_1 \triangleq \beta_1 q = (0.954)(\$80) = \$76.32$ and at $x = a_1 \triangleq \alpha_1 q = (2.9023)(\$80) = \$232.184$.

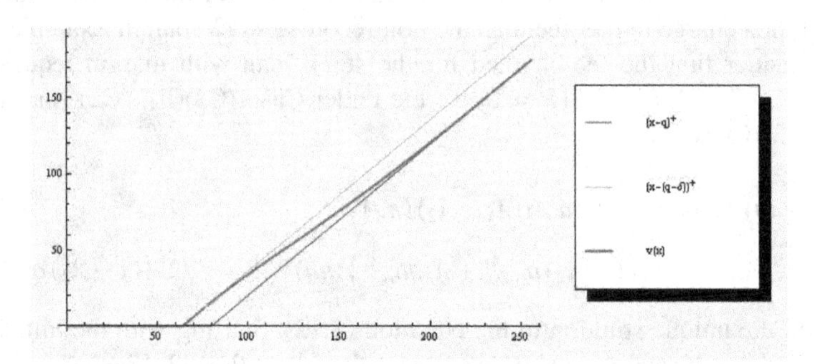

Figure 2. : Graphical illustration of the price of the initial value of the callable stock loan (v) with respect to initial stock price (x) under the set of the model parameters: $r = 0.05$, $\kappa = 0.15$, $\gamma = 0.02$, $d = 0.015$, $\pi = 0.8$, $q = 80$, and $\delta = 64$.

5. Proof of 3.1 and Theorem 3.1

In this section, we shall provide the proof of Lemma 3.1 and Theorem 3.1. We shall first show that Equation (3.12) and (3.11) admit a unique solution which will then be used to prove Lemma 3.1. Next, we shall rigorously prove the existence of the threshold dividend rate d^* defined in (3.9). Finally, we show that Equation (3.14) admits a unique solution under (C3), which will then be used to prove Theorem 3.1.

Recall the two solutions $\psi(x)$ and $\varphi(x)$ of the differential equation $\mathcal{A}f = 0$ given in Definition 3.6, and their ratio $F(x) = \frac{\psi}{\varphi}(x)$ given in (3.8) which was chosen to be an increasing function. For a given function f, we consider the change of variable $y = F(x)$ and the transformation

$$\widehat{f}(y) = \frac{f}{\varphi}\left(F^{-1}(y)\right). \tag{5.1}$$

The above transformation is first introduced in [7] as a tool of transforming an optimal stopping problem for a general diffusion to that of the standard Brownian motion. To further motivate the use of this transformation, let

$$s(x) = \int_1^x \exp\left\{-\int_1^y \frac{2(\widetilde{r}-d)z}{(\kappa z)^2}dz\right\}dy \tag{5.2}$$

be the *scale function* of the underlying diffusion X_t, and let

$$W = W(\psi,\varphi) = \frac{\psi'(x)}{s'(x)}\varphi(x) - \frac{\varphi'(x)}{s'(x)}\psi(x)$$

be the *Wronskian determinant* of $\psi(\cdot)$ and $\varphi(\cdot)$, which is a positive constant. If a mapping $f(x)$ is twice differentiable, then a simple differentiation yields

$$\widehat{f}''(F(x))\,WF'(x) = \mathcal{A}f(x)\,\varphi(x)\,m'(x) \tag{5.3}$$

where $m'(x) = \frac{2}{(\kappa x)^2 s'(x)}$ denotes the density of the *speed measure* of the underlying diffusion X_t. Equation (5.3) implies that f is $\mathcal{A}$-superharmonic (respectively, $\mathcal{A}$-subharmonic) on a interval $\mathcal{I}$ if and only if the transformed function $\widehat{f}$ is concave (respectively, convex) on the transformed interval $F(\mathcal{I})$. In particular, if f is $\mathcal{A}$-harmonic then $\widehat{f}$ is linear[4]. Therefore, by applying this transformation to both payoff functions of borrower and lender and the value function, it helps visualizing the form of the value function and making the mathematical proofs simpler and more transparent.

Moreover, direct computation yields that Equations (3.12) and (3.11) can be expressed in terms of $\widehat{g}_1$ as:

$$\frac{\widehat{g}_1(y_0)}{y_0 - F(\pi q)} = \widehat{g}_1'(y_0) \tag{5.4}$$

[4]When $\widetilde{r} \geq 0$, these results are implied by the standard one-dimensional diffusion theory (see [15]). For $\widetilde{r} < 0$, the validity of (5.3) cannot be covered by the same general theory. Nevertheless, for the infinitesimal generator A considered in this paper, explicit calculation ensures the same result to hold. It is interesting to study to what extent the same result remains true when $\widetilde{r} < 0$.

with $y_0 = F(a_0)$, and similarly, the system of equations (3.14) can be rewritten as

$$\begin{cases} \frac{\widehat{g}_1(y_1)-\widehat{g}_2(z_1)}{y_1-z_1} = \widehat{g}_1'(y_1) \\ \frac{\widehat{g}_1(y_1)-\widehat{g}_2(z_1)}{y_1-z_1} = \widehat{g}_2'(z_1) \end{cases} \tag{5.5}$$

with $y_1 = F(a_1)$ and $z_1 = F(b_1)$. While the explicit expression of $\widehat{g}_1$ and $\widehat{g}_2$ can be given, all we need in this section are the sign of $\widehat{g}_1''$ and $\widehat{g}_2''$, which are recorded below for further references:

	(C1)	(C2)(i)-(iii)	(C2)(iv) and (C3)
$sign\left(\widehat{g}_1''\right)$	$+$ on $(F(q), \infty)$	$-$ on $(F(q), \infty)$	$+$ on $\left(F(q), F\left(\frac{\overline{r}}{d}q\right)\right)$ and $-$ on $\left(F\left(\frac{\overline{r}}{d}q\right), \infty\right)$

Table1. Sign of $\widehat{g}_1''$

	(C1)	(C2)(i)-(iii)	(C2)(iv) and (C3)
$sign\left(\widehat{g}_2''\right)$	$+$ on $(F(\pi q), \infty)$	$-$ on $(F(\pi q), \infty)$	$+$ on $\left(F(\pi q), F\left(\frac{\overline{r}}{d}\pi q\right)\right)$ and $-$ on $\left(F\left(\frac{\overline{r}}{d}\pi q\right), \infty\right)$

Table2. Sign of $\widehat{g}_2''$

The following lemma shows that, under (C2) and (C3), (5.4) admits a unique solution.

Lemma 5.1. *Define* $S(y) \triangleq \frac{\widehat{g}_1(y)}{y-F(\pi q)}$.

1. *Under (C2)(i)-(iii), on $(F(q), \infty)$, $S(y)$ is first increasing, then decreasing and admits a unique maximum point y_0, which is also the unique solution to (5.4) on $(F(q), \infty)$.*

2. *Under (C2)(iv) and (C3), on $\left(F\left(\frac{\overline{r}}{d}q\right), \infty\right)$, $S(y)$ is first increasing, then decreasing and admits a unique maximum point y_0, which also uniquely solves (5.4) on $\left(F\left(\frac{\overline{r}}{d}q\right), \infty\right)$.*

   Moreover, if $\lambda_1 > 1$ then $y_0 < F\left(\frac{\lambda_1}{\lambda_1-1}q\right)$.

Proof. *Note that $S(y)$ gives the slope of the line connecting $(y, \widehat{g}_1(y))$ and $(F(\pi q), 0)$, and equation (5.4) says that there exists a line starting from $(F(\pi q), 0)$ which is also tangent to $\widehat{g}_1$. This follows from the concavity of $\widehat{g}_1$ on $(F(q), \infty)$ for cases (C2)(i)-(iii) and the concavity of $\widehat{g}_1$ on $\left(F\left(\frac{\overline{r}}{d}q\right), \infty\right)$ for cases (C2)(iv) and (C3). A graphical illustration for case 1 and case 2 are given below, and a formal proof is provided in the appendix.* □

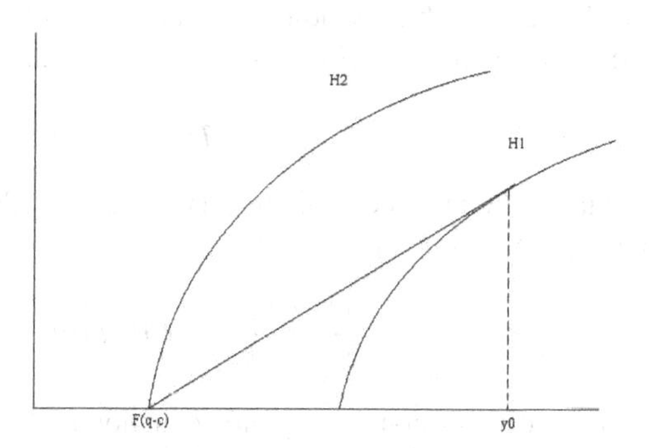

Figure 3. : A graphical illustration of the existence of a line starting at $(F(\pi q), 0)$ and is tangential to $\widehat{g}_1$ for cases (C2)(i)-(iii).

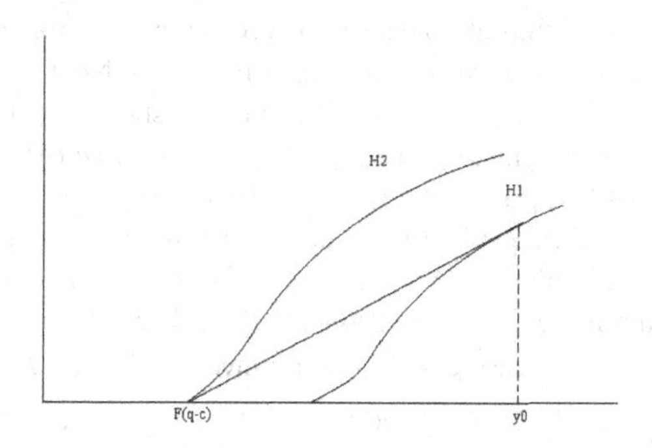

Figure 4. : A graphical illustration of the existence of a line starting at $(F(\pi q), 0)$ and is tangential to $\widehat{g}_1$ for cases (C2)(iv) and (C3).

Using the existence result proved in above lemma, we now provide a rigorous proof for Lemma 3.1.

Proof. [**Proof of Lemma 3.1**] Let v_m be as given in (3.4),

1. This is well known and can be found in [25].

2. We propose that the function v_m solves the following free boundary problem:

$$\mathcal{A}v_m(x) = \widetilde{r}v_m(x) \text{ for } x \in (\pi q, a_0), v_m(\pi q) = 0, v_m(x) = x - q \text{ for } x \geq a_0$$

together with the smooth-fit condition $v_m'(a_0) = g_1'(a) = 1$. Writing $y_0 = F(a_0)$, above free boundary value problem is equivalent to

$$\widehat{v}_m''(y) = 0 \text{ for } y \in (F(\pi q), y_0), \widehat{v}_m(F(\pi q)) = 0, \widehat{v}_m(y) = \widehat{g}_1(y) \text{ for } y \geq y_0$$

together with the smooth-fit condition $\widehat{v}_m'(y_0) = \widehat{g}_1'(y_0)$. Solving this boundary value problem we obtain

$$\widehat{v}_m(y) = \widehat{g}_1(y_0)\frac{y - F(\pi q)}{y_0 - F(\pi q)} \text{ for } F(\pi q) \leq y \leq y_0.$$

Invoking the smooth fit condition yields equation (5.4), which by Lemma 5.1 uniquely determines the boundary y_0. From the transformation $v_m(x) = \widehat{v}_m(F(x))\varphi(x)$ we see that v_m is given by the right hand side of (3.10). To verify v_m obtained above is indeed the solution to the optimal stopping problem (3.4), we need to show that v_m is $(\mathcal{A} - \widetilde{r})$-superharmonic and dominates g_1 on $[\pi q, \infty)$. Note that $(\mathcal{A} - \widetilde{r})v_m(x) = 0$ for $x \in [\pi q, a_0]$, $(\mathcal{A} - \widetilde{r})v_m(x) = rq - dx \leq 0$ for $x \geq a_0$ (since $a_0 \geq q \vee \frac{\widetilde{r}}{d}q$), and v_m is continuously differentiable at a_0. Therefore by a simple application of Ito's lemma v_m is $(\mathcal{A} - \widetilde{r})$-superharmonic on $[\pi q, \infty)$. Next, since $v_m \geq 0 = g_1$ on $[\pi q, q]$, and $v_m = g_1$ on $[a_0, \infty)$, it suffices to show $v_m(x) \geq g_1(x)$ on $[q, a_0]$, which is equivalent to show $\widehat{v}_m(y) \geq \widehat{g}_1(y)$ on $[F(q), y_0]$. Recall that $\widehat{v}_m$ is linear on $[F(q), y_0]$ and $(\widehat{g}_1 - \widehat{v}_m)(y_0) = (\widehat{g}_1 - \widehat{v}_m)'(y_0) = 0$. First consider cases (C2)(i)-(iii). In this case $\widehat{g}_1(y)$ is concave on $[F(q), y_0]$, hence $\widehat{g}_1 - \widehat{v}_m$ is concave on $[F(q), y_0]$ with $(\widehat{g}_1 - \widehat{v}_m)(y_0) = (\widehat{g}_1 - \widehat{v}_m)'(y_0) = 0$, which implies $\widehat{g}_1 - \widehat{v}_m \leq 0$ on $[F(q), y_0]$. Next consider cases (C2)(iv) and (C3). In this case $\widehat{g}_1(y)$ is first convex on $\left[F(q), F\left(\frac{\widetilde{r}}{d}q\right)\right]$ then concave on $\left[F\left(\frac{\widetilde{r}}{d}q\right), y_0\right]$. Then by the same token as above, we have $\widehat{g}_1 - \widehat{v}_m \leq 0$ on $\left[F\left(\frac{\widetilde{r}}{d}q\right), y_0\right]$. Therefore on $\left[F(q), F\left(\frac{\widetilde{r}}{d}q\right)\right]$, $\widehat{g}_1 - \widehat{v}_m$ is a convex function taking negative value at both end points, which immediately implies that $\widehat{g}_1 - \widehat{v}_m \leq 0$ on $\left[F(q), F\left(\frac{\widetilde{r}}{d}q\right)\right]$. □

From now on, let us write $v_m(x, d)$ to denote the value of a collateralized loan with margin requirement on a rate of dividend payment d. The following lemma shows that, the threshold dividend rate d^* defined in (3.9) is well-defined and is bounded above by $\widetilde{r}$.

Lemma 5.2. *For $\widetilde{r} > 0$, $d \longmapsto \frac{\partial}{\partial x}v_m(\pi q+, d)$ is nonincreasing with $\frac{\partial}{\partial x}v_m(\pi q+, \widetilde{r}) < 1$ and $\lim_{d\to 0}\frac{\partial}{\partial x}v_m(\pi q+, d) > 1$. This implies that d^* defined in (3.9) is well-defined and $0 < d^* < \widetilde{r}$.*

Proof. *Note that the mapping $d \longmapsto X_t$ is nonincreasing, from the definition of v_m it is clear $d \longmapsto v_m(x, d)$ is nonincreasing for each $x > \pi q$. By definition*

$\frac{\partial}{\partial x}v_m((\pi q)+,d) = \lim_{x\to(\pi q)+}\frac{v_m(x,d)}{x-(\pi q)}$, therefore $d \longmapsto \frac{\partial}{\partial x}v_m((\pi q)+,d)$ is also non-increasing. By (3.10), we get

$$\frac{\partial}{\partial x}v_m((\pi q)+,d) = \frac{\widehat{g}_1(y_0)}{y_0 - F(\pi q)}(\lambda_1-\lambda_2)(\pi q)^{\lambda_1-1}. \tag{5.6}$$

First let $d = \widetilde{r}$, in which case, $\lambda_1 > 1 > \lambda_2$. Since y_0 solves (5.4) and $\widehat{g}_1''(y) < 0$ for $y > F(q)$, (5.6) reads as

$$\begin{aligned}\frac{\partial}{\partial x}v_m((\pi q)+,d) &= \widehat{g}_1'(y_0)(\lambda_1-\lambda_2)(\pi q)^{\lambda_1-1}\\ &< \widehat{g}_1'(F(q))(\lambda_1-\lambda_2)(\pi q)^{\lambda_1-1} = \left(\frac{\pi q}{q}\right)^{\lambda_1-1} < 1,\end{aligned}$$

Next let $d < \widetilde{r}$. Since by Lemma 5.1, $S(y) = \frac{\widehat{g}_1(y)}{y-F(\pi q)}$ is increasing on $\left[F\left(\frac{\widetilde{r}}{d}q\right), y_0\right]$, (5.6) becomes

$$\begin{aligned}\frac{\partial}{\partial x}v_m((\pi q)+,d) &\geq \frac{\widehat{g}_1\left(F\left(\frac{\widetilde{r}}{d}q\right)\right)}{F\left(\frac{\widetilde{r}}{d}q\right) - F(\pi q)}(\lambda_1-\lambda_2)(\pi q)^{\lambda_1-1}\\ &= \frac{\widetilde{r}^{1-\lambda_2} - \widetilde{r}^{-\lambda_2}d}{\widetilde{r}^{\lambda_1-\lambda_2}d^{1-\lambda_1} - \left(\frac{\pi q}{q}\right)^{\lambda_1-\lambda_2}}(\lambda_1-\lambda_2)\left(\frac{\pi q}{q}\right)^{\lambda_1-1}.\end{aligned}$$

When $d \to 0$, $\lambda_1 \to 1$ and $\lambda_2 \to -\frac{2\widetilde{r}}{\kappa^2} < 0$, and using elementary calculus one can verify that $\lim_{d\to0} d^{\lambda_1-1} = 1$, therefore

$$\begin{aligned}&\lim_{d\to0}\frac{\widetilde{r}^{1-\lambda_2} - \widetilde{r}^{-\lambda_2}d}{\widetilde{r}^{\lambda_1-\lambda_2}d^{1-\lambda_1} - \left(\frac{\pi q}{q}\right)^{\lambda_1-\lambda_2}d^{1-\lambda_2}}(\lambda_1-\lambda_2)\left(\frac{\pi q}{q}\right)^{\lambda_1-1}\\ &= \lim_{d\to0} d^{\lambda_1-1}\left(1+\frac{2\widetilde{r}}{\kappa^2}\right) = 1+\frac{2\widetilde{r}}{\kappa^2}.\end{aligned}$$

Since $\tilde{r} > 0$, we can conclude that $\lim_{d\to0}\frac{\partial}{\partial x}v_m((\pi q)+,d) > 1$. □

Above lemma in particular implies that, under (C2)(iv) we have $\frac{\partial}{\partial x}v_m(\pi q+) \leq 1$, and under (C3) we have $\frac{\partial}{\partial x}v_m(\pi q+) > 1$. From Lemma 3.1, we know that $(\pi q, a_0)$ is the continuous region for v_m. Therefore for all $x \in (\pi q, a_0)$,

$$\frac{\kappa^2}{2}x^2 v_m''(x) + (\widetilde{r} - d)xv_m'(x) - \widetilde{r}v_m(x) = 0. \tag{5.7}$$

Under Cases (C2)(i)-(iii), we have $d \geq \widetilde{r}$, then above equation suggests that $v_m(x)$ is strictly convex in the continuous region. From Lemma 3.1 we also know that

$v'_m(a_0) = 1$ (actually this is how a_0 is determined). Combined with the strict convexity of v_m, this implies that under (C2)(i)-(iii), we also have $\frac{\partial}{\partial x} v_m(\pi q+) < 1$. Finally, for Case (C1), using the closed form expression given in Lemma 3.1, we have $\frac{\partial}{\partial x} v_m(\pi q+) = 1 + \frac{2\bar{r}}{\kappa^2} > 1$. Combining these arguments, we see that Case (C2) precisely corresponds to the case $\frac{\partial}{\partial x} v_m(\pi q+) \le 1$, and (C1) and (C3) together correspond to the case $\frac{\partial}{\partial x} v_m(\pi q+) > 1$. The next lemma shows that, under (C3), the system of transcendental equations (5.5) admits a unique pair of solutions.

Lemma 5.3. *Under (C3), (5.5) admits a unique pair of solutions (z_1, y_1) on the interval*
$\left(F(\pi q), F\left(\frac{\bar{r}}{d}\pi q\right)\right) \times \left(F\left(\frac{\bar{r}}{d}q\right), \infty\right)$. *Moreover, $y_1 < y_0 < F\left(\frac{\lambda_1}{\lambda_1 - 1}q\right)$.*

Proof. *Equation (5.5) says there exists a line being tangent to both $\widehat{g}_1$ and $\widehat{g}_2$. This follows from the convexity of $\widehat{g}_2$ on $\left(F(\pi q), F\left(\frac{\bar{r}}{d}\pi q\right)\right)$ and the concavity of $\widehat{g}_1$ on $\left(F\left(\frac{\bar{r}}{d}q\right), \infty\right)$. A graphical illustration is given below, and a formal proof is given in the appendix. We emphsize that, the condition $\frac{\partial}{\partial x} v_m(\pi q+) > 1$ plays an important role in establishing the existence result.* □

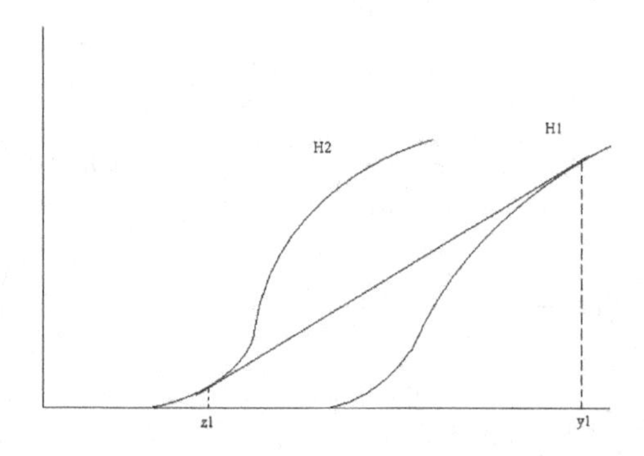

Figure 5. : A graphical illustration of the existence of a line tangential to both $\widehat{g}_1$ and $\widehat{g}_2$.

We are now in a position to prove our main theorem.

Proof. [**Proof of Theorem 3.1**] Let v be as defined in (3.5). In each of the cases below let $\tau^* = \inf\{t > 0 : v(X_t) = g_1(X_t)\}$ and $\sigma^* = \inf\{t > 0 : v(X_t) = g_2(X_t)\}$. As in the proof of part (ii) of theorem 2 in Kyprianou [21], if we can verify that $e^{-\bar{r}t\wedge\sigma^*} v(X_{t\wedge\sigma^*})$ is a supermartingale, $e^{-\bar{r}t\wedge\tau^*} v(X_{t\wedge\tau^*})$ is a submartingale and $g_1(x) \le v(x) \le g_2(x)$ for all x with $v(X_{\tau^*}) = g_1(X_{\tau^*})\, a.s.$ and $v(X_{\sigma^*}) =$

$g_2(X_{\sigma^*})\, a.s.$, then v is indeed the value of the game and (σ^*, τ^*) is the saddle point of the game.

1. Under (C1), $\widetilde{r} \geq 0$ and $d = 0$. In this case $v(x) = g_2(x) = (x - \pi q)^+$ and $(\mathcal{A} - \widetilde{r})\, v(x) = \widetilde{r}\pi q \geq 0$. Let $\tau^* = \inf\{t > 0 : X_t \leq \pi q\}$ and $\sigma^* = 0$. Then it is straightforward to check that $e^{-\widetilde{r}t\wedge\sigma^*} v(X_{t\wedge\sigma^*})$ is a supermartingale, $e^{-\widetilde{r}t\wedge\tau^*} v(X_{t\wedge\tau^*})$ is a submartingale and $g_1(x) \leq v(x) \leq g_2(x)$ for all x with $v(X_{\tau^*}) = g_1(X_{\tau^*})$ and $v(X_{\sigma^*}) = g_2(X_{\sigma^*})$.

2. Under (C2), we have (i) $\widetilde{r} = -\frac{\kappa^2}{2}$ and $d = 0$, or (ii) $0 > \widetilde{r} > -\frac{\kappa^2}{2}$ and $d = 0$, (iii) $\{d \geq \widetilde{r}\} / \left\{0 \geq \widetilde{r} \geq -\frac{\kappa^2}{2}, d = 0\right\}$ and (iv) $\widetilde{r} > 0$ and $\widetilde{r} > d \geq d^*$. In this case $v = v_m$, $\tau^* \triangleq \inf\{t > 0 : X_t \geq a_0\}$ and $\sigma^* = \inf\{t > 0 : X_t \leq \pi q\}$. In the proof of Lemma 3.1, we already show that v_m is $(\mathcal{A} - \widetilde{r})$-superharmonic on $[\pi q, \infty)$ and $g_1(x) \leq v_m(x)$ for x. And the two equalities $v(X_{\tau^*}) = g_1(X_{\tau^*})\, a.s.$ and $v(X_{\sigma^*}) = g_2(X_{\sigma^*})\, a.s.$ are clearly true. Therefore it remains to show that $e^{-\widetilde{r}t\wedge\tau^*} v_m(X_{t\wedge\tau^*})$ is a submartingale and $v_m(x) \leq g_2(x)$ for all $x \geq \pi q$. From (3.10) it is clear that $(\mathcal{A} - \widetilde{r})\, v_m(x) = 0$ on $(0, \pi q) \cup (\pi q, a_0)$ and $v_m'(\pi q+) > 0 = v_m'(\pi q-)$, therefore if we let $L_t^{\pi q}$ be the local time of X at πq, then by a generalized Ito formula (see e.g., problem 6.24, page 215 in [18]),

$$e^{-\widetilde{r}t\wedge\tau^*} v_m(X_{t\wedge\tau^*}) = v_m(x) + \int_0^{t\wedge\tau^*} e^{-\widetilde{r}s} v_m'(X_s)\, \kappa X_s dW_s + v_m'(\pi q+) \int_0^{t\wedge\tau^*} e^{-\widetilde{r}s} dL_s^{\pi q},$$

which implies $e^{-\widetilde{r}t\wedge\tau^*} v_m(X_{t\wedge\tau^*})$ is indeed a submartingale. Next, since $v_m = g_2 = 0$ on $(0, \pi q]$ and $v_m = g_1 \leq g_2$ on $[a_0, \infty)$, it suffices to show $v_m(x) \leq g_2(x)$ on $[\pi q, a_0]$, which is equivalent to show $\widehat{v}_m(y) \leq \widehat{g}_2(y)$ on $[F(\pi q), y_0]$. Recall that $\widehat{v}_m$ is linear on $[F(\pi q), y_0]$, $\widehat{v}_m(F(\pi q)) = \widehat{g}_2(F(\pi q)) = 0$ and $\widehat{v}_m(y_0) < \widehat{g}_2(y_0)$. Let us first consider the case (C2)(i)-(iii). In this case $\widehat{g}_2$ is concave on $[F(\pi q), y_0]$. Therefore on $[F(\pi q), y_0]$ $\widehat{g}_2 - \widehat{v}_m$ is a concave function taking nonnegative value at both end points, which implies that $\widehat{g}_2 - \widehat{v}_m \geq 0$ on $[F(\pi q), y_0]$. Next consider the case (C2)(iv). In this case $\widehat{g}_2(y)$ is first convex on $\left[F(\pi q), F\left(\frac{\widetilde{r}}{d}\pi q\right)\right]$ then concave on $\left[F\left(\frac{\widetilde{r}}{d}\pi q\right), y_0\right]$. Since $d \geq d^*$, $v_m'((\pi q)+) \leq 1$, which is equivalent to the condition $\widehat{v}_m'(F(\pi q)+) \leq \widehat{g}_2'(F(\pi q)+)$. Therefore on $\left[F(\pi q), F\left(\frac{\widetilde{r}}{d}\pi q\right)\right]$ $\widehat{g}_2 - \widehat{v}_m$ is a convex function with $(\widehat{g}_2 - \widehat{v}_m)(F(\pi q)) = 0$ and $(\widehat{g}_2 - \widehat{v}_m)'(F(\pi q)) \geq 0$, which implies that $\widehat{g}_2 \geq \widehat{v}_m$ on $\left[F(\pi q), F\left(\frac{\widetilde{r}}{d}\pi q\right)\right]$. In particular this implies $(\widehat{g}_2 - \widehat{v}_m)\left(F\left(\frac{\widetilde{r}}{d}\pi q\right)\right) \geq 0$. Finally on $\left[F\left(\frac{\widetilde{r}}{d}\pi q\right), y_0\right]$ one can apply the same argument as above to conclude that $\widehat{g}_2 \geq \widehat{v}_m$ on $\left[F\left(\frac{\widetilde{r}}{d}\pi q\right), y_0\right]$.

3. Under (C3), we have $\rho > 0$ and $d^* > d > 0$. In this case, let us first note that since $d < d^*$, $v_m'((\pi q)+) > 1 = g_2'((\pi q)+)$, therefore v_m is larger than g_2 in

a right neighborhood of πq. Therefore v_m can not be the value of the game. Given this observation, we propose that the function v solves the following free boundary problem:

$$\mathcal{A}v(x) = \widetilde{r}v(x) \text{ for } x \in (b_1, a_1), v(x) = g_2(x) \text{ for } x \leq b_1, v(x) = g_1(x) \text{ for } x \geq a_1$$

together with smooth fit conditions $v'(b_1) = v'(a_1) = 1$. Writing $z_1 = F(b_1)$ and $y_1 = F(a_1)$, above free boundary problem is equivalent to

$$\widehat{v}''(y) = 0 \text{ for } y \in (z_1, y_1), \widehat{v}(y) = \widehat{g}_2(y) \text{ for } y \leq z_1, \widehat{v}(y) = \widehat{g}_1(y) \text{ for } y \geq y_1$$

together with smooth fit conditions $\widehat{v}'(z_1) = \widehat{g}_2'(z_1)$ and $\widehat{v}'(y_1) = \widehat{g}_1'(y_1)$. Solving this boundary value problem we obtain

$$\widehat{v}(y) = \frac{y - z_1}{y_1 - z_1}\widehat{g}_1(y_1) + \frac{y_1 - y}{y_1 - z_1}\widehat{g}_2(z_1).$$

Invoking the two smooth fit conditions yields the system of equations (5.5), which by Lemma 5.3 uniquely determins the pair of boundaries (z_1, y_1). And from the transformation $v(x) = \widehat{v}(F(x))\varphi(x)$ we see that v is given by the right hand side of (3.13). Then in this case $\tau^* = \inf\{t > 0 : X_t \geq a_1\}$ and $\sigma^* = \inf\{t > 0 : X_t \leq b_1\}$, and we need to verify that v is indeed the value of the game. Firsly the two equalities $v(X_{\tau^*}) = g_1(X_{\tau^*})\, a.s.$ and $v(X_{\sigma^*}) = g_2(X_{\sigma^*})\, a.s.$ are clearly true. Then since $(\mathcal{A} - \widetilde{r})v(x) = 0$ for $x \in (b_1, a_1)$, $(\mathcal{A} - \widetilde{r})v(x) = \widetilde{r}q - dx \leq 0$ for $x > a_1$ (since $a_1 \geq \frac{\widetilde{r}}{d}q$), and v is continuously differentiable at a_1. Therefore by Ito's lemma $e^{-\widetilde{r}t\wedge\sigma^*}v(X_{t\wedge\sigma^*})$ is a supermartingale. Moreover, Since $(\mathcal{A} - \widetilde{r})v(x) = 0$ for $x \in (0, \pi q) \cup (b_1, a_1)$, $(\mathcal{A} - \widetilde{r})v(x) = \widetilde{r}\pi q - dx \geq 0$ for $x \in (\pi q, b_1)$ (since $b_1 \leq \frac{\widetilde{r}}{d}\pi q$), v is continuously differentiable at b_1 and $v'(\pi q+) > 0 = v'(\pi q-)$, by Ito's lemma

$$\begin{aligned} e^{-\widetilde{r}t\wedge\tau^*}v(X_{t\wedge\tau^*}) &= v(x) + \int_0^{t\wedge\tau^*} e^{-\widetilde{r}s}(\widetilde{r}\pi q - dX_s)\mathbf{1}_{(\pi q, b_1)}(X_s)\,ds \\ &+ \int_0^{t\wedge\tau^*} e^{-\widetilde{r}s}v'(X_s)\kappa X_s dW_s + v_m'(\pi q+)\int_0^{t\wedge\tau^*} e^{-\widetilde{r}s}dL_s^{\pi q}, \end{aligned}$$

from which we see that $e^{-\widetilde{r}t\wedge\tau^*}v(X_{t\wedge\tau^*})$ is a submartingale. To show $g_1(x) \leq v(x) \leq g_2(x)$ for all x, it is equivalent to show that $\widehat{g}_1(y) \leq \widehat{v}(y) \leq \widehat{g}_2(y)$ for all y. Since $\widehat{v} = \widehat{g}_1 \leq \widehat{g}_2$ on $[y_1, \infty)$ and $\widehat{v} = \widehat{g}_2 \geq \widehat{g}_1$ on $(0, z_1]$, we only need to show $\widehat{g}_1 \leq \widehat{v} \leq \widehat{g}_2$ on $[z_1, y_1]$. Note that $\widehat{v}$ is linear on $[z_1, y_1]$, while $\widehat{g}_1$ is first convex on $\left[z_1, F\left(\frac{\widetilde{r}}{d}q\right)\right]$ then concave on $\left[F\left(\frac{\widetilde{r}}{d}q\right), y_1\right]$. Then $\widehat{g}_1 - \widehat{v}$ is concave on $\left[F\left(\frac{\widetilde{r}}{d}q\right), y_1\right]$ with $(\widehat{g}_1 - \widehat{v})(y_1) = (\widehat{g}_1 - \widehat{v})'(y_1) = 0$, which implies $\widehat{g}_1 - \widehat{v} \leq 0$ on $\left[F\left(\frac{\widetilde{r}}{d}q\right), y_1\right]$. And on $\left[z_1, F\left(\frac{\widetilde{r}}{d}q\right)\right]$ $\widehat{g}_1 - \widehat{v}$ is a convex function taking negative value at both end points, which immediately implies that

$\widehat{g}_1 - \widehat{v} \le 0$ on $\left[z_1, F\left(\frac{\bar{r}}{d}q\right)\right]$. Next, note that $\widehat{g}_2$ is first convex on $\left[z_1, F\left(\frac{\bar{r}}{d}\pi q\right)\right]$ then concave on $\left[F\left(\frac{\bar{r}}{d}\pi q\right), y_1\right]$. Then $\widehat{g}_2 - \widehat{v}$ is convex on $\left[z_1, F\left(\frac{\bar{r}}{d}\pi q\right)\right]$ with $(\widehat{g}_2 - \widehat{v})(z_1) = (\widehat{g}_2 - \widehat{v})'(z_1) = 0$, which implies $\widehat{g}_2 - \widehat{v} \ge 0$ on $\left[z_1, F\left(\frac{\bar{r}}{d}\pi q\right)\right]$. And on $\left[F\left(\frac{\bar{r}}{d}\pi q\right), y_1\right]$ $\widehat{g}_2 - \widehat{v}$ is a concave function taking positive value at both end points, which immediately implies that $\widehat{g}_2 - \widehat{v} \ge 0$ on $\left[F\left(\frac{\bar{r}}{d}\pi q\right), y_1\right]$. This completes the proof. □

6. Conclusion and future work

In the present paper, we examine the impact of the margin requirement and callable feature on the value and service fee of a collateralized loan, and study which one should a loan manager choose to include in the contract. By first identifying respective cases when the smooth-fit condition holds or fails across the lower exercising boundary, and then in each case solving for the corresponding free boundary value problems with appropriate boundary conditions, we rigorously identified the closed-form solution to a perpetual zero-sum Dynkin game as the fair value of a callable collateralized loan. The explicit solution is then applied to classify the ranges of the model parameters so that: 1) including the callable feature is more beneficial; 2) including margin requirement is more preferable; 3) both options are indifferent to each other. As in the case of the non-recourse collateralized loan, dividend payment from the asset as collateral plays an indispensable role in creating business opportunity in the cases of the collateralized loans with callable feature and margin requirement, without which the business opportunities may be significantly diminished. The explicit values of the collateralized loans with callable feature and margin requirement in this paper provide a transparent and systematic guideline to the loan manager in choosing these collateralized loans for his clients, offering clients more selections on the loan features than the case of the non-recourse collateralized loans.

The closed-form solution of the callable collateralized loan provides an efficient channel to analyze the scope of business opportunities generated from the call protection against that from the non-recourse collateralized loan. An important implication drawn from this comparative analysis is that embedding callable feature in the collateralized loan certainly induces incentive for the lender to carry out business for the callable feature provides protection from the eroded market value of the collateral. Yet, the same call protection may also stimulate the borrower to increase the size of the loan than he would in the non-recourse collateralized loan. This competing nature between lender and borrower can only be resolved under specific model and this paper provides a complete analysis under the Black-Scholes economy.

In the present paper, we discuss different types of collateralized loans under the complete, frictionless market, in which every contingent claim can be replicated by trading in the stock and money market. In reality, the market is not frictionless market as assumed by the Black-Scholes framework. In this case, the valuation of a collateralized loan can treated by (utility-)indifference pricing arguments from the point of view of the borrowers, as discussed in [14] for the non-recourse stock loans under the incomplete markets. It would be very interesting to see whether the present work could be extended to the setting of incomplete market. In particular, it is of interest to see how a callable collateralized loan can be evaluated by using indifference pricing arguments; whether a closed-form solution would still be available; and what the preference of the lender is over the adoption of margin requirement or callable feature in the collateralized loan contract.

7. Acknowledgement

The authors are grateful for the valuable comments and his insightful suggestions on an earlier version of this paper from Alain Bensoussan. We are also grateful to many other people who offered insights into this work, including seminar participants at National University of Singapore, The University of Hong Kong, and conference participants at The World Congress of Bachelier Finance Society. Phillip Yam acknowledges the financial support from The Hong Kong RGC GRF 404012. Chi Chung Siu acknowledges Professor Masaaki Kijima, Professor Takashi Shibata, and Professor Yukio Muromachi for their kind invitation to the Tokyo Metropolitan University (TMU) Finance Workshop 2014.

8. Appendix

Proof. [Proof of Lemma 5.1] Using the explicit expression of ψ and φ a direct computation yields that, for $y > F(q)$,

$$\widehat{g}_1(y) = \begin{cases} 1 - qe^{-y} & \text{for Case (C2)(i)} \\ y^{\frac{1-\lambda_2}{\lambda_1-\lambda_2}} - qy^{-\frac{\lambda_2}{\lambda_1-\lambda_2}} & \text{for Cases (C2)(ii)-(iv) and (C3)} \end{cases}, \tag{8.1}$$

and

$$\widehat{g}_1'(y) = \begin{cases} qe^{-y} & \text{for Case (C2)(i)} \\ \frac{(1-\lambda_2)}{\lambda_1-\lambda_2} y^{\frac{1-\lambda_1}{\lambda_1-\lambda_2}} + \frac{\lambda_2}{\lambda_1-\lambda_2} qy^{-\frac{\lambda_1}{\lambda_1-\lambda_2}} & \text{for Cases (C2)(ii)-(iv) and (C3)} \end{cases}, \tag{8.2}$$

Let us also define $R(y) = \widehat{g}_1'(y)(y - F(\pi q)) - \widehat{g}_1(y)$, then $S'(y) = \frac{R(y)}{(y-F(\pi q))^2}$.

1. Consider cases (C2)(i)-(iii). Since $\widehat{g}_1(F(q)) = 0$, $R(F(q)) = \widehat{g}_1'(F(q))(F(q) - F(\pi q)) - \widehat{g}_1(F(q)) > 0$. From Table 1 we see that $R'(y) = \widehat{g}_1''(y)(y - F(\pi q)) < 0$ for $y > F(q)$. In each case below we will prove that, $\lim_{y\to\infty} R(y) < 0$. Then the function $R(y)$ decreases from a positive number to a negative number, and hence has a unique zero y_0, i.e., y_0 is the unique solution to $\frac{\widehat{g}_1(y)}{(y-F(\pi q))} = \widehat{g}_1'(y)$ on $[F(q), \infty)$. Therefore $S(y)$ is

first increasing then decreasing, with a unique maximum point at y_0 over $[F(q), \infty)$.

Case (C2)(i): $d = 0$ and $\widetilde{r} = -\frac{\kappa^2}{2}$. In this case for $y > F(q)$, $\widehat{g}_1'(y) = qe^{-y}$. Thus

$$\lim_{y\to\infty} R(y) = \lim_{y\to\infty} \left[qe^{-y}(y - F(\pi q)) - (1 - qe^{-y})\right] = -1 < 0.$$

Case (C2)(ii): $d = 0$ and $0 > \widetilde{r} > -\frac{\kappa^2}{2}$. In this case $\lambda_1 = 1 > \lambda_2 > 0$. By (8.1) and (8.2), we have $\lim_{y\to\infty} \widehat{g}_1'(y) = 1$, hence

$$\lim_{y\to\infty} R(y) = \lim_{y\to\infty} R(y) = \lim_{y\to\infty} \frac{1}{1-\lambda_2} y^{-\frac{\lambda_2}{1-\lambda_2}} - F(\pi q) = -F(\pi q) < 0.$$

Case (C2)(iii): $\{d \geq \widetilde{r}\} / \left\{0 \geq \widetilde{r} \geq -\frac{\kappa^2}{2}, d = 0\right\}$. In this case we have $\lambda_1 > 1 > \lambda_2$, then by (8.1) and (8.2) we have $\lim_{y\to\infty} \widehat{g}_1'(y) = 0$ and hence $\lim_{y\to\infty} R(y) = \lim_{y\to\infty} \frac{1-\lambda_1}{\lambda_1-\lambda_2} y^{\frac{1-\lambda_2}{\lambda_1-\lambda_2}} + \frac{\lambda_1}{\lambda_1-\lambda_2} y^{-\frac{\lambda_2}{\lambda_1-\lambda_2}} = -\infty$.

2. Consider cases (C2)(iv) and (C3). In this case $\lambda_1 > 1 > \lambda_2$, then use (8.1) and (8.2) one can verify that $\lim_{y\to\infty} \widehat{g}_1'(y) = 0$. Hence

$$\lim_{y\to\infty} R(y) = \lim_{y\to\infty} \frac{1-\lambda_1}{\lambda_1-\lambda_2} y^{\frac{1-\lambda_2}{\lambda_1-\lambda_2}} + \frac{\lambda_1}{\lambda_1-\lambda_2} y^{-\frac{\lambda_2}{\lambda_1-\lambda_2}} = -\infty. \tag{8.3}$$

By Table1, $\widehat{g}_1''(y) < 0$ for $y > F\left(\frac{\widetilde{r}}{d} q\right)$, thus

$$R'(y) = \widehat{g}_1''(y)(y - F(\pi q)) < 0 \text{ for } y > F(\pi q). \tag{8.4}$$

Also from Table 1, $\widehat{g}_1''(y) > 0$ for $F(q) < y < F\left(\frac{\widetilde{r}}{d} q\right)$, and since $\widehat{g}_1(F(q)) = 0$

$$\begin{aligned} &R\left(F\left(\frac{\widetilde{r}}{d} q\right)\right) \\ &= \widehat{g}_1'\left(F\left(\frac{\widetilde{r}}{d} q\right)\right)\left(F\left(\frac{\widetilde{r}}{d} q\right) - F(\pi q)\right) - \widehat{g}_1'(\zeta)\left(F\left(\frac{\widetilde{r}}{d} q\right) - F(q)\right) > 0, \end{aligned} \tag{8.5}$$

where in the equality we applied mean value theorem. Combining (8.3), (8.4) and (8.5), we see that the function $R(y)$ has a unique zero y_0 on $\left[F\left(\frac{\widetilde{r}}{d} q\right), \infty\right)$, i.e., y_0 is the unique solution to $\frac{\widehat{g}_1(y)}{(y-F(\pi q))} = \widehat{g}_1'(y)$ on $\left[F\left(\frac{\widetilde{r}}{d} q\right), \infty\right)$. Therefore $S(y)$ is first increasing then decreasing, with a unique maximum point at y_0 over $\left[F\left(\frac{\widetilde{r}}{d} q\right), \infty\right)$.

Finally, note that if $\lambda > 1$, $F\left(\frac{\lambda_1}{\lambda_1 - 1}q\right)$ solves the equation $\widehat{g}_1'(y)\,y - \widehat{g}_1(y) = 0$, which implies $R\left(F\left(\frac{\lambda_1}{\lambda_1-1}q\right)\right) < 0$, therefore $y_0 < F\left(\frac{\lambda_1}{\lambda_1-1}q\right)$. □

Proof. [Proof of Lemma 5.3] Define $R_1(y,z) = \widehat{g}_1'(y)(y - z) - (\widehat{g}_1(y) - \widehat{g}_2(z))$. And write $z_l = F(\pi q)$ and $z_r = F\left(\frac{\widetilde{r}}{d}(\pi q)\right)$. By writing down explicite expression for $\widehat{g}_1$ one can check that $\lim_{y\to\infty} \widehat{g}_1'(y) = 0$ and hence $\lim_{y\to\infty} R_1(y,z) = -\infty$ for each z. According to Table 1, we also have $\frac{\partial}{\partial y}R_1(y,z) = \widehat{g}_1''(y)(y-z) < 0$ for $y > F\left(\frac{\widetilde{r}}{d}q\right)$, and

$$\begin{aligned} R_1\left(F\left(\frac{\widetilde{r}}{d}q\right), z\right) &> \widehat{g}_1'\left(F\left(\frac{\widetilde{r}}{d}q\right)\right)\left(F\left(\frac{\widetilde{r}}{d}q\right) - z\right) - \left(\widehat{g}_1\left(F\left(\frac{\widetilde{r}}{d}q\right)\right) - \widehat{g}_1(z)\right) \\ &\geq \widehat{g}_1'\left(F\left(\frac{\widetilde{r}}{d}q\right)\right)\left(F\left(\frac{\widetilde{r}}{d}q\right) - z\right) - \widehat{g}_1'\left(F\left(\frac{\widetilde{r}}{d}q\right)\right)\left(F\left(\frac{\widetilde{r}}{d}q\right) - z\right) = 0, \end{aligned}$$

where the first inequality holds since $\widehat{g}_2(z) > \widehat{g}_1(z)$ for all $z > z_l$, the second inequality comes from the convexity of $\widehat{g}_1$ on $\left(0, F\left(\frac{\widetilde{r}}{d}q\right)\right)$. Thus for each $z \in [z_l, z_r]$, $y \longmapsto R_1(y,z)$ decreases from a positive value to negative infinity, and hence admits a unique zero $y(z)$ on $\left(F\left(\frac{\widetilde{r}}{d}q\right), \infty\right)$. Moreover, by Implicit Function Theorem the mapping $z \longmapsto y(z)$ is C^1 with $y'(z) = -\left.\frac{\frac{\partial}{\partial z}R_1(y,z)}{\frac{\partial}{\partial y}R_1(y,z)}\right|_{y=y(z)} = -\frac{\widehat{g}_2'(z) - \widehat{g}_1'(y(z))}{\widehat{g}_1''(y(z))(y(z)-z)}$. We remark that when $z = z_l$, $y(z_l) = y_0$ with y_0 given in Lemma 5.1. Next let us define $R_2(y,z) = \widehat{g}_2'(z)(y - z) - (\widehat{g}_1(y) - \widehat{g}_2(z))$, and $R(z) = R_2(y(z), z)$. We want to show that $R(z)$ has a unique zero in the interval(z_l, z_r). First note that for $z \in (z_l, z_r)$

$$\begin{aligned} R'(z) &= \frac{\partial}{\partial y} R_2(y,z)|_{y=y(z)} \cdot y'(z) + \frac{\partial}{\partial z} R_2(y,z)|_{y=y(z)} \\ &= -\frac{\left(\widehat{g}_2'(z) - \widehat{g}_1'(y(z))\right)^2}{\widehat{g}_1''(y(z))(y(z) - z)} + \widehat{g}_2''(z)(y(z) - z) > 0 \end{aligned} \tag{8.6}$$

since according to Table 1 and 2 $\widehat{g}_1''(y(z)) < 0$ as $y(z) > F(q)$, and $\widehat{g}_2''(z) > 0$ as $z \in (z_l, z_r)$. Under (C3), $g_2'(\pi q) = 1 < \frac{\partial}{\partial x}v_m(\pi q+)$, which implies $\widehat{g}_2'(F(\pi q)) < \widehat{v}_m'(F(\pi q)) = \widehat{v}_m'(y_0) = \widehat{g}_1'(y_0)$. Hence

$$\begin{aligned} R(z_l) &= R_2(y_0, F(\pi q)) \\ &= \widehat{g}_2'(F(\pi q))(y_0 - F(\pi q)) - \widehat{g}_1(y_0) \\ &< \widehat{g}_1'(y_0)(y_0 - F(\pi q)) - \widehat{g}_1(y_0) = 0. \end{aligned} \tag{8.7}$$

For $z > z_l$,

$$\widehat{g}_2'(z) = \frac{z^{-\frac{\lambda_1}{\lambda_1-\lambda_2}}}{\lambda_1 - \lambda_2}\left((1-\lambda_2)z^{\frac{1}{\lambda_1-\lambda_2}} + \lambda_2(\pi q)\right), \tag{8.8}$$

By comparing (8.8) and (8.2), we see that $\widehat{g}_2'(y) > \widehat{g}_1'(y)$ for all $y > F(q)$. And by Table 2, $\widehat{g}_2''(y) < 0$ for $y > z_r$, hence

$$\begin{aligned} R(z_r) &= R_2(y(z_r), z_r) \\ &= \widehat{g}_2'(z_r)(y(z_r) - z_r) - (\widehat{g}_1(y(z_r)) - \widehat{g}_2(z_r)) \\ &> \widehat{g}_2'(y(z_r))(y(z_r) - z_r) - (\widehat{g}_1(y(z_r)) - \widehat{g}_2(z_r)) \\ &> \widehat{g}_1'(y(z_r))(y(z_r) - z_r) - (\widehat{g}_1(y(z_r)) - \widehat{g}_2(z_r)) = 0. \end{aligned} \tag{8.9}$$

Combining (8.6), (8.7) and (8.9) we conclude that on the interval $[z_l, z_r]$, $R(z)$ increases from a negative number to a positive number, and hencc has a unique zero, which we shall call z_1. So if we define $y_1 \triangleq y(z_1)$, then (z_1, y_1) is the unique pair of solutions solving the system of equations

$$\begin{cases} R_1(y, z) = 0 \\ R_2(y, z) = 0 \end{cases},$$

which is clearly equivalent to equations (5.5). To prove our last assertion, first note that

$$\begin{aligned} y'(z) &= -\frac{\widehat{g}_2'(z) - \widehat{g}_1'(y(z))}{\widehat{g}_1''(y(z))(y(z) - z)} \\ &= \frac{\widehat{g}_2'(z)(y(z) - z) - (\widehat{g}_1(y(z)) - \widehat{g}_2(z))}{-\widehat{g}_1''(y(z))(y(z) - z)^2} = \frac{R(z)}{-\widehat{g}_1''(y(z))(y(z) - z)^2}. \end{aligned}$$

Therefore $z \longmapsto y(z)$ decreases for $z \in [z_l, z_1]$ and increases for $z \in [z_1, z_r]$, which implies that $y(z_1) < y(z_l) = y_0$. □

References

1. Alvarez, L. H. R. (2008). A class of solvable stopping games. *Applied Mathematics & Optimization* **58 (3)**, 291–314.
2. Alvarez, L. H. R. (2010). Minimum guaranteed payments and costly cancellation rights: A stopping game perspective. *Mathematical Finance* **20 (4)**, 733-751.
3. Bensoussan, A., Friedman, A. (1974). Nonlinear variational inequalities and differential games with stopping times. *Journal of Functional Analysis* **16 (3)**, 305–352.
4. Bensoussan, A., Friedman, A. (1977). Non-zero sum stochastic differential games with stopping times and free-boundary problems. *Transactions of the American Mathematical Society* **231 (2)**, 275–327.
5. Cai, N., Sun, L. (2014). Valuation of stock loans with jump risk. *Journal of Economic Dynamics and Control* **40 (3)**, 213–241.
6. Dai, M., Xu, Z. Q., (2011).Optimal redeeming strategy of stock loans with finite maturity. *Mathematical Finance* **21 (4)**, 775–793.

7. Dayanik, S., Karatzas, I. (2003). On the optimal stopping problem for one-dimensional diffusions. *Stochastic Processes and Their Applications* **107 (2)**, 172–212.
8. Dynkin, E. B. (1969). Game variant of a problem on optimal stopping. *Soviet Mathematics Doklady* **10**, 270–274.
9. Egami, M., Leung, T., Yamazaki, K. (2013). Default swap games driven by spectrally negative Lévy processes *Stochastic Processes and Their Applications* **123 (2)**, 347-384.
10. Ekström, E., Peskir, G. (2008). Optimal stopping games for Markov processes. *SIAM Journal of Control and Optimizations* **47**, 684–702.
11. Ekström, E., Wanntorp, H. (2008). Margin call stock loans. *Working Paper.*
12. Ekström, E., Villeneuve, S. (2006). On the values of the optimal stopping games. *Annals of Applied Probability* **16 (3)** , 1576-1596.
13. Gapeev, P. V. (2005). The spread option optimal stopping game. In Kyprianou, A., Wim, S., and Wilmott, P., editors, *In Exotic Option Pricing and Advanced Lévy Models,* 293–305. John-Wiley.
14. Grasselli, M. R., Velez, C. G. (2013). Stock loans in incomplete markets. *Applied Mathematical Finance* **20(2)**, 118–136.
15. Itô, K., McKean, H. P. (1974). *Diffusion Processes and Their Sample Paths.* Chapman & Hall.
16. Jones, R. A., Nickerson, D. (2002). Mortgage contract, strategic options and stochastic collateral. *Journal of Real Estate Finance and Economics* **24 (1/2)**, 35–58.
17. Karatzas, I. (1988). On the pricing of American options. *Applied Mathematics & Optimization* **17**, 37–60.
18. Karatzas, I., Shreve, S. (1988). *Brownian Motion and Stochastic Calculus.* Springer, New York.
19. Kifer, Y. (2000). Game options. *Finance and Stochastics* **4**, 443–463.
20. Kifer, Y. (2013). Dynkin's games and Israeli's options. *ISRN Probability and Statistics* **2013**, 1–17.
21. Kyprianou, A. E. (2004). Some calculations for Israeli's options. *Finance and Stochastics* **8**, 73–86.
22. Liang, Z., Wu, W., Jiang, S. (2010). Stock loan with automatic termination clause, cap, and margin. *Computers and Mathematics with Applications* **60 (12)**, 3160–3176.
23. Liang, Z., Wu, W. (2012). Variational inequalities in stock loan models. *Optimization and Engineering* **13 (3)**, 459–470.
24. Liu, G., Xu, Y. (2010). Capped stock loans. *Computers and Mathematics with Applications* **59 (11)**, 3458–3558.
25. Merton, R. C. (1973). Theory of rational option pricing. *Bell Journal of Economics and Management Science* **4**, 7141–183.
26. Neveu, J. (1975). *Discrete-Parameter Martingales*, North-Holland, Amsterdam.
27. Peskir, G. (2007). Principle of smooth fit and diffusions with angles. *Stochastics* **79**, 293–302.
28. Peskir, G., Shiryaev, A. (2006). *Optimal Stopping and Free-Boundary Problems.*, Lectures in Mathematics, ETH Zurich, Birhauser.
29. Wong, T. W., Wong, H. Y. (2012). Stochastic volatility asymptotics of stock loans. *Journal of Mathematical Analysis and Applications* **394 (1)**, 337–346.

30. Wong, T. W., Wong, H. Y. (2013). Valuation of stock loans using exponential phase-type Lévy models. *Applied Mathematics and Computation* **222** (**1**), 275–289.
31. Xia, J., Zhou, X. Y. (2007). Stock loans. *Mathematical Finance* **17**, 307–317.
32. Yam, S. C. P., Yung, S. P., Zhou, W. (2014). Game call options revisited. *Mathematical Finance* **24** (**1**), 173-206.
33. Zhang, Q., Zhou, X. Y. (2009). Valuation of stock loans with regime switching. *SIAM Journal on Control and Optimizations* **48** (**3**), 1229–1250.

Cash Management and Control Band Policies for Spectrally One-sided Lévy Processes*

Kazutoshi Yamazaki

Faculty of Engineering Science, Kansai University, 3-3-35 Yamate-cho, Suita-shi, Osaka 564-8680, Japan. Email: kyamazak@kansai-u.ac.jp

We study the control band policy arising in the context of cash balance management. A policy is specified by four parameters (d, D, U, u). The controller pushes the process up to D as soon as it goes below d and pushes down to U as soon as it goes above u, while he does not intervene whenever it is within the set (d, u). We focus on the case when the underlying process is a spectrally one-sided Lévy process and obtain the expected fixed and proportional controlling costs as well as the holding costs under the band policy.

Key words: cash balance management; impulse control; Lévy processes; scale functions

1. Introduction

In a cash balance management problem, one continuously monitors and modifies the cash balance that fluctuates stochastically over time. In a most general model, a controller is allowed, at a cost, to both increase and decrease the balance so as to prevent the excess and shortage. The excess and shortage costs, collectively called the *holding costs*, are modeled by (typically a convex) function of the balance integrated over time. The *controlling costs* consist of fixed and proportional costs, where the former is incurred at each adjustment whereas the latter is proportional to the adjustment quantity. The objective is to minimize the sum of expected values of these costs.

---

*Send all correspondence to Kazutoshi Yamazaki, Department of Mathematics, Faculty of Engineering Science, Kansai University, 3-3-35 Yamate-cho, Suita-shi, Osaka 564-8680, Japan. Email: kyamazak@kansai-u.ac.jp. The author thanks the anonymous referee for constructive comments and suggestions. K. Yamazaki is in part supported by MEXT KAKENHI grant number 26800092, the Inamori foundation research grant, and the Kansai University subsidy for supporting young scholars 2014.

In most of the existing literature, the common goal is to show the optimality of the *band policy* that is specified by four parameters (d, D, U, u) such that $d < u$ and $D, U \in (d, u)$: the controller pushes the balance up to D as soon as it goes below d and pushes down to U as soon as it exceeds u; he does not intervene whenever it is within the set $[d, u]$. To our best knowledge, the existing optimality results are limited only for the Brownian motion (with a drift) case. In particular, Constantinides and Richard [9], Harrison and Taylor [12], Harrison et al. [11] solve for the linear holding cost case; Buckley and Korn [7] solve for the quadratic holding cost case.

In this paper, we study the band policy of the same form by generalizing the underlying process to a class of spectrally negative Lévy processes; namely, the cash balance, in the absence of control, follows a general Lévy process with only negative jumps. We obtain the associated net present values (NPV) of the total discounted controlling costs as well as those of the holding costs. While it is out of scope of this paper, its potential application lies in obtaining the solution to the cash management problem by choosing appropriately the values of (d, D, U, u) and show the quasi-variational inequalities (QVI) of Bensoussan and Lions [5].

While the inclusion of jumps makes the problem significantly harder, there have recently been several results on related stochastic control problems. In particular, there are two special cases of the cash balance management problem that have been solved analytically for a general spectrally negative Lévy process. First, under the additional constraint that the process can only be augmented, a two-parameter band policy, known as the (s, S)-policy, has been shown to be optimal by Yamazaki [18] (as a generalization of the previous results by [4, 6] for processes with compound Poisson jumps). Second, in the absence of fixed controlling costs, Baurdoux and Yamazaki [2] show the optimality of another two-parameter band policy where the optimally controlled process becomes a doubly reflected Lévy process of [1, 17]. For other stochastic control problems where the optimal policy is characterized by two parameters, we refer the reader to [3, 16] for optimal dividend problems with fixed transaction costs and [10, 13] for two-player stochastic games.

The objective of this paper is to obtain semi-analytical expressions of the NPV's of the total discounted costs associated with the band policy. Following the same paths of the above mentioned papers, we use the scale function to efficiently write these quantities. We expect these expressions to be beneficial in solving the cash management problem; the forms written in terms of the scale function can potentially help one to analyze the smoothness of the value function and to verify the optimality of a candidate band policy.

The rest of the paper is organized as follows. Section 2 reviews the spectrally negative Lévy process, the band policy, and the scale function. Sections 3 and 4 obtain, using the scale function, the NPV's of the controlling and holding costs, respectively. Section 5 concludes the paper with discussions on its contributions as well as potential challenges in its application in cash management problems.

2. Mathematical Formulation

Let $(\Omega, \mathcal{F}, \mathbb{P})$ be a probability space hosting a *spectrally negative Lévy process* $X = \{X_t; t \geq 0\}$ whose *Laplace exponent* is given by

(1)
$$\psi(s) := \log \mathbb{E}\left[e^{sX_1}\right] = cs + \frac{1}{2}\sigma^2 s^2 + \int_{(-\infty,0)} (e^{sz} - 1 - sz1_{\{-1<z<0\}})\nu(\mathrm{d}z), \quad s \geq 0,$$

where ν is a Lévy measure with the support $(-\infty, 0)$ that satisfies the integrability condition $\int_{(-\infty,0)}(1 \wedge z^2)\nu(\mathrm{d}z) < \infty$. It has paths of bounded variation if and only if $\sigma = 0$ and $\int_{(-1,0)} |z|\, \nu(\mathrm{d}z) < \infty$; in this case, we write (1) as

$$\psi(s) = \delta s + \int_{(-\infty,0)} (e^{sz} - 1)\nu(\mathrm{d}z), \quad s \geq 0,$$

with $\delta := c - \int_{(-1,0)} z\, \nu(\mathrm{d}z)$. We exclude the case in which X is the negative of a subordinator (i.e., X has monotone paths a.s.). This assumption implies that $\delta > 0$ when X is of bounded variation. Let $\mathbb{P}_x$ be the conditional probability under which $X_0 = x$ (also let $\mathbb{P} \equiv \mathbb{P}_0$), and let $\mathbb{F} := \{\mathcal{F}_t : t \geq 0\}$ be the filtration generated by X.

Fix (d, D, U, u) such that $d < u$ and $D, U \in (d, u)$. We consider adjusting the process X by adding and subtracting the processes $R \equiv R(d, D, U, u)$ and $L \equiv L(d, D, U, u)$, respectively; the resulting controlled process becomes:

$$A_t = A_t(d, D, U, u) := X_t + R_t - L_t, \quad t \geq 0.$$

The process R pushes the process up to D as soon as it goes below d while the process L pushes it down to U as soon as it goes above u. We consider the right-continuous versions for R and L. For the sake of completeness, we construct the processes as follows. In doing so, we also define an auxiliary process

$$\tilde{A}_t := A_{t-} + \Delta X_t, = A_t - (\Delta R_t - \Delta L_t), \quad t \geq 0,$$

which can be understood as the *pre-controlled* process that does not reflect at t the adjustments made by the processes R_t and L_t. Here and throughout, let $\Delta\xi_t := \xi_t - \xi_{t-}$, for any right-continuous process ξ.

Construction of the processes A, $\tilde{A}$, L and R

Step 1 Set $A_{0-} = \tilde{A}_0 = x$ and $L_{0-} = R_{0-} = 0$.

Step 1-1 If $d \leq x \leq u$, set

$$A_0 = x \quad \text{and} \quad L_0 = R_0 = 0.$$

If $x < d$, set

$$A_0 = D, \quad L_0 = 0, \quad \text{and} \quad R_0 = D - x.$$

If $x > u$, set

$$A_0 = U, \quad L_0 = x - U, \quad \text{and} \quad R_0 = 0.$$

Step 1-2 Set $n = 0$ and define $T^{(0)} = 0$.

Step 2 **Step 2-1** Set

$$\tilde{A}_t = A_{T^{(n)}} + (X_t - X_{T^{(n)}}), \quad T^{(n)} < t \leq T^{(n+1)} := T_u^{(n+1)+} \wedge T_d^{(n+1)-}$$

where we define

$$T_u^{(n+1)+} := \inf\left\{t \geq T^{(n)} : \tilde{A}_t > u\right\},$$
$$T_d^{(n+1)-} := \inf\left\{t \geq T^{(n)} : \tilde{A}_t < d\right\}.$$

Step 2-2 Set $A_t = \tilde{A}_t$, $R_t = R_{T^{(n)}}$ and $L_t = L_{T^{(n)}}$ for $T^{(n)} < t < T^{(n+1)}$ and

$$A_{T^{(n+1)}} = \begin{cases} U, & \text{if } T^{(n+1)} = T_u^{(n+1)+}, \\ D, & \text{if } T^{(n+1)} = T_d^{(n+1)-}, \end{cases}$$

$$R_{T^{(n+1)}} = \begin{cases} R_{T^{(n)}}, & \text{if } T^{(n+1)} = T_u^{(n+1)+}, \\ R_{T^{(n)}} + (D - \tilde{A}_{T^{(n+1)}}), & \text{if } T^{(n+1)} = T_d^{(n+1)-}, \end{cases}$$

$$L_{T^{(n+1)}} = \begin{cases} L_{T^{(n)}} + (u - U), & \text{if } T^{(n+1)} = T_u^{(n+1)+}, \\ L_{T^{(n)}}, & \text{if } T^{(n+1)} = T_d^{(n+1)-}. \end{cases}$$

Step 2-3 Increment the value of n by 1 and go back to **Step 2-1**.

In the algorithm above, the processes are first initialized in **Step 1**. In the constructions in **Step 2**, the process R_t (resp. L_t) stays constant while the pre-controlled process $\tilde{A}_t$ remains on $[d, \infty)$ (resp. $(-\infty, u]$), and it increases by $D - \tilde{A}_t$ (resp. $\tilde{A}_t - U$) as soon as $\tilde{A}$ enters $(-\infty, d)$ (resp. (u, ∞)). By construction, R and L are non-decreasing a.s. and the controlled process A_t always remains on the interval $[d, u]$. It is easy to see that these processes are $\mathbb{F}$-adapted; in particular, the processes A and $\tilde{A}$ are strong Markov processes.

2.1 Scale functions

We conclude this section with a brief review on the scale function.

Fix $q > 0$. For any spectrally negative Lévy process, there exists a function called the q-scale function

$$W^{(q)} : \mathbb{R} \to [0, \infty),$$

which is zero on $(-\infty, 0)$, continuous and strictly increasing on $[0, \infty)$, and is characterized by the Laplace transform:

$$\int_0^\infty e^{-sx} W^{(q)}(x) \mathrm{d}x = \frac{1}{\psi(s) - q}, \qquad s > \Phi(q),$$

where

$$\Phi(q) := \sup\{\lambda \geq 0 : \psi(\lambda) = q\}.$$

Here, the Laplace exponent ψ in (1) is known to be zero at the origin and convex on $[0, \infty)$; therefore $\Phi(q)$ is well defined and is strictly positive as $q > 0$. We also define, for $x \in \mathbb{R}$,

$$\begin{aligned}
\overline{W}^{(q)}(x) &:= \int_0^x W^{(q)}(y) \mathrm{d}y, \\
Z^{(q)}(x) &:= 1 + q\overline{W}^{(q)}(x), \\
\overline{Z}^{(q)}(x) &:= \int_0^x Z^{(q)}(z) \mathrm{d}z = x + q \int_0^x \int_0^z W^{(q)}(w) \mathrm{d}w \mathrm{d}z.
\end{aligned}$$

Because $W^{(q)}$ is uniformly zero on the negative half line, we have $Z^{(q)}(x) = 1$ and $\overline{Z}^{(q)}(x) = x$ for $x \leq 0$.

Let us define the first down- and up-crossing times, respectively, of X by

$$\tau_b^- := \inf\{t \geq 0 : X_t < b\} \quad \text{and} \quad \tau_b^+ := \inf\{t \geq 0 : X_t > b\}, \quad b \in \mathbb{R}. \tag{2}$$

Then, for any $b > 0$ and $x \leq b$,

(3)

$$\mathbb{E}_x\left[e^{-q\tau_b^+} 1_{\{\tau_b^+ < \tau_0^-\}}\right] = \frac{W^{(q)}(x)}{W^{(q)}(b)} \quad \text{and} \quad \mathbb{E}_x\left[e^{-q\tau_0^-} 1_{\{\tau_b^+ > \tau_0^-\}}\right] = Z^{(q)}(x) - Z^{(q)}(b)\frac{W^{(q)}(x)}{W^{(q)}(b)}.$$

In addition, as in Theorem 8.7 of [14], for any measurable function f bounded on $[d, u]$, we have

$$\mathbb{E}_x\left[\int_0^{\tau_d^- \wedge \tau_u^+} e^{-qt} f(X_t) \mathrm{d}t\right] = \varphi_d(u; f)\frac{W^{(q)}(x-d)}{W^{(q)}(u-d)} - \varphi_d(x; f), \tag{4}$$

where

$$\varphi_{d'}(x'; f) := \int_{d'}^{x'} W^{(q)}(x' - y) f(y) \mathrm{d}y, \quad d', x' \in \mathbb{R}. \tag{5}$$

Remark 2.1.

1. If X is of unbounded variation or the Lévy measure is atomless, it is known that $W^{(q)}$ is $C^1(\mathbb{R}\backslash\{0\})$; see, e.g., [8]. Hence,
   (a) $Z^{(q)}$ is $C^1(\mathbb{R}\backslash\{0\})$ and $C^0(\mathbb{R})$ for the bounded variation case, while it is $C^2(\mathbb{R}\backslash\{0\})$ and $C^1(\mathbb{R})$ for the unbounded variation case, and
   (b) $\overline{Z}^{(q)}$ is $C^2(\mathbb{R}\backslash\{0\})$ and $C^1(\mathbb{R})$ for the bounded variation case, while it is $C^3(\mathbb{R}\backslash\{0\})$ and $C^2(\mathbb{R})$ for the unbounded variation case.
2. Regarding the asymptotic behavior near zero, as in Lemmas 4.3 and 4.4 of [15],

$$W^{(q)}(0) = \begin{cases} 0, & \text{if } X \text{ is of unbounded variation,} \\ \frac{1}{\delta}, & \text{if } X \text{ is of bounded variation,} \end{cases} \tag{6}$$

$$W^{(q)\prime}(0+) := \lim_{x \downarrow 0} W^{(q)\prime}(x) = \begin{cases} \frac{2}{\sigma^2}, & \text{if } \sigma > 0, \\ \infty, & \text{if } \sigma = 0 \text{ and } \nu(-\infty, 0) = \infty, \\ \frac{q + \nu(-\infty,0)}{\delta^2}, & \text{if } \sigma = 0 \text{ and } \nu(-\infty, 0) < \infty. \end{cases} \tag{7}$$

3. As in (8.18) and Lemma 8.2 of [14],

$$\frac{W^{(q)\prime}(y+)}{W^{(q)}(y)} \leq \frac{W^{(q)\prime}(x+)}{W^{(q)}(x)}, \quad y > x > 0.$$

   In all cases, $W^{(q)\prime}(x-) \geq W^{(q)\prime}(x+)$ for all $x > 0$.

3. Controlling costs

In this section, we compute the controlling costs given by, for all $x \in \mathbb{R}$,

$$v_L(x) := \mathbb{E}_x\Big[\sum_{0 \leq s \leq t} e^{-qs} (\Delta L_s + k_L) 1_{\{\Delta L_s > 0\}} \Big], \tag{8}$$

$$v_R(x) := \mathbb{E}_x\Big[\sum_{0 \leq s \leq t} e^{-qs} (\Delta R_s + k_R) 1_{\{\Delta R_s > 0\}} \Big], \tag{9}$$

for given constants $k_L, k_R \in \mathbb{R}$. Throughout, we fix (d, D, U, u) such that $d < u$ and $D, U \in (d, u)$.

We shall write these in terms of the scale function as reviewed above. Because both $W^{(q)}$ and $\overline{W}^{(q)}$ are nondecreasing, we can define the measures $W^{(q)}(\mathrm{d}x)$ and $\overline{W}^{(q)}(\mathrm{d}x)$ such that, for any $y > x > 0$,

$$W^{(q)}(x, y) = W^{(q)}(y) - W^{(q)}(x) \quad \text{and} \quad \overline{W}^{(q)}(x, y) = \overline{W}^{(q)}(y) - \overline{W}^{(q)}(x).$$

Let us also define

$$\Xi(d, D, U, u) := \overline{W}^{(q)}(U - d, u - d)W^{(q)}(D - d) - W^{(q)}(U - d, u - d)\overline{W}^{(q)}(D - d).$$

We first obtain the expression for (8).

Proposition 3.1. *Let*

$$\epsilon_L := (u - U) + k_L.$$

1. *For all $d \le x \le u$,*

$$v_L(x) = \frac{\epsilon_L}{\Xi(d, D, U, u)}\left[Z^{(q)}(x - d)\frac{W^{(q)}(D - d)}{q} - W^{(q)}(x - d)\overline{W}^{(q)}(D - d)\right].$$

2. *For all $x > u$,*

$$\begin{aligned} v_L(x) &= (x - U) + k_L + v_L(U) \\ &= (x - U) + k_L + \frac{\epsilon_L}{q}\frac{W^{(q)}(D - d)Z^{(q)}(U - d) - qW^{(q)}(U - d)\overline{W}^{(q)}(D - d)}{\Xi(d, D, U, u)}. \end{aligned}$$

3. *For all $x < d$,*

$$v_L(x) = v_L(D) = \frac{\epsilon_L}{q}\frac{W^{(q)}(D - d)}{\Xi(d, D, U, u)}.$$

Proof. Fix $d \le x \le u$. Suppose

$$T_b^+ := \inf\left\{t \ge 0 : \tilde{A}_t > b\right\} \quad \text{and} \quad T_b^- := \inf\left\{t \ge 0 : \tilde{A}_t < b\right\}, \quad b \in \mathbb{R}. \tag{10}$$

Because the law of $\{\tilde{A}_t; t \le T_u^+ \wedge T_d^-\}$ and that of $\{X_t; t \le \tau_u^+ \wedge \tau_d^-\}$ are the same (see the above construction of the process $\tilde{A}$), the strong Markov property and $\tilde{A}_{T_u^+} = u$ on $\{T_u^+ < \infty\}$ (due to the fact that X has no positive jumps) gives

$$\begin{aligned} v_L(x) &= \mathbb{E}_x\left[e^{-qT_u^+}1_{\{T_u^+ < T_d^-\}}\right](v_L(U) + \epsilon_L) + \mathbb{E}_x\left[e^{-qT_d^-}1_{\{T_u^+ > T_d^-\}}\right]v_L(D) \\ &= \mathbb{E}_x\left[e^{-q\tau_u^+}1_{\{\tau_u^+ < \tau_d^-\}}\right](v_L(U) + \epsilon_L) + \mathbb{E}_x\left[e^{-q\tau_d^-}1_{\{\tau_u^+ > \tau_d^-\}}\right]v_L(D). \end{aligned}$$

Hence, by (3),

(11)

$$\begin{aligned} v_L(x) &= \frac{W^{(q)}(x - d)}{W^{(q)}(u - d)}[v_L(U) + \epsilon_L] + \left[Z^{(q)}(x - d) - Z^{(q)}(u - d)\frac{W^{(q)}(x - d)}{W^{(q)}(u - d)}\right]v_L(D) \\ &= \frac{W^{(q)}(x - d)}{W^{(q)}(u - d)}\left[v_L(U) + \epsilon_L - Z^{(q)}(u - d)v_L(D)\right] + Z^{(q)}(x - d)v_L(D). \end{aligned}$$

In particular, by substituting $x = U, D$, we obtain

$$v_L(U) = \frac{W^{(q)}(U-d)}{W^{(q)}(u-d)} \left[v_L(U) + \epsilon_L - Z^{(q)}(u-d) v_L(D) \right] + Z^{(q)}(U-d) v_L(D),$$

$$v_L(D) = \frac{W^{(q)}(D-d)}{W^{(q)}(u-d)} \left[v_L(U) + \epsilon_L - Z^{(q)}(u-d) v_L(D) \right] + Z^{(q)}(D-d) v_L(D). \tag{12}$$

By computing $v_L(U) - v_L(D) W^{(q)}(U-d)/W^{(q)}(D-d)$, we attain the relation:

$$v_L(U) = v_L(D) \left[Z^{(q)}(U-d) - \frac{W^{(q)}(U-d)}{W^{(q)}(D-d)} q \overline{W}^{(q)}(D-d) \right]. \tag{13}$$

Substituting this back in (12) and solving for $v_L(D)$, we obtain

$$v_L(D) = \frac{\epsilon_L}{q} \frac{W^{(q)}(D-d)}{\Xi(d, D, U, u)}. \tag{14}$$

In addition, substituting this in (13) gives

$$v_L(U) = \frac{\epsilon_L}{q} \frac{W^{(q)}(D-d) Z^{(q)}(U-d) - q W^{(q)}(U-d) \overline{W}^{(q)}(D-d)}{\Xi(d, D, U, u)}. \tag{15}$$

These together with (11) complete the proof of (1). The proofs of (2) and (3) are immediate by the construction of the process L and by (14) and (15). □

We now move on to obtaining the expression for (9). Toward this end, we assume that the first moment of X_t is finite.

Assumption 3.1. *Suppose* $\mu := \mathbb{E}[X_1] = \psi'(0+) \in (-\infty, \infty)$.

We define the following short-hand notations:

$$\epsilon_R := (D-d) + k_R,$$

$$Y^{(q)}(y) \equiv Y^{(q)}(y; \epsilon_R) := \overline{Z}^{(q)}(y) + \frac{\mu}{q} - \left(\frac{\mu}{q} + \epsilon_R\right) Z^{(q)}(y), \quad y \in \mathbb{R}.$$

Proposition 3.2. *Suppose Assumption 3.1 holds.*

1. *For* $d \le x \le u$,

$$\begin{aligned} &v_R(x) \\ &= \frac{Z^{(q)}(x-d)}{q} \frac{W^{(q)}(D-d)[Y^{(q)}(u-d) - Y^{(q)}(U-d)] - Y^{(q)}(D-d) W^{(q)}(U-d, }{\Xi(d, D, U, u)} \\ &- Y^{(q)}(x-d) \\ &+ W^{(q)}(x-d) \frac{Y^{(q)}(D-d) \overline{W}^{(q)}(U-d, u-d) - \overline{W}^{(q)}(D-d)[Y^{(q)}(u-d) - Y^{(q)}(U}{\Xi(d, D, U, u)} \end{aligned}$$

2. *For all $x > u$,*

$$\begin{aligned} v_R(x) &= v_R(U) \\ &= \left[Z^{(q)}(U-d) - \frac{W^{(q)}(U-d)}{W^{(q)}(D-d)} q\overline{W}^{(q)}(D-d)\right] \\ &\times \frac{[Y^{(q)}(u-d) - Y^{(q)}(U-d)]W^{(q)}(D-d) - W^{(q)}(U-d, u-d)Y^{(q)}(D-d)}{q\Xi(d, D, U, u)} \\ &+ \frac{W^{(q)}(U-d)}{W^{(q)}(D-d)} Y^{(q)}(D-d) - Y^{(q)}(U-d). \end{aligned}$$

3. *For all $x < d$,*

$$\begin{aligned} v_R(x) &= (D-x) + k_R + v_R(D) \\ &= (D-x) + k_R \\ &+ \frac{[Y^{(q)}(u-d) - Y^{(q)}(U-d)]W^{(q)}(D-d) - W^{(q)}(U-d, u-d)Y^{(q)}(D-d)}{q\Xi(d, D, U, u)}. \end{aligned}$$

Proof. Fix $d \le x \le u$. Because the law of $\{\tilde{A}_t; t \le T_u^+ \wedge T_d^-\}$ and that of $\{X_t; t \le \tau_u^+ \wedge \tau_d^-\}$ are the same (see (10)), the strong Markov property gives

$$\begin{aligned} v_R(x) &= \mathbb{E}_x\left[e^{-qT_u^+} 1_{\{T_u^+ < T_d^-\}}\right] v_R(U) + \mathbb{E}_x\left[e^{-qT_d^-} 1_{\{T_u^+ > T_d^-\}}\right] v_R(D) \\ &+ \mathbb{E}_x\left[e^{-qT_d^-} 1_{\{T_d^- < T_u^+\}}(d - \tilde{A}_{T_d^-} + \epsilon_R)\right] \\ &= \mathbb{E}_x\left[e^{-q\tau_u^+} 1_{\{\tau_u^+ < \tau_d^-\}}\right] v_R(U) + \mathbb{E}_x\left[e^{-q\tau_d^-} 1_{\{\tau_u^+ > \tau_d^-\}}\right] v_R(D) \\ &+ \mathbb{E}_x\left[e^{-q\tau_d^-} 1_{\{\tau_d^- < \tau_u^+\}}(d - X_{\tau_d^-} + \epsilon_R)\right]. \end{aligned}$$

Here, Lemma 3.1 of [3] and (3) give

$$\mathbb{E}_x\left[e^{-q\tau_d^-} 1_{\{\tau_d^- < \tau_u^+\}}(d - X_{\tau_d^-} + \epsilon_R)\right] = -Y^{(q)}(x-d) + Y^{(q)}(u-d)\frac{W^{(q)}(x-d)}{W^{(q)}(u-d)}.$$

Substituting this and (3),

$$\begin{aligned} v_R(x) &= \frac{W^{(q)}(x-d)}{W^{(q)}(u-d)} v_R(U) + \left[Z^{(q)}(x-d) - Z^{(q)}(u-d)\frac{W^{(q)}(x-d)}{W^{(q)}(u-d)}\right] v_R(D) \\ &- Y^{(q)}(x-d) + Y^{(q)}(u-d)\frac{W^{(q)}(x-d)}{W^{(q)}(u-d)} \\ &= \left[Z^{(q)}(x-d) - Z^{(q)}(u-d)\frac{W^{(q)}(x-d)}{W^{(q)}(u-d)}\right] v_R(D) - Y^{(q)}(x-d) \\ &+ \left[Y^{(q)}(u-d) + v_R(U)\right]\frac{W^{(q)}(x-d)}{W^{(q)}(u-d)}. \end{aligned} \tag{16}$$

In particular, by setting $x = D, U$, we obtain

$$
\begin{aligned}
v_R(D) &= \left[Z^{(q)}(D-d) - Z^{(q)}(u-d)\frac{W^{(q)}(D-d)}{W^{(q)}(u-d)}\right]v_R(D) - Y^{(q)}(D-d) \\
&\quad + \left[Y^{(q)}(u-d) + v_R(U)\right]\frac{W^{(q)}(D-d)}{W^{(q)}(u-d)}, \\
v_R(U) &= \left[Z^{(q)}(U-d) - Z^{(q)}(u-d)\frac{W^{(q)}(U-d)}{W^{(q)}(u-d)}\right]v_R(D) - Y^{(q)}(U-d) \\
&\quad + \left[Y^{(q)}(u-d) + v_R(U)\right]\frac{W^{(q)}(U-d)}{W^{(q)}(u-d)}.
\end{aligned} \tag{17}
$$

In order to solve this system of equations, we compute

$$
\begin{aligned}
v_R(U) - v_R(D)\frac{W^{(q)}(U-d)}{W^{(q)}(D-d)} &= \left[Z^{(q)}(U-d) - Z^{(q)}(u-d)\frac{W^{(q)}(U-d)}{W^{(q)}(u-d)}\right]v_R(D) \\
&\quad - \frac{W^{(q)}(U-d)}{W^{(q)}(D-d)}\left[Z^{(q)}(D-d) - Z^{(q)}(u-d)\frac{W^{(q)}(D-d)}{W^{(q)}(u-d)}\right]v_R(\\
&\quad + \frac{W^{(q)}(U-d)}{W^{(q)}(D-d)}Y^{(q)}(D-d) - Y^{(q)}(U-d) \\
&= \left[Z^{(q)}(U-d) - \frac{W^{(q)}(U-d)}{W^{(q)}(D-d)}Z^{(q)}(D-d)\right]v_R(D) \\
&\quad + \frac{W^{(q)}(U-d)}{W^{(q)}(D-d)}Y^{(q)}(D-d) - Y^{(q)}(U-d),
\end{aligned}
$$

and therefore

$$
\begin{aligned}
v_R(U) &= \left[Z^{(q)}(U-d) - \frac{W^{(q)}(U-d)}{W^{(q)}(D-d)}q\overline{W}^{(q)}(D-d)\right]v_R(D) \\
&\quad + \frac{W^{(q)}(U-d)}{W^{(q)}(D-d)}Y^{(q)}(D-d) - Y^{(q)}(U-d).
\end{aligned} \tag{18}
$$

Substituting this in (17) and solving for $v_R(D)$ gives

(19)

$$
v_R(D) = \frac{[Y^{(q)}(u-d) - Y^{(q)}(U-d)]W^{(q)}(D-d) - W^{(q)}(U-d, u-d)Y^{(q)}(D-d)}{q\Xi(d, D, U, u)},
$$

and hence

(20)

$$v_R(U) = \left[Z^{(q)}(U-d) - \frac{W^{(q)}(U-d)}{W^{(q)}(D-d)} q\overline{W}^{(q)}(D-d)\right]$$
$$\times \frac{[Y^{(q)}(u-d) - Y^{(q)}(U-d)]W^{(q)}(D-d) - W^{(q)}(U-d,u-d)Y^{(q)}(D-d)}{q\Xi(d,D,U,u)}$$
$$+ \frac{W^{(q)}(U-d)}{W^{(q)}(D-d)} Y^{(q)}(D-d) - Y^{(q)}(U-d).$$

By (16) and (18),

$$v_R(x) = Z^{(q)}(x-d)v_R(D) - Y^{(q)}(x-d) + B(d,D,U,u)\frac{W^{(q)}(x-d)}{W^{(q)}(u-d)}, \tag{21}$$

where

$$B(d,D,U,u) := -q\left[\overline{W}^{(q)}(U-d,u-d) + \frac{W^{(q)}(U-d)}{W^{(q)}(D-d)}\overline{W}^{(q)}(D-d)\right]v_R(D)$$
$$+ \frac{W^{(q)}(U-d)}{W^{(q)}(D-d)} Y^{(q)}(D-d) + Y^{(q)}(u-d) - Y^{(q)}(U-d).$$

Here in particular

$$\left[\overline{W}^{(q)}(U-d,u-d) + \frac{W^{(q)}(U-d)}{W^{(q)}(D-d)}\overline{W}^{(q)}(D-d)\right]v_R(D)$$
$$= \left[\frac{W^{(q)}(u-d)}{W^{(q)}(D-d)}\overline{W}^{(q)}(D-d) + \frac{\Xi(d,D,U,u)}{W^{(q)}(D-d)}\right]v_R(D)$$
$$= \frac{W^{(q)}(u-d)}{W^{(q)}(D-d)}\overline{W}^{(q)}(D-d)v_R(D)$$
$$+ \frac{1}{q}\left[Y^{(q)}(u-d) - Y^{(q)}(U-d) - \frac{W^{(q)}(U-d,u-d)}{W^{(q)}(D-d)} Y^{(q)}(D-d)\right].$$

Hence,

$$B(d,D,U,u) = \frac{W^{(q)}(u-d)}{W^{(q)}(D-d)}\left[Y^{(q)}(D-d) - q\overline{W}^{(q)}(D-d)v_R(D)\right]$$
$$= W^{(q)}(u-d)\frac{Y^{(q)}(D-d)\overline{W}^{(q)}(U-d,u-d) - \overline{W}^{(q)}(D-d)[Y^{(q)}(u-d) - Y^{(q)}(U-d)]}{\Xi(d,D,U,u)}.$$

Substituting this and (19) in (21), the proof of (1) is complete. The proofs of (2) and (3) are immediate by the construction of the process R and by (19) and (20)□

4. Holding costs

Fix (d, D, U, u) such that $d < u$ and $D, U \in (d, u)$ and define

$$w(x) = w(x; f) := \mathbb{E}_x \left[\int_0^\infty e^{-qt} f(A_t) \mathrm{d}t \right],$$

for any measurable function f bounded on $[d, u]$. We define

$$\Theta(d, D, U, u; f) := W^{(q)}(D - d)[\varphi_d(u; f) - \varphi_d(U; f)] - W^{(q)}(U - d, u - d)\varphi_d(D; f).$$

Proposition 4.1.

1. *For any* $d \leq x \leq u$,

$$w(x) = \frac{W^{(q)}(x - d)}{W^{(q)}(D - d)} \varphi_d(D; f) - \varphi_d(x; f) + \left[\frac{Z^{(q)}(x - d)}{q} - W^{(q)}(x - d) \frac{\overline{W}^{(q)}(D - d)}{W^{(q)}(D - d)} \right] \frac{\Theta(d, D, U, u; f)}{\Xi(d, D, U, u)}.$$

2. *For* $x < d$,

$$w(x) = w(D) = \frac{\Theta(d, D, U, u; f)}{q\Xi(d, D, U, u)}.$$

3. *For* $x > u$,

$$\begin{aligned} w(x) &= w(U) \\ &= \frac{\Theta(d, D, U, u; f)}{q\Xi(d, D, U, u)} \left[Z^{(q)}(U - d) - q \frac{W^{(q)}(U - d)}{W^{(q)}(D - d)} \overline{W}^{(q)}(D - d) \right] \\ &\quad - \varphi_d(U; f) + \frac{W^{(q)}(U - d)}{W^{(q)}(D - d)} \varphi_d(D; f). \end{aligned}$$

Proof. Fix $d \leq x \leq u$. Again, because the law of $\{\tilde{A}_t; t \leq T_u^+ \wedge T_d^-\}$ and that of $\{X_t; t \leq \tau_u^+ \wedge \tau_d^-\}$ are the same, the strong Markov property gives

$$w(x) = \mathbb{E}_x \left[e^{-q\tau_u^+} 1_{\{\tau_u^+ < \tau_d^-\}} \right] w(U) + \mathbb{E}_x \left[e^{-q\tau_d^-} 1_{\{\tau_u^+ > \tau_d^-\}} \right] w(D) + \mathbb{E}_x \left[\int_0^{\tau_d^- \wedge \tau_u^+} e^{-qt} f(X_t) \mathrm{d}t \right].$$

By (3) and (4),

(22)

$$w(x) = \frac{W^{(q)}(x - d)}{W^{(q)}(u - d)} \left[w(U) - Z^{(q)}(u - d) w(D) + \varphi_d(u; f) \right] + Z^{(q)}(x - d) w(D) - \varphi_d(x; f).$$

In particular, by setting $x = U, D$,

$$w(U) = \frac{W^{(q)}(U-d)}{W^{(q)}(u-d)}\left[w(U) - Z^{(q)}(u-d)w(D) + \varphi_d(u;f)\right] + Z^{(q)}(U-d)w(D) - \varphi_d(U;f),$$

(23)

$$w(D) = \frac{W^{(q)}(D-d)}{W^{(q)}(u-d)}\left[w(U) - Z^{(q)}(u-d)w(D) + \varphi_d(u;f)\right] + Z^{(q)}(D-d)w(D) - \varphi_d(D;f).$$

Hence by computing $w(U) - w(D)W^{(q)}(U-d)/W^{(q)}(D-d)$, we obtain

$$w(U) = w(D)\left[Z^{(q)}(U-d) - q\frac{W^{(q)}(U-d)}{W^{(q)}(D-d)}\overline{W}^{(q)}(D-d)\right] - \varphi_d(U;f) + \frac{W^{(q)}(U-d)}{W^{(q)}(D-d)}\varphi_d(D;f).$$

Substituting this in (23),

$$w(D) = \frac{qw(D)}{W^{(q)}(u-d)}\left[-W^{(q)}(D-d)\overline{\overline{W}}^{(q)}(U-d,u-d) - W^{(q)}(U-d)\overline{W}^{(q)}(D-d)\right] + \frac{W^{(q)}(U-d)}{W^{(q)}(u-d)}\varphi_d(D;f) + \frac{W^{(q)}(D-d)}{W^{(q)}(u-d)}(\varphi_d(u;f) - \varphi_d(U;f)) + Z^{(q)}(D-d)w(D) - \varphi_d(D;f).$$

Solving this, we have

$$w(D) = \frac{\Theta(d,D,U,u;f)}{q\Xi(d,D,U,u)}. \tag{24}$$

Substituting this in (22),

$$w(x) = \frac{W^{(q)}(x-d)}{W^{(q)}(u-d)}\frac{\Theta(d,D,U,u;f)}{\Xi(d,D,U,u)}\left[-\overline{\overline{W}}^{(q)}(U-d,u-d) - \frac{W^{(q)}(U-d)}{W^{(q)}(D-d)}\overline{W}^{(q)}(D-d)\right] + \frac{W^{(q)}(x-d)}{W^{(q)}(u-d)}\left[\varphi_d(u;f) - \varphi_d(U;f) + \frac{W^{(q)}(U-d)}{W^{(q)}(D-d)}\varphi_d(D;f)\right] + Z^{(q)}(x-d)\frac{\Theta(d,D,U,u;f)}{q\Xi(d,D,U,u)} - \varphi_d(x;f).$$

In order to simplify this, note that

$$\frac{\Theta(d,D,U,u;f)}{\Xi(d,D,U,u)}\Big[-\overline{W}^{(q)}(U-d,u-d)-\frac{W^{(q)}(U-d)}{W^{(q)}(D-d)}\overline{W}^{(q)}(D-d)\Big]$$

$$=-\frac{1}{W^{(q)}(D-d)}\frac{\Theta(d,D,U,u;f)}{\Xi(d,D,U,u)}\Big[\Xi(d,D,U,u)+W^{(q)}(u-d)\overline{W}^{(q)}(D-d)\Big]$$

$$=-(\varphi_d(u;f)-\varphi_d(U;f))+\frac{W^{(q)}(U-d,u-d)\varphi_d(D;f)}{W^{(q)}(D-d)}$$

$$-\frac{W^{(q)}(u-d)\overline{W}^{(q)}(D-d)}{W^{(q)}(D-d)}\frac{\Theta(d,D,U,u;f)}{\Xi(d,D,U,u)}.$$

Substituting this,

$$w(x)=-\frac{W^{(q)}(x-d)}{W^{(q)}(u-d)}\Big[\varphi_d(u;f)-\varphi_d(U;f)-\frac{W^{(q)}(U-d,u-d)}{W^{(q)}(D-d)}\varphi_d(D;f)\Big]$$

$$-W^{(q)}(x-d)\frac{\overline{W}^{(q)}(D-d)}{W^{(q)}(D-d)}\frac{\Theta(d,D,U,u;f)}{\Xi(d,D,U,u)}$$

$$+\frac{W^{(q)}(x-d)}{W^{(q)}(u-d)}\Big[\varphi_d(u;f)-\varphi_d(U;f)+\frac{W^{(q)}(U-d)}{W^{(q)}(D-d)}\varphi_d(D;f)\Big]$$

$$+Z^{(q)}(x-d)\frac{\Theta(d,D,U,u;f)}{q\Xi(d,D,U,u)}-\varphi_d(x;f)$$

$$=\frac{W^{(q)}(x-d)}{W^{(q)}(D-d)}\varphi_d(D;f)-\varphi_d(x;f)$$

$$+\left[\frac{Z^{(q)}(x-d)}{q}-W^{(q)}(x-d)\frac{\overline{W}^{(q)}(D-d)}{W^{(q)}(D-d)}\right]\frac{\Theta(d,D,U,u;f)}{\Xi(d,D,U,u)},$$

which completes the proof of (1). The proofs for (2) and (3) are also immediate by the construction of A and by (24). □

5. Concluding Remarks

We have studied the band policy with parameters (d, D, U, u) and its associated NPV's of the controlling and holding costs. We focused on the case that is driven by a general spectrally negative Lévy process. Using the fluctuation theory, we expressed the NPV's using the scale function. Here, we conclude this paper with its contributions as well as challenges in applying to solve the cash management problem where one wants to minimize the total NPV of the costs over the set of impulse controls.

In a cash management problem, an admissible policy is given by a set of nondecreasing processes $\pi := \{R^\pi, L^\pi\}$ that are $\mathbb{F}$-adapted and increase only with jumps. The objective is to minimize the sum of holding and controlling costs given by

$$V^\pi(x) := \mathbb{E}_x\Big[\int_0^\infty e^{-qt} f(A_t^\pi)\mathrm{d}t + \sum_{0\le t<\infty} e^{-qt}[c_L(\Delta L_t^\pi + k_L)1_{\{\Delta L_t^\pi>0\}} + c_R(\Delta R_t^\pi + k_R)1_{\{\Delta R_t^\pi>0\}}]\Big],$$

where $c_L, c_R \in \mathbb{R}$ and $A_t^\pi := X_t + R_t^\pi - L_t^\pi$ is the resulting process controlled by the policy π.

It is clear that the band policies studied in this paper are admissible, and it is naturally conjectured that, under a certain (for instance, convexity) assumption on the holding cost function f, the optimal strategy is given by a band policy for a suitable choice of the parameters (d, D, U, u).

From the well-known existing results on impulse control, the candidate values of (d, D, U, u) are first chosen so that the value function becomes continuous/smooth at the levels d and u, and its slopes at D and U equal, respective, the negative of the unit proportional cost for R^π and the unit proportional cost for L^π. More precisely, if V^* is the value function, it is expected to satisfy the following:

$$\begin{aligned} V^{*'}(d-) &= V^{*'}(d+), \\ V^{*'}(D) &= -c_R, \\ V^{*'}(U) &= c_L, \\ V^{*'}(u-) &= V^{*'}(u+). \end{aligned} \tag{25}$$

Here, for the case X is of bounded variation, because of irregularity of the lower half-line (see, e.g., page 142 of [14]), the first smooth fit condition is replaced with the continuous fit condition: $V^*(d-) = V^*(d+)$.

Using the analytical expressions of the NPV's under the band policy, these four equations can be written concisely in terms of the scale function. In particular, the asymptotic behaviors of the scale function near zero as summarized in Remark 2.1(2) are expected to be helpful in simplifying these. In turn, the problem reduces to identifying the four parameters (d, D, U, u) as a solution to the system of four equations. Unfortunately, however, this is likely to become a big hurdle. Because the equations turn out to be nonlinear and somewhat complicated, even the existence/uniqueness of a solution is expected to be difficult to show. With regard to this, we refer the reader to [2, 3, 10, 13, 16, 18] for simpler cases where two (instead of four) parameters are sought.

After the four parameters (d, D, U, u) that satisfy (25) are identified, the last step is to verify the optimality. This is equivalent to showing that the candidate value function solves the QVI of [5]. This is indeed the most challenging part of the problem. However, there are several benefits about having the semi-explicit expressions written in terms of the scale function. First, the harmonicity on (d,u) can be proven easily thanks to the smoothness of the scale function and because the processes $e^{-q(t\wedge\tau_0^-\wedge\tau_b^+)}W^{(q)}(X_{t\wedge\tau_0^-\wedge\tau_b^+})$, $e^{-q(t\wedge\tau_0^-\wedge\tau_b^+)}Z^{(q)}(X_{t\wedge\tau_0^-\wedge\tau_b^+})$, $e^{-q(t\wedge\tau_0^-\wedge\tau_b^+)}(\overline{Z}^{(q)}(X_{t\wedge\tau_0^-\wedge\tau_b^+})+\mu/q)$, $t\geq 0$, for any fixed $b>0$ are martingales. In addition, the property given as Remark 2.1(3) has been shown to be useful in the verification as in the existing results [2, 3, 10, 13, 18].

Overall, the cash management problem of this form is conjectured to be challenging to solve. However, the results obtained in this paper would certainly be helpful and potentially lead to an efficient way of solving the problem.

References

1. Avram, F., Palmowski, Z. and Pistorius, M.R. (2007), "On the optimal dividend problem for a spectrally negative Lévy process," *Ann. Appl. Probab.*, **17(1)**, 156–180.
2. Baurdoux, E. and Yamazaki, K. (2015), "Optimality of doubly reflected Lévy processes in singular control," *Stochastic Process. Appl.*, **125(7)**, 2727–2751.
3. Bayraktar, E., Kyprianou, A.E. and Yamazaki, K. (2014), "Optimal dividends in the dual model under transaction costs," *Insurance Math. Econom.*, **54**, 133-143.
4. Benkherouf, L. and Bensoussan, A. (2009), "Optimality of an (s,S) policy with compound Poisson and diffusion demands: a quasi-variational inequalities approach," *SIAM J. Control Optim.*, **48(2)**, 756–762.
5. Bensoussan, A. and Lions, J.-L. (1984), *Impulse control and quasi-variational inequalities*, John Wiley & Sons Ltd.
6. Bensoussan, A., Liu, R.H. and Sethi, S.P. (2005), "Optimality of an (s,S) policy with compound Poisson and diffusion demands: a quasi-variational inequalities approach," *SIAM J. Control Optim.*, **44(5)**, 1650–1676.
7. Buckley, I. and Korn, R. (1998), "Optimal index tracking under transaction costs and impulse control," *Int. J. Theoretical Appl. Finance*, **1(3)**, 315–330.
8. Chan, T., Kyprianou, A.E. and Savov, M. (2011), "Smoothness of scale functions for spectrally negative Lévy processes," *Probab. Theory Relat. Fields*, **150**, 691–708.
9. Constantinides, G.M. and Richard, S.F. (1978) "Existence of optimal simple policies for discounted-cost inventory and cash management in continuous time," *Oper. Res.*, **26(4)**, 620–636.
10. Egami, M., Leung, T. and Yamazaki, K. (2013), "Default swap games driven by spectrally negative Lévy processes," *Stochastic Process. Appl.*, **123(2)**, 347-384.
11. Harrison, J.M., Sellke, T.M. and Taylor, A.J. (1983), "Impulse control of Brownian motion," *Math. Oper. Res.*, **8(3)**, 454–466.
12. Harrison, J.M. and Taylor, A.J. (1977/78), "Optimal control of a Brownian storage system," *Stochastic Processes Appl.*, **6(2)**, 454–466.
13. Hernández-Hernández, D. and Yamazaki, K. (2015), "Games of singular control and stopping driven by spectrally one-sided Lévy processes," *Stochastic Process. Appl.*, **125(1)**, 1–38.
14. Kyprianou, A.E. (2006), *Introductory lectures on fluctuations of Lévy processes with applications*, Springer-Verlag.
15. Kyprianou, A.E. and Surya, B.A. (2007), "Principles of smooth and continuous fit in the determination of endogenous bankruptcy levels," *Finance Stoch.*, **11(1)**, 131–152.

16. Loeffen, R.L. (2009), "An optimal dividends problem with transaction costs for spectrally negative Lévy processes," *Insurance Math. Econom.*, **45(1)**, 41–48.
17. Pistorius, M.R. (2003), "On doubly reflected completely asymmetric Lévy processes," *Stochastic Process. Appl.*, **107(1)**, 131–143.
18. Yamazaki, K. (2013), "Inventory control for spectrally positive Lévy demand processes," *arXiv*, 1303.5163.

A Second-order Monotone Modification of the Sharpe Ratio

Mikhail Zhitlukhin

Department of Probability and Statistics, Steklov Mathematical Institute, 8 Gubkina St., Moscow, Russia,
Laboratory of Quantitative Finance, Higher School of Economics, 31 Shabolovka St., Moscow, Russia. Email: mikhailzh@mi.ras.ru

The Sharpe ratio is not monotone: an investment strategy which yields a higher return may have a lower Sharpe ratio. This may lead to suboptimal investment decisions.
In this paper we present a modification of the Sharpe ratio, which is monotone with respect to second-order stochastic dominance. We study its properties and obtain a representation which allows to compute it in an efficient way.

Key words: Sharpe ratio, performance measure, second order stochastic dominance

1. Introduction

This paper concerns the problem of evaluating performance of investment strategies. By performance we mean a numerical characteristic which can be used to describe the quality of a strategy. The central object of the paper is the Sharpe ratio and its modification we propose, the second-order monotone Sharpe ratio.

The Sharpe ratio, which was introduced by W. F. Sharpe [13, 14] and originally called the "reward-to-variability ratio", has become one the most widely known tools to evaluate performance of investments, used by both practitioners and academicians. Its success is largely due to its simplicity and clear interpretation. Nevertheless, it has also been criticized extensively and many modifications have been proposed to overcome its drawbacks. An extensive overview of different performance measures can be found in e. g. [5, 6, 15].

This paper focuses on the issue that the Sharpe ratio is not monotone, namely that an investment strategy with a lower return can have a higher Sharpe ratio. This effect can lead to suboptimal investment decisions (see e.g. [1]) and, in general, contradicts with the natural understanding of what risk is (see e.g. [2]).

We propose a modification of the Sharpe ratio which is defined as the maximum Sharpe ratio of investment strategies yielding a return not higher than the given strategy. Equivalently, it is the minimal second-order monotone functional dominating the standard Sharpe ratio. It will be shown that the new performance measure is quasi-concave, continuous and second-order monotone. Then, by considering a dual optimization problem, we provide a representation of the new measure, which allows to compute it in an simple way.

Section 2 contains a review of basic concepts, in particular the Sharpe ratio and second-order stochastic monotonicity. Section 3 introduces the second-order monotone modification of the Sharpe and formulates the main results about it. The proofs are in Section 4.

2. The Sharpe ratio and second-order monotonicity

Let $(\Omega, \mathcal{F}, \mathbb{P})$ be a probability space and L^2 denote the space of all square integrable random variables. We consider a one-step model of an economy and interpret random variables $X \in L^2$ as possible excessive[1] returns from different investment plans after some fixed time (for example, 1 year).

In order to compare different investment plans, a natural approach is to assign a number to each of them, which shows its investment quality, and then compare these numbers. We can formalize this by saying that a *performance measure* is a map $\rho\colon L^2 \to \overline{\mathbb{R}}$, which assigns a number $\rho(X)$, possibly $\pm\infty$, to the strategy X representing its investment quality.

In this paper we will consider a performance measure, which is a modification of the Sharpe ratio. Recall, that the Sharpe ratio of an investment X can be defined as the ratio of the expected excessive return X to its standard deviation,

$$S(X) = \frac{\mathbb{E}X}{\sqrt{\operatorname{Var} X}}.$$

Strictly speaking, we consider here only the *ex-ante* Sharpe ratio, i.e. we measure the future performance of an investment strategy with known probability distribution. In the *ex-post* setting, on the other hand, one evaluates the past performance from a sequence of realized returns. Although mathematically the problem of evaluating the ex-post ratio can be reduced to the ex-ante setting by considering the empirical distribution of past returns, this will include additional uncertainty due to the estimation of distribution parameters. This issue will not be covered in the paper, and we will always assume the distribution of returns is given.

As it was noted in the introduction, the Sharpe ratio is not monotone: a higher return may result in a lower Sharpe ratio. A simple example shows this: let X be normally distributed with mean 1 and variance 1, and take $Y = \min(X, 2)$. Then $Y \le X$ a.s., but straightforward computation shows that $S(Y) > 1.05 > S(X)$.

[1] Compared to some benchmark strategy.

We will modify the Sharpe ratio to make it a monotone performance measure, which still has a clear meaning and can be easily computed and used, for example, for portfolio optimization.

Let us formalize the notion of monotonicity that will be used. We will consider functionals on the space L^2 monotone with respect to second-order stochastic dominance, which have a natural interpretation in the language of preferences of risk-averse investors. Recall that a random variable X is said to dominate a random variable Y in the second stochastic order, which we denote in this paper by $X \succcurlyeq Y$, if for any increasing concave function U such that both $\mathbb{E}U(X)$ and $\mathbb{E}U(Y)$ exist it holds that $\mathbb{E}U(X) \geq \mathbb{E}U(Y)$. Then, a functional $\rho\colon L^2 \to \overline{\mathbb{R}}$ is called monotone with respect to second-order stochastic dominance if $\rho(X) \geq \rho(Y)$ for any $X \succcurlyeq Y$ from the domain of ρ.

Second-order stochastic dominance has a clear interpretation: if $X \succcurlyeq Y$ then X is preferred to Y by any risk-averse investor, whose utility function is an arbitrary concave increasing function U. Strassen [16], Rothschild and Stiglitz [11, 12], Kellerer [9] provided a fundamental connection of such dominance with characterization of variability (see also Ewald and Yor [8]). They showed that $X \succcurlyeq Y$ if and only if there exists a random variable Z such that

$$Y \stackrel{d}{=} X + Z \quad \text{and} \quad \mathbb{E}(Z \mid X) \leq 0 \text{ a.s.},$$

where "$\stackrel{d}{=}$" denotes the equality of the distributions (it may be necessary to consider a copy of X on another probability space, where such Z exists). In other words, Y bears additional risk Z compared to X.

Second-order stochastic dominance is weaker than almost sure dominance, i.e. if $\mathbb{P}(X \geq Y) = 1$, then $X \succcurlyeq Y$. Consequently, as the example from the introduction shows, the Sharpe ratio is not second-order monotone. From an investor's point of view, this means that underreporting profit may result in a higher Sharpe ratio.

3. The second-order monotone Sharpe ratio and its properties

The above consideration leads us to the following idea of defining a modification of the Sharpe ratio.

Definition. The second-order monotone Sharpe ratio of an investment $X \in L^2$ is defined by

$$\mathcal{MS}(X) = \sup\{\mathcal{S}(Y) \mid Y \in L^2,\ Y \leq X \text{ a.s.}\},$$

i. e. it is the maximum Sharpe ratio among all investment strategies yielding a return not higher than X (the proof that it is indeed a second-order monotone performance measure will be given below).

If the underlying probability space is sufficiently rich, the definition implies that $\mathcal{MS}(X) \in [0, +\infty]$ because $\mathcal{MS}(X) = 0$ if $\mathbb{E}X \leq 0$, and $\mathcal{MS}(X) = +\infty$ if $X \geq 0$ a.s. and $\mathbb{P}(X > 0) > 0$. To see why the first claim holds, take a sequence $Z_n \geq 0$ such that $\mathbb{E}Z_n$ are bounded but $\operatorname{Var} Z_n \to \infty$, and let $Y_n = X - Z_n$ for which

$\mathcal{S}(Y_n) \to 0$. The second claim follows by considering $Y_n = n^{-1}\mathbb{I}(X \geq n^{-1})$ since $\mathcal{S}(Y_n) \to \infty$.

In other words, all investment strategies with non-positive average excessive returns are bad, and all those yielding sure profit are exceptionally good. Following Cherny and Madan [4], the latter situation means that the performance of such an *arbitrage opportunity* is infinite. These two observations can be considered as limitations of the monotone Sharpe ratio; this is the price we have to pay for the monotonicity. On the other hand, the most interesting case in practice is, of course, the case of profitable but risky investments, i.e. when $\mathbb{E}X > 0$ and $\mathbb{P}(X < 0) > 0$, where the monotone Sharpe ratio is non-trivial. Moreover, one should keep in mind that, for example, the standard Sharpe ratio also lacks a clear financial interpretation when $\mathbb{E}X < 0$.

The two theorems below contain the main results of the paper about the second-order monotone Sharpe ratio. In view of the first theorem, recall that a function f on a vector space V is called quasiconcave if the set $\{x : f(x) \geq a\}$ is convex for any a, or, equivalently, $f(\lambda x + (1 - \lambda)y) \geq \min\{f(x), f(y)\}$ for any $x, y \in V$ and $\lambda \in [0, 1]$.

Theorem 3.1. *The second-order monotone Sharpe ratio is a quasiconcave functional on L^2 continuous at any $X \in L^2$ such that $\mathbb{E}X > 0$ and $\mathbb{P}(X < 0) > 0$, and is monotone with respect to second-order stochastic dominance.*

Moreover, it is the minimal second-order monotone functional which dominates the Sharpe ratio[2].

For a random variable $X \in L^2$ introduce the function $g_X : \mathbb{R} \to \mathbb{R}$ defined by

$$g_X(x) = \mathbb{E}(X \cdot (x - X)^+)$$

with the notation $(x - X)^+ = \max\{x - X, 0\}$.

Theorem 3.2. *For any $X \in L^2$ such that $\mathbb{E}X > 0$ and $\mathbb{P}(X < 0) > 0$ the equation $g_X(x) = 0$ has a unique positive solution $x_* = x_*(X)$ and following representation is valid:*

$$\mathcal{MS}(X) = \sqrt{\frac{x_*}{\mathbb{E}(x_* - X)^+} - 1}.$$

The first theorem establishes natural properties one can expect a performance measure should satisfy. Quasi-concavity means that a performance measure favors diversification. Continuity means that small changes in an investment strategy do not largely affect the performance. The interpretation of second-order monotonicity was discussed above.

[2]This means that for any other $\rho \colon L^2 \to \overline{\mathbb{R}}$ which is second-order monotone and such that $\rho(X) \geq \mathcal{S}(X)$ for any $X \in L^2$, it holds that $\rho(X) \geq \mathcal{MS}(X)$ for any $X \in L^2$.

One general system of properties (axioms) a performance measure should satisfy was proposed by Cherny and Madan [4]. In addition to the above properties[3] they required scale invariance, i.e. $\rho(\lambda X) = \rho(X)$ for a performance measure ρ and any constant $\lambda > 0$, which is also satisfied by the monotone Sharpe ratio. A performance measure satisfying these four properties is called an acceptability index. Additional properties of acceptability indices emphasized in [4] are law invariance ($X \overset{d}{=} Y$ implies $\rho(X) = \rho(Y)$; observe that it follows from second-order monotonicity), expectation consistency (if $\mathbb{E}X \leq 0$ then $\rho(X) = 0$; if $\mathbb{E}X > 0$ then $\rho(X) > 0$) and arbitrage consistency ($\rho(X) = +\infty$ if and only if $X \geq 0$ a.s. and $\mathbb{P}(X > 0) > 0$). Cherny and Madan, and Eberlein and Madan [7] provided interesting examples of performance measures satisfying all these properties. The monotone Sharpe ratio also satisfies them, so it fits the classical Sharpe ratio into the modern framework of performance and risk analysis.

From practical point of view, continuity and quasi-concavity are also crucial for performing portfolio optimization (combining given investment strategies X_i to maximize the performance of a portfolio $X = \sum_i \lambda_i X_i$), as there exist efficient algorithms for quasi-convex and quasi-concave optimization. However, without a fast method of computing the monotone Sharpe ratio (since its definition is implicit), efficient optimization would be problematic. This is where the second theorem becomes useful.

In particular, if X has a continuous density, the function $g(x)$ is continuously differentiable with the derivative $g'(x) = \mathbb{E}[X\mathbb{I}(X < x)]$ and unimodal: it does not increase on $(-\infty, x_0]$ and does not decrease on $[x_0, +\infty)$ where x_0 is a root of the equation $g'(x) = 0$. Thus, x_* can be easily found numerically.

If X is a discrete random variable taking a finite number of values, the solution of the equation $g(x) = 0$ can be found by computing $g(x_k)$ starting from the smallest value x_1 to the largest value x_n using that $g(x_k) = x_k\mathbb{E}[X\mathbb{I}(X < x_k)] - \mathbb{E}[X^2\mathbb{I}(X < x_k)]$ and recursively computing $\mathbb{E}[X\mathbb{I}(X < x_k)]$ and $\mathbb{E}[X^2\mathbb{I}(X < x_k)]$.

4. Proofs

To prove the theorems, we first establish an auxiliary lemma. Everywhere in this section we assume that all random variables are from L^2.

Lemma 4.1. *1) If $\mathbb{E}X > 0$ and $\mathbb{P}(X < 0) > 0$, the supremum in the definition of $\mathcal{MS}(X)$ is attained, i.e. there exists $\hat{Y} \in L^2$ such that $\hat{Y} \leq X$ a.s. and $\mathcal{MS}(X) = \mathcal{S}(\hat{Y})$.*

2) It is possible to find $\hat{Y}$ from the previous statement such that it is $\sigma(X)$-measurable, $\hat{Y}(\omega) = X(\omega)$ on $\{\omega : X(\omega) \leq 0\}$ and $\hat{Y}(\omega) \geq 0$ on $\{\omega : X(\omega) > 0\}$.

[3]Note that Cherny and Madan considered performance measures for L^∞, which are monotone with respect to dominance with probability one, and satisfy the so-called Fatou property instead of continuity. However, most of their results can be carried over to our setting.

3) The monotone Sharpe ratio is law-invariant, i.e if $Law(X) = Law(X')$, then $\mathcal{MS}(X) = \mathcal{MS}(X')$.

Proof. For a fixed $X \in L^2$ define the set $\mathcal{Y}(X) \subset L^2$ of random variables Y such that $Y = X$ whenever $\{X \le 0\}$ and $0 \le Y \le X$ whenever $\{X > 0\}$. If $\mathbb{E}X > 0$ and $\mathbb{P}(X < 0) > 0$ the maximization in the definition of $\mathcal{MS}(X)$ can be carried over only $\mathcal{Y}(X)$, i.e.

$$\mathcal{MS}(X) = \sup_{Y \in \mathcal{Y}(X)} \mathcal{S}(Y).$$

Indeed, if $Y \le X$ and $\mathbb{E}Y > 0$, taking $Y' = X\mathbb{I}(X \le 0) + Y\mathbb{I}(X > 0)$ we get $Y' \le X$ and $\mathcal{S}(Y') \ge \mathcal{S}(Y)$.

The set $\mathcal{Y}(X)$ is totally bounded and complete in L^2, so it is compact. Moreover, if $\mathbb{P}(X < 0) > 0$ then $0 \notin \mathcal{Y}(X)$ and because the Sharpe ratio $\mathcal{S}(Y)$ is continuous at any $Y \in L^2$ except $Y \equiv 0$, it attains the maximum on $\mathcal{Y}(X)$. This proves statement 1.

The first part of the second statement follows from that if $Y \le X$, then for $\overline{Y} = \mathbb{E}(Y \mid X)$ it holds that $\overline{Y} \le X$, $\mathbb{E}\overline{Y} = \mathbb{E}Y$, $\operatorname{Var}\overline{Y} \le \operatorname{Var} Y$, hence[4] $\mathcal{S}(\overline{Y}) \ge \mathcal{S}(Y)$. The second part follows from that the supremum in $\mathcal{MS}(X)$ is attained at some $\hat{Y} \in \mathcal{Y}(X)$.

Statement 3 follows from that if $Law(X)$ is such that $\mathbb{E}X > 0$ and $\mathbb{P}(X < 0) > 0$ then $\hat{Y}$ is $\sigma(X)$-measurable, so it can be represented as a function $\hat{Y} = f(X)$. If $\mathbb{E}X \le 0$ then $\mathcal{MS}(X) = 0$ and if $X \ge 0$ a.s. and $\mathbb{P}(X \ge 0) > 0$, then $\mathcal{MS}(X) = +\infty$. □

Proof. [Proof of Theorem 1] The Sharpe ratio is continuous at any non-constant $X \in L^2$. Hence if $\mathbb{E}X > 0$ and $\mathbb{P}(X < 0) > 0$ the Sharpe ratio is continuous on any set $\mathcal{Y}(X')$ for X' such that $\|X - X'\| \le \min\{\mathbb{E}X, \|X^-\|\}/2$ as this guarantees $\mathbb{E}X' > 0$ and $\mathbb{P}(X' < 0) > 0$, so $\mathbb{P}(Y \ge 0) > 0$ and $\mathbb{P}(Y < 0) > 0$ for any $Y \in \mathcal{Y}(X')$, and therefore $\operatorname{Var} Y > 0$.

The correspondence $\mathcal{Y}\colon L^2 \twoheadrightarrow L^2$ is continuous, i.e. upper hemicontinuous and lower hemicontinuous. Indeed, for any L^2-neighborhood U of $\mathcal{Y}(X)$ there exists $\epsilon > 0$ such that $X' \in U$ whenever $\|X - X'\| \le \epsilon$ because $X \in \mathcal{Y}(X)$. Then if $\|X - X'\| \le \epsilon$ we also have $\mathcal{Y}(X') \subset U$, which proves the upper hemicontinuity. To establish the lower hemicontinuity, take $X_n \to X$ and some $Y \in \mathcal{Y}(X)$. Then $Y_n = \min\{Y, X_n\} \cdot \mathbb{I}(X_n \ge 0, Y \ge 0) + X_n\mathbb{I}(X_n < 0)$ belongs to $\mathcal{Y}(X_n)$ and converges to Y, which is equivalent to the lower hemicontinuity of $\mathcal{Y}$ at X (Theorem 17.21 in [3]). Now we can apply Berge's maximum theorem which implies that the monotone Sharpe ratio is continuous at any X such that $\mathbb{E}X > 0$ and $\mathbb{P}(X < 0) > 0$.

---

[4] $\operatorname{Var}\overline{Y} = \mathbb{E}(\mathbb{E}(Y \mid X))^2 - (\mathbb{E}Y)^2 \le \mathbb{E}(\mathbb{E}(Y^2 \mid X)) - (\mathbb{E}Y)^2 = \operatorname{Var} Y$, where conditional Jensen's inequality is applied.

To prove the quasiconcavity, first observe that the Sharpe ratio is quasiconcave. Indeed, if $\mathcal{S}(Y_1) \geq y$ and $\mathcal{S}(Y_2) \geq y$, then for any $\lambda \in [0,1]$ we have

$$\mathcal{S}(\lambda Y_1 + (1-\lambda)Y_2) \geq \frac{\lambda \mathbb{E}Y_1 + (1-\lambda)\mathbb{E}Y_2}{\lambda\sqrt{\operatorname{Var} Y_1} + (1-\lambda)\sqrt{\operatorname{Var} Y_2}} \geq y$$

where in the first inequality we used the linearity of the expectation and the convexity of the standard deviation. If $0 < x \leq \mathcal{MS}(X_i) < \infty$ for some $X_1, X_2 \in L^2$, take Y_i such that $\mathcal{S}(Y_i) = \mathcal{MS}(X_i)$. Then

$$\mathcal{MS}(\lambda X_1 + (1-\lambda)X_2) \geq \mathcal{S}(\lambda Y_1 + (1-\lambda)Y_2) \geq x$$

where the first inequality follows from that $\lambda Y_1 + (1-\lambda)Y_2 \leq \lambda X_1 + (1-\lambda)X_2$.

Finally, we prove the second-order monotonicity. Suppose $X \succcurlyeq X'$. We will assume $0 < \mathcal{MS}(X), \mathcal{MS}(X') < \infty$ (i.e. $\mathbb{E}X > 0$ and $\mathbb{P}(X < 0) > 0$ and the same for X'); the other cases are easy.

As it was noted above, there exists Z such that $X' \overset{d}{=} X + Z$ and $\mathbb{E}(Z \mid X) \leq 0$. Using the law-invariance of the monotone Sharpe ratio, we may assume that X, X', Z are defined on the same probability space and the equality $X' = X + Z$ holds with probability one.

Take $Y \in L^2$ such that $Y \leq X'$ and $\mathcal{MS}(X') = \mathcal{S}(Y)$. Then for $\overline{Y} = \mathbb{E}(Y \mid X)$ we have $\mathcal{MS}(X') \leq \mathcal{S}(\overline{Y})$ because $\mathbb{E}\overline{Y} = \mathbb{E}Y$, and $\operatorname{Var}\overline{Y} \leq \operatorname{Var} Y$. Since $X = X' - Z \geq Y - Z$, it holds that $X \geq \mathbb{E}(Y \mid X) - \mathbb{E}(Z \mid X) \geq \overline{Y}$, so $\mathcal{MS}(X) \geq \mathcal{S}(\overline{Y}) \geq \mathcal{MS}(X')$.□

Proof. [Proof of Theorem 2] Suppose $\mathbb{E}X > 0$, $\mathbb{P}(X < 0) > 0$. We will assume $\mathbb{E}X = 1$ and then the case of arbitrary $\mathbb{E}X > 0$ will be reduced to this one. The definition of $\mathcal{MS}(X)$,

$$\mathcal{MS}(X) = \sup_{Y \in L^2} \left\{ \frac{\mathbb{E}Y}{\sqrt{\operatorname{Var} Y}} \;\middle|\; Y \leq X \text{ a.s.} \right\},$$

can be equivalently written in the form

$$\frac{1}{(\mathcal{MS}(X))^2} + 1 = \inf_{Z,a} \left\{ \mathbb{E}Z^2 \mid Z \leq aX \text{ a.s., } \mathbb{E}Z = 1 \right\},$$

where we minimize over $Z \in L^2$ and $a \in \mathbb{R}$. Here Z corresponds to $Y/\mathbb{E}Y$ and a corresponds to $1/\mathbb{E}Y$ in the definition of $\mathcal{MS}(X)$. Note the constraints automatically imply $a > 0$, because otherwise they cannot be satisfied simultaneously since $\mathbb{E}Z \leq a\mathbb{E}X = a$ according to our assumption that $\mathbb{E}X = 1$.

Consider the dual Lagrangian

$$L_d(p,q) = \inf_{Z,a} \left\{ \mathbb{E}Z^2 + \mathbb{E}[p(Z - aX)] + q(\mathbb{E}Z - 1) \right\},$$

which is a function of variables $p \in L^2$, $p \geq 0$ a.s., and $q \in \mathbb{R}$. A necessary condition for $L_d(p,q) > -\infty$ (the dual problem is feasible) is $\mathbb{E}(pX) = 0$. In that case we have

$$L_d(p,q) = \inf_{Z \in L^2} \mathbb{E}[Z^2 + (p+q)Z - q] = -\frac{1}{4}\mathbb{E}(p+q)^2 - q,$$

where we excluded Z by choosing it to minimize the expression under the expectation with probability one.

Consider the dual problem

$$V_d = \sup_{p,q}\{L_d(p,q) \mid p \geq 0 \text{ a.s., } \mathbb{E}(pX) = 0\}$$

and introduce the value function of the primal problem $\phi \colon L^2 \times \mathbb{R} \to \overline{\mathbb{R}}$,

$$\phi(u,v) = \inf_{Z,a} F(Z,a,u,v),$$

where

$$F(Z,a,u,v) = \begin{cases} \mathbb{E}Z^2, & \text{if } Z \leq aX + u \text{ a.s., } \mathbb{E}Z = 1 + v, \\ +\infty, & \text{otherwise.} \end{cases}$$

The function $\phi(u,v)$ is continuous at zero, which implies the strong duality (see Theorem 7 in [10]), i.e. that the value of the dual problem is equal to the value of the primal problem: $V_d = (\mathcal{MS}(X))^{-2} + 1$.

Maximizing the dual Lagrangian over q, we obtain

$$\begin{aligned} \frac{1}{(\mathcal{MS}(X))^2} &= \sup_p \left\{\mathbb{E}p - \tfrac{1}{4}\operatorname{Var} p \mid p \geq 0 \text{ a.s., } \mathbb{E}(pX) = 0\right\} \\ &= \sup_{r,s}\left\{s - \tfrac{1}{4}s^2 \operatorname{Var} r \mid r \geq 0 \text{ a.s., } s \geq 0,\ \mathbb{E}(rX) = 0,\ \mathbb{E}r = 1\right\}, \end{aligned}$$

where the new variables $r \in L^2$, $s \in \mathbb{R}$ correspond to $p/\mathbb{E}p$ and $\mathbb{E}p$ respectively. Maximizing over s we obtain

$$(\mathcal{MS}(X))^2 + 1 = \inf_r \left\{\mathbb{E}r^2 \mid r \geq 0 \text{ a.s., } \mathbb{E}(rX) \leq 0,\ \mathbb{E}r = 1\right\}.$$

Observe that we replaced the equality constraint $\mathbb{E}(rX) = 0$ by the inequality constraint. This does not change the solution of the problem, because if $\mathbb{E}(rX) < 0$ we can take $r' = (r - \mathbb{E}(rX))/(1 - \mathbb{E}(rX))$ which satisfies the constraints of the problem and gives a smaller value of the functional being minimized.

Next we pass to the dual problem again:

$$\begin{aligned} (\mathcal{MS}(X))^2 + 1 &= \sup_{\alpha,\beta,\gamma}\left\{\inf_r\left\{\mathbb{E}r^2 - \mathbb{E}(\alpha r) + \beta\mathbb{E}(rX) + \gamma(\mathbb{E}r - 1)\right\}\right\} \\ &= -\inf_{\alpha,\beta,\gamma}\left\{\tfrac{1}{4}\mathbb{E}(\beta X - \alpha + \gamma)^2 + \gamma \mid \alpha \geq 0 \text{ a.s., } \beta \geq 0\right\} \end{aligned}$$

where $\alpha \in L^2$, $\beta, \gamma \in \mathbb{R}$ and in the second equality we chose r minimizing the expression under the expectation with probability one.

Using that the infimum is negative (the left-hand side of the equation is positive), we can restrict γ to $\gamma < 0$, and β to $\beta > 0$. Observe that if β and γ are fixed, the infimum will be attained at $\alpha = (\beta X + \gamma) \cdot \mathbb{I}(\beta X + \gamma \geq 0)$. Denoting $x = -\gamma/\beta$ we obtain

$$(\mathcal{MS}(X))^2 + 1 = -\inf_{\beta,x} \left\{ \tfrac{1}{4}\beta^2 \mathbb{E}[(X-x)^2 \mathbb{I}(X < x)] - \beta x \mid \beta < 0,\ x > 0 \right\}$$

$$= \sup_{x>0} \left[\frac{x^2}{\mathbb{E}[(X-x)^2 \mathbb{I}(X < x)]} \right].$$

The function $f(x)$ defined for $x > 0$ by

$$f(x) = \mathbb{E}[(X/x - 1)^2 \cdot \mathbb{I}(X/x < 1))]$$

is differentiable (because the function $z \mapsto (z-1)^2 \cdot \mathbb{I}(z < 1)$ is differentiable) and its derivative $f'(x) = 2g(x)/x^3$ is continuous, where $g(x)$ is defined in Section 3; the continuity follows from the Lebesgue convergence theorem. Moreover, $g(x) < 0$ for $x \leq 0$ and $g(x) \to +\infty$ as $x \to +\infty$, hence the equation $g(x) = 0$ has a solution.

To prove that the positive solution is unique, suppose $x_* > 0$ is the minimal solution. Then $\mathbb{E}[X \cdot \mathbb{I}(X < x_*)] > 0$ and for any $x_* \leq x_1 < x_2$ we have $g(x_1) - g(x_2) \geq (x_2 - x_1) \cdot \mathbb{E}[X \cdot \mathbb{I}(X < x_1)] > 0$, so the solution is indeed unique. Consequently, x_* minimizes $f(x)$, and thus $\mathcal{MS}(X) = \sqrt{1/f(x_*) - 1}$. From $g(x_*) = 0$, we get $f(x_*) = \mathbb{E}(x_* - X)^+/x_*$, which proves the representation of the monotone Sharpe ratio in the case $\mathbb{E}X = 1$.

Finally, observe that we have $x_*(X) = x_*(X/\mathbb{E}X) \cdot \mathbb{E}X$, where $x_*(X)$ and $x_*(X/\mathbb{E}X)$ are the solutions of the equation $g(x) = 0$ for X and $X' = X/\mathbb{E}X$ respectively. Using that the monotone Sharpe ratio is invariant under the multiplication of X by a positive constant, one can see that the representation also holds for arbitrary $\mathbb{E}X > 0$. □

Acknowledgement. I thank Igor Evstigneev for drawing my attention to the problem and valuable discussion and comments. The research is supported by RFBR grants No. 14-01-31468, 14-01-00739.

References

1. Abdulali, A. and Weinstein, E. (2002), "Hedge fund transparency: quantifying valuation bias for illiquid assets," *Risk* **15**, 25–28.
2. Aumann, R. J. and Serrano, R. (2008), "An Economic Index of Riskiness," *Journal of Political Economy* **116**, 810–836.
3. Aliprantis, C. D. and Border, K. C. (2007), *Infinite Dimensional Analysis: A Hitchhiker's Guide*, Springer.
4. Cherny, A. and Madan, D. (2009), "New measures for performance evaluation," *Review of Financial Studies* **22**, 2571–2606.

5. Cogneau, P. and Huber, G. (2009). "The (more than) 100 Ways to Measure Portfolio Performance – Part 1: Standardized Risk-Adjusted Measures," *Journal of Performance Measurement* **13**, 56–71.
6. Cogneau, P. and Huber, G. (2009). "The (more than) 100 Ways to Measure Portfolio Performance – Part 2: Special Measures and Comparison," *Journal of Performance Measurement* **14**, 56–69.
7. Eberlein, E. and Madan, D. B. (2009), "Hedge fund performance: sources and measures," *International Journal of Theoretical and Applied Finance* **12**, 267–282.
8. Ewald, C.-O. and Yor, M. (2014), "On Increasing Risk, Inequality and Poverty Measures: Peacocks and Lyrebirds," available at SSRN: http://ssrn.com/abstract=2461595.
9. Kellerer, H. G. (1972), "Markov-Komposition und eine Anwendung auf Martingale," *Mathematische Annalen* **198**, 99–122.
10. Rockafellar, R. T. (1974), *Conjugate Duality and Optimization*, SIAM, Philadelphia.
11. Rothschild, M. and Stiglitz, J. (1970), "Increasing Risk: I. A Definition," *Journal of Economic Theory* **2**, 225–243.
12. Rothschild, M. and Stiglitz, J. (1971), "Increasing Risk: II. Its Economic Consequences," *Journal of Economic Theory* **3**, 66–84.
13. Sharpe, W. F. (1966), "Mutual Fund Performance," *Journal of Business* **39**, 119–138.
14. Sharpe, W. F. (1994), "The Sharpe Ratio," *The Journal of Portfolio Management* **21**, 49–58.
15. Le Sourd, V. (2007), *Performance measurement for traditional investment*, EDHEC Risk and Asset Management Research Centre.
16. Strassen, V. (1965), "The existence of probability measures with given marginals," *The Annals of Mathematical Statistics* **36**, 423–439.

Printed in the United States
By Bookmasters